Advances in Geospatial Technologies (AGT) Book Series

ISSN:2327-5715
EISSN:2327-5723

Editor-in-Chief: Nilanjan Dey, Techno India College of Technology, India

MISSION

The geospatial technology field is a fast-paced, high growth industry that is involved in a variety of fields, including military planning, public health, land use, environmental protection, and Google Earth mapping. With such a diverse body of applications, the research in geospatial technologies is always evolving and new theories, methodologies, tools, and applications are being developed. **Advances in Geospatial Technologies (AGT) Book Series** is a reference source and outlet for research that discusses all aspects of geographic information, including areas such as geomatics, geodesy, GIS, cartography, remote sensing, and other areas. Because geospatial technologies are so pervasive in such a variety of areas, AGT also includes books that address interdisciplinary applications of the technologies.

COVERAGE

- Global Positioning Systems
- Cartography
- Surveying Technologies
- Public Health Applications of Geospatial Technologies
- Spatial Reference Systems
- Public Sector Use of Geospatial Technologies
- Geographical information systems
- Digital Terrain Modeling
- Disaster Informatics
- Geomatics

IGI Global is currently accepting manuscripts for publication within this series. To submit a proposal for a volume in this series, please contact our Acquisition Editors at Acquisitions@igi-global.com or visit: http://www.igi-global.com/publish/.

The Advances in Geospatial Technologies (AGT) Book Series (ISSN 2327-5715) is published by IGI Global, 701 E. Chocolate Avenue, Hershey, PA 17033-1240, USA, www.igi-global.com. This series is composed of titles available for purchase individually; each title is edited to be contextually exclusive from any other title within the series. For pricing and ordering information please visit http://www.igi-global.com/book-series/advances-geospatial-technologies/73686. Postmaster: Send all address changes to above address.

Titles in this Series

For a list of additional titles in this series, please visit:
http://www.igi-global.com/book-series/advances-geospatial-technologies/73686

Advanced Geospatial Practices in Natural Environment Resource Management
Rubeena Vohra (Bharati Vidyapeeth's College of Engineering, India) and Ashish Kumar (Bennett Universit, India)
Engineering Science Reference • © 2024 • 292pp • H/C (ISBN: 9798369313961) • US $275.00

Emerging Trends, Techniques, and Applications in Geospatial Data Science
Loveleen Gaur (Amity University, India & Taylor's University, Malaysia & University of the South Pacific, Fiji) and P.K. Garg (Indian Institute of Technology, Roorke, India)
Engineering Science Reference • © 2023 • 309pp • H/C (ISBN: 9781668473191) • US $240.00

Addressing Environmental Challenges Through Spatial Planning
Athar Hussain (Netaji Subhas University of Technology, India) Kailash Chandra Tiwari (Delhi Technological University, India) and Alpana Gupta (Indian Institute of Technology, Delhi, India)
Engineering Science Reference • © 2022 • 329pp • H/C (ISBN: 9781799883319) • US $215.00

Methods and Applications of Geospatial Technology in Sustainable Urbanism
José António Tenedório (Universidade NOVA de Lisboa, Portugal) Rossana Estanqueiro (Universidade NOVA de Lisboa, Portugal) and Cristina Delgado Henriques (Universidade de Lisboa, Portugal)
Engineering Science Reference • © 2021 • 684pp • H/C (ISBN: 9781799822493) • US $215.00

Interdisciplinary Approaches to Spatial Optimization Issues
Sami Faiz (University of Tunis El Manar, Tunis, Tunisia) and Soumaya Elhosni (University of Tunis El Manar, Tunis, Tunisia)
Engineering Science Reference • © 2021 • 271pp • H/C (ISBN: 9781799819547) • US $215.00

For an entire list of titles in this series, please visit:
http://www.igi-global.com/book-series/advances-geospatial-technologies/73686

IGI Global
PUBLISHER of TIMELY KNOWLEDGE

701 East Chocolate Avenue, Hershey, PA 17033, USA
Tel: 717-533-8845 x100 • Fax: 717-533-8661
E-Mail: cust@igi-global.com • www.igi-global.com

Geospatial Application Development Using Python Programming

Mohammad Gouse Galety
Samarkand International University of Technology, Uzbekistan

Arul Kumar Natarajan
Samarkand International University of Technology, Uzbekistan

Tesfaye Fufa Gedefa
Space Science and Geospatial Institute, Ethiopia

Tsegaye Demsis Lemma
Space Science and Geospatial Institute, Ethiopia

A volume in the Advances in Geospatial Technologies (AGT) Book Series

IGI Global
PUBLISHER of TIMELY KNOWLEDGE

Published in the United States of America by
IGI Global
Engineering Science Reference (an imprint of IGI Global)
701 E. Chocolate Avenue
Hershey PA, USA 17033
Tel: 717-533-8845
Fax: 717-533-8661
E-mail: cust@igi-global.com
Web site: http://www.igi-global.com

Library of Congress Cataloging-in-Publication Data

Names: Galety, Mohammad Gouse, 1976- editor. | Natarajan, Arulkumar, 1986- editor. | Gedefa, Tesfaye, 1990- editor. | Lemma, Tsegaye, 1990- editor.

Title: Geospatial application development using Python programming / edited by Mohammad Galety, Arulkumar Natarajan, Tesfaye Gedefa, Tsegaye Lemma.
Description: Hershey, PA : Engineering Science Reference, 2024. | Includes bibliographical references and index. | Summary: "This book provides a comprehensive and cutting-edge resource for scholars, researchers, and professionals in geospatial science and computer programming. This book seeks to bridge the gap between geospatial technologies and Python programming by offering in-depth insights into developing geospatial applications through Python-based solutions"-- Provided by publisher.
Identifiers: LCCN 2024009392 (print) | LCCN 2024009393 (ebook) | ISBN 9798369317549 (hardcover) | ISBN 9798369317556 (ebook)
Subjects: LCSH: Geospatial data--Computer processing. | Python (Computer program language)
Classification: LCC G70.217.G46 G434 2024 (print) | LCC G70.217.G46 (ebook) | DDC 910.285/5133--dc23/eng/20240323
LC record available at https://lccn.loc.gov/2024009392
LC ebook record available at https://lccn.loc.gov/2024009393

This book is published in the IGI Global book series Advances in Geospatial Technologies (AGT) (ISSN: 2327-5715; eISSN: 2327-5723)

British Cataloguing in Publication Data
A Cataloguing in Publication record for this book is available from the British Library.
All work contributed to this book is new, previously-unpublished material.

For electronic access to this publication, please contact: eresources@igi-global.com.

Table of Contents

Detailed Table of Contents

Geospatial solutions represent a pivotal toolset for analyzing, interpreting, and visualizing spatial data across diverse domains, facilitating informed decision-making and fostering innovation. This book chapter provides a comprehensive overview of geospatial solutions, emphasizing their critical role in addressing spatially explicit challenges and driving efficiency, productivity, and innovation across various sectors. Furthermore, it explores the integration of Python programming in geospatial applications, highlighting its versatility and extensive ecosystem of libraries and tools tailored for spatial data analysis and visualization. The fundamentals of web mapping are discussed in depth, elucidating spatial representation, technologies, and tools commonly employed in web mapping applications. Also, the chapter explores Python's role in retrieving geospatial data with Python, visualization methods, and interactive web mapping.

Remote sensing is a crucial technique in environmental and spatial investigations, and Python is a popular programming language for analyzing this data. This chapter provides a comprehensive guide to using Python for remote sensing data analysis, covering various data types, attributes, and practical implementations. It introduces Python and its data processing libraries, discusses preprocessing operations like data conversion and import, geometric rectification, and radiometric correction, and covers image enhancement techniques like edge detection, contrast enhancement, and filtering. It also covers image analysis techniques like band mathematics, indices, classification, and segmentation. The chapter also covers exporting data and generating visualization maps and charts. Python's application in remote sensing data analysis is illustrated through case studies.

This chapter aims to explore the power of Python in spatial data visualization. Spatial data visualization is the process of representing spatial information visually, enabling one to explore and communicate patterns, distributions, and relationships within the data. An informative spatial data visualization with Python effectively represents and communicates spatial information using visual elements, enabling users to gain insights and make informed decisions related to geospatial data. Python provides many sets of libraries and tools for handling and visualizing geospatial data to enhance understanding, facilitate exploration, and present geographic patterns and relationships clearly and intuitively. The chapter demonstrates the capabilities of Python for spatial data visualization by showcasing various techniques, spatial data formats, and tools with Geopandas, Matplotlib, Plotly, and Folium libraries. Examples and code snippets are provided for the readers to gain solid knowledge about spatial data visualization using Python.

The field of spatial data visualization in Python is propelled by the significance of data science and geospatial analysis. Python is pivotal in providing a flexible toolkit for visualizing intricate spatial data. This guide explores the fundamentals, techniques, and practical applications of spatial data visualization, tracing its historical roots alongside the ascent of data science and geospatial analytics. Examining libraries like Matplotlib, Geopandas, Folium, and Plotly, the guide equips readers with essential tools for geospatial data visualization in diverse fields, including urban planning, environmental science, and healthcare. Emphasizing dynamic and interactive visualizations, it explores spatial data's pivotal role in space exploration, virtual and augmented reality, satellite navigation, industry accuracy, and pandemic response. A focus lies on integrating IP geolocation for enhanced marketing accuracy, security, and compliance, showcasing how geospatial visualization improves public experiences in navigation, weather forecasting, public safety, healthcare.

The chapter uses Python to explore vector and raster data formats within geospatial visualization. It highlights their pivotal role across diverse environmental science, urban planning, and natural resource management domains. A nuanced comprehension of these formats is deemed essential for proficient geospatial visualization in Python, as they facilitate the storage and manipulation of spatial data. Vector data formats accurately represent points, lines, and polygons within a coordinate system. In contrast, raster data formats are tailored to depict continuous surfaces or grids of data. An array of libraries and tools are outlined for exploring and visualizing these geospatial data formats in Python, each serving distinct functionalities ranging from data manipulation to visualization. The chapter systematically introduces the concept of geospatial visualization, elucidates the disparities and application scenarios of vector and raster data formats, and subsequently elucidates various Python libraries and tools conducive to geospatial data manipulation and visualization.

Geospatial data visualization is a powerful tool for understanding and analyzing spatial patterns and relationships. This chapter dives into the world of geospatial data visualization using Folium geospatial Python library. It discloses the significance and versatility of Folium for creating dynamic and accessible maps. This chapter provides hands-on experience of geospatial data visualization using Python geospatial library Folium. This chapter provides practical examples. Readers will gain insights into the power of Folium for creating interactive and visually compelling maps. Thereby, making geospatial data accessible and engaging

This chapter addresses the challenge of road route selection by employing a systematic geospatial data integration approach during the design phase. The methodology revolves around the acquisition and analysis of datasets, such as satellite imagery, DEMs, GIS, and other important data sets. Leveraging Python programming, the chapter endeavors to develop a tailored geospatial framework for optimizing road route selection, with a focus on evaluating socio-economic, engineering and environmental impacts and aligning with sustainability objectives. Resulting in a robust methodology determining optimal alignments based on terrain, environmental, and socioeconomic factors using MCA. The chapter extend to providing data-driven recommendations for the development of resilient infrastructure, ensuring that road design aligns with sustainability principles to foster economic prosperity, environmental conservation, and societal well-being a paradigm shift in road expansion strategies.

This study employs the Python library for road network analysis in the Debre Berhan Metropolitan City, focusing on determining the best route, shortest path, road types, length in km, and speed limit in km/h. Geopandas is utilized to visualize the road network data, enabling the calculation of the shortest path and extraction of road type and speed limit information. Network analysis libraries like,folium, NetworkX convert the data into a graph representation, using algorithms like Dijkstra's or A* to find the shortest path based on distance or travel time. Geopandas then overlays road segments on a map, highlighting the shortest path and indicating road type, length, and speed limit, offering a robust framework for assessing and optimizing the road network in urban areas and contributing to the effective management and enhancement of transportation infrastructure.

(MAPE) of 237%, and a root mean square error (RMSE) of 518 for GRU, which demonstrates the effectiveness of the deep learning approach in predicting corn yield.

Foreword

In today's rapidly evolving technological landscape, the fusion of geospatial data analysis with Python programming has emerged as a powerful toolset, revolutionizing how we understand, interpret, and interact with our world. "Geospatial Application Development Using Python Programming" navigates this dynamic intersection, offering readers a comprehensive journey into geospatial applications empowered by Python.

As our societies become increasingly interconnected and data-driven, harnessing the potential of geospatial data has become indispensable across various industries and disciplines. From urban planning and environmental monitoring to agriculture and transportation logistics, the insights derived from geospatial analysis facilitate informed decision-making and drive innovation and sustainable development.

This book is a guiding beacon for novice learners and seasoned practitioners alike, offering a structured approach to mastering the essential concepts and techniques of geospatial application development using Python. Through a carefully curated collection of chapters, readers embark on a progressive expedition, starting with fundamental principles and gradually advancing toward cutting-edge applications and methodologies.

Each chapter within this compendium represents the boundless possibilities that emerge when the precision of geospatial data meets the flexibility and scalability of Python programming. From spatial data visualization and remote sensing to deep learning-driven predictive analytics, the diverse array of topics covered herein reflects the multifaceted nature of geospatial applications and the versatility of Python as a programming language.

Moreover, this book transcends mere theoretical discourse, providing readers with practical insights and hands-on examples that empower them to translate knowledge into action. Whether individuals are GIS professionals seeking to enhance their skill set, aspiring developers venturing into the world of geospatial technology, or academic researchers exploring new frontiers, *Geospatial Application Development Using Python Programming* is an indispensable companion on this exhilarating journey.

In an era defined by rapid technological innovation and global interconnectedness, the ability to harness the power of geospatial data is not merely a competitive advantage - it is a strategic imperative. As we embark on this transformative voyage together, may this book ignite curiosity, inspire creativity, and empower individuals to unlock the full potential of geospatial application development using Python programming.

R Thamarai Selvi
Department of Computer Applications (MCA), Bishop Heber College (Autonomous), India

Preface

In an era marked by the relentless march of technological progress, the unique synergy between geospatial data analysis and Python programming is a testament to the transformative power of interdisciplinary collaboration. Our book, *Geospatial Application Development Using Python Programming*, is a pioneering effort that explores and illuminates this symbiotic relationship. It offers readers a comprehensive roadmap uniquely tailored to navigate the complexities of geospatial applications in the modern world.

The subject matter of this book transcends traditional disciplinary boundaries, encompassing the realms of geography, data science, computer programming, and beyond. At its core, it delves into the art and science of leveraging Python programming to manipulate, visualize, and analyze geospatial data, unlocking many opportunities for innovation and insight.

In today's interconnected world, the relevance of geospatial technology extends far beyond the confines of academic discourse. From urban planning and disaster management to environmental conservation and precision agriculture, geospatial analysis applications are theoretical concepts and practical tools that permeate virtually every aspect of our daily lives. By providing readers with the requisite knowledge and skills to harness the power of Python programming in this context, our book equips them to confront the challenges and capitalize on the opportunities of the digital age.

The target audience for this book is diverse and inclusive, encompassing a broad spectrum of individuals with varying levels of expertise and experience. GIS professionals seeking to enhance their technical proficiency, aspiring developers looking to expand their skill set, academic researchers exploring new avenues of inquiry, and students embarking on their educational journey will find value in the insights and resources offered within these pages.

Each chapter submission in this volume represents a valuable contribution to the collective understanding of geospatial application development using Python programming. Let us briefly delve into the importance and relevance of each chapter:

1. **Introduction to Geospatial Data and Python Programming:** This foundational chapter provides a comprehensive overview of geospatial data analysis and Python programming concepts and techniques. Laying the groundwork for subsequent discussions serves as a springboard for deeper exploration into the subject matter.
2. **Python Programming for Geospatial Applications: Web Mapping, Interactive Visualization, and Beyond:** Building upon the foundational knowledge established in the preceding chapter, this chapter delves into the practical applications of Python programming for geospatial analysis. From web mapping to interactive visualization, readers gain insights into the diverse tools and techniques available.
3. **Python-Powered Remote Sensing Data:** Remote sensing is a cornerstone of geospatial analysis, providing invaluable insights into our environment from a distance. This chapter explores how Python programming can be leveraged to process, analyze, and interpret remote sensing data, enabling readers to extract meaningful information from satellite imagery and other sources.
4. **Towards Spatial Data Visualization with Python: Unveiling Geographic Patterns and Trends:** Visualization is a powerful tool for understanding complex geospatial data. This chapter delves into the principles and techniques of spatial data visualization using Python, equipping readers with the skills to unveil geographic patterns and trends hidden within their datasets.
5. **Spatial Data Visualization with Python: Techniques, Tools, and Real-World Applications:** Building upon the foundations laid in the previous chapter, this chapter explores advanced techniques and methodologies for spatial data visualization using Python. From static maps to dynamic visualizations, readers gain insights into the diverse array of visualization tools and libraries in the Python ecosystem.
6. **Exploring Vector and Raster Data Formats for Geospatial Visualization with Python:** Data formats play a crucial role in geospatial analysis, shaping how we store, access, and manipulate spatial data. This chapter provides readers with a comprehensive overview of vector and raster data formats, equipping them with the knowledge to work with different types of geospatial data effectively.
7. **Geospatial Data Visualization with Folium:** Folium is a powerful Python library for interactive geospatial data visualization. This chapter explores how Folium can be leveraged to create dynamic, interactive maps for web-based applications, enabling readers to harness the full potential of geospatial visualization in their projects.
8. **Applying Geospatial Data to Choose the Best Route During the Road Design Stage:** Transportation infrastructure is critical in shaping the urban

landscape. This chapter explores how geospatial data analysis can be applied to optimize route selection during the road design stage, thereby improving efficiency and safety in transportation planning.

9. **Python Library for Road Network Analysis In the case of Debre Berhan City:** Building upon the principles established in the preceding chapter, this chapter focuses on a specific case study: road network analysis in the city of Debre Berhan. By examining real-world applications of geospatial data analysis, readers gain insights into the practical implications of their newfound knowledge.
10. **Impacts of Climate Changes on Traffic Flows Using Geospatial Data Analysis: Shifting the Traffic Flow to the Next Level:** Climate change poses significant challenges to transportation infrastructure and unpredictable traffic flows and patterns. This chapter explores how geospatial data analysis can be used to understand and mitigate the impacts of climate change on traffic flows, thereby improving resilience and adaptability in transportation planning.
11. **Deep Learning Approach to Estimate the Maize Yield Prediction using Data from Cameroon: Shifting the Maize Yield Production to the Next Level:** Deep learning represents a cutting-edge approach to geospatial analysis, offering unparalleled insights into complex phenomena. This chapter explores how deep learning techniques can be applied to predict maize yield using data from Cameroon, highlighting the transformative potential of advanced analytics in agriculture.

In conclusion, *Geospatial Application Development Using Python Programming* represents a seminal contribution to geospatial technology and data science. By providing readers with a comprehensive overview of the principles, techniques, and applications of geospatial analysis using Python programming, this book equips them to confront the challenges and capitalize on the opportunities of the digital age. As the boundaries between geography, data science, and computer programming continue to blur, the insights and resources offered within these pages will serve as an indispensable guide for navigating the complexities of the modern world.

Let this book serve as a compass as individuals embark on their journey into the fascinating world of geospatial application development using Python programming.

Mohammad Gouse Galety
Samarkand International University of Technology, Uzbekistan

Arul Kumar Natarajan
Samarkand International University of Technology, Uzbekistan

Tesfaye Fufa Gedefa
Ethiopian Space Science and Geospatial Institute (SSGI), Ethiopia

Tsegaye Demsis Lemma
Ethiopian Space Science and Geospatial Institute (SSGI), Ethiopia

Chapter 1
Introduction to Geospatial Data and Python Programming

D. Shanmugapriyaa
Coimbatore Institute of Technology, India

V. Sachin Kumar
Coimbatore Institute of Technology, India

M. Sujithra
Coimbatore Institute of Technology, India

Ram Sanjai
Coimbatore Institute of Technology, India

B. Senthilkumar
https://orcid.org/0000-0001-8117-3314
Kumaraguru College of Technology, India

ABSTRACT

Utilising geographical linkages, predictive modelling, and problem-solving techniques, geospatial analysis is an essential discipline in many businesses. This chapter explores the deep meaning of geographical data and how machine learning can benefit from it. Geospatial data, derived from sources such as GPS devices and satellite pictures, serves as the basis for comprehending infrastructure, topography, weather, and population dynamics. Crowdsourced information and open data projects enhance the data set available for research. By addressing issues like dimensionality reduction and missing data, spatial data preparation ensures the quality and diversity of data sources. Geospatial analysis is improved for a variety of applications using machine learning algorithms, which include supervised, unsupervised, and reinforcement learning techniques. Temporal dynamics, as examined by methods such as ARIMA and LSTM networks, track variations within regions. Ethical considerations, efficient data visualization, and data fusion techniques all contribute to thorough.

DOI: 10.4018/979-8-3693-1754-9.ch001

GETTING STARTED WITH GEOSPATIAL DATA AND PYTHON

The advent of the digital age has ushered in an era where vast amounts of data are at our fingertips, providing unprecedented opportunities to unravel intricate patterns within our surroundings. In this data-rich landscape, geospatial data emerges as a cornerstone, offering a unique lens through which we can understand and interpret the world. This introduction sets the stage for our exploration into the realms of geospatial data analysis, focusing on its definition, characteristics, and the role of Python as a powerful tool in this dynamic field. In today's era of data-driven decision making, integrating geospatial data analysis into powerful programming languages is essential. This article aims to provide a comprehensive understanding of the interaction between Python and geospatial data, illuminating the reasons for Python's prominence, the nuances, and practical applications of geospatial data through real case studies.

Understanding Geospatial Data

Geospatial data refers to information that holds a geographic or spatial component, linking it to specific locations on Earth. It encapsulates a diverse array of data types, ranging from the simple coordinates of a point to complex spatial patterns and relationships. The unique characteristic of geospatial data lies in its ability to provide context to information, adding a spatial dimension that is crucial for understanding relationships, distributions, and patterns. Geospatial data contains different types of data that are categorized into different types. Vector data, which includes points, lines, and polygons, is ideal for representing discrete objects such as roads and administrative boundaries. Raster data, represented as a cell grid, is suitable for continuous phenomena such as elevation and satellite images. This section provides a foundation and guides readers through the basic structures that define geospatial data.

In exploring the sources of spatial data, this text examines satellite imagery, GPS data, and remote. sensing technologies. Understanding the provenance of data allows researchers to assess its reliability and accuracy. In addition, we highlight the various applications of geospatial information, from urban planning and disaster management to agriculture and epidemiology. This research lays the foundation for identifying the broad impact of geospatial information in decision-making processes.

Why Python for Geospatial Analysis

Python is a high-level, interpreted programming language known for its simplicity, readability, and versatility. Created by Guido van Rossum and first released in 1991,

Python has grown to become one of the most popular programming languages globally. Its design philosophy emphasizes code readability and ease of use, making it an excellent choice for beginners and experienced

Figure 1. Logo of Python language

developers alike. Python is chosen for geospatial data analysis due to its simplicity, versatility, and powerful libraries. Its clear syntax makes it beginner-friendly, while libraries like GDAL/OGR, NumPy, and Matplotlib offer robust tools for reading, processing, and visualizing geospatial data. Python's active community ensures continuous support and improvement. Its integration with GIS tools, cross-platform compatibility, and role in data science contribute to its popularity, making Python an ideal language for both beginners and professionals in geospatial analysis.

- ***Clear and Readable Syntax:*** Python's syntax is designed to be clear and readable, resembling plain English. This readability not only makes it easier for developers to write code but also enhances collaboration and code maintenance.
- ***Interpreted and Interactive***: Python is an interpreted language, allowing developers to run code line by line, making it interactive and facilitating rapid development and debugging. This feature contributes to Python's popularity in various fields, from web development to data science.
- ***Extensive Standard Library***: Python comes with a comprehensive standard library that includes modules and packages for various tasks. This extensive library simplifies many common programming tasks, providing developers with pre-built functionality to save time and effort.

- ***Community Support:*** Python has a large and active community of developers, contributing to an abundance of resources, documentation, and third-party libraries. The supportive community is a key factor in Python's continuous growth and adaptation to emerging technologies.
- ***Cross-Platform Compatibility:*** Python is a cross-platform language, meaning that code written in Python can run on different operating systems without modification. This portability enhances the flexibility of Python applications and facilitates collaboration across diverse environments.
- ***Versatility in Application:*** Python is a general-purpose language that can be used for a wide range of applications. From web development (Django, Flask) to data analysis (Pandas, NumPy), machine learning (TensorFlow, PyTorch), and scientific computing, Python has become a go-to language in various domains.
- ***Dynamically Typed and Object-Oriented:*** Python is dynamically typed, allowing for flexibility in variable assignment. It also supports object-oriented programming principles, enabling developers to create reusable and modular code with classes and objects.
- ***Rapid Development:*** The simplicity and readability of Python, combined with its extensive libraries, contribute to rapid development. Python's focus on code simplicity and expressiveness makes it an ideal choice for both small scripts and large-scale projects.

EXPLORING GEOSPATIAL DATA TYPES AND SOURCES

In the vast landscape of geospatial data, the intricacies of Earth's features are captured through various data types, each offering a unique perspective on our surroundings. Among these, Vector Data and Raster Data stand as pillars, shaping the foundation of geospatial analysis and interpretation.

Types Of Geospatial Data

Geospatial data, often referred to as spatial data, brings a unique perspective to information by incorporating spatial context, allowing us to analyse and visualize data within its geographical framework. This dimension is fundamental in addressing questions related to location, proximity, and spatial relationships, providing a holistic understanding of our surroundings.

Vector Data

Vector data, a cornerstone in geospatial analysis, represents the world through points, lines, and polygons. This type of data excels in capturing discrete features with precision, making it indispensable for representing structures, boundaries, and intricate urban layouts. As we explore vector data, we uncover its applications in urban planning, cadastral mapping, and network analysis.

Raster Data

Figure 2. Classification of spatial datatypes

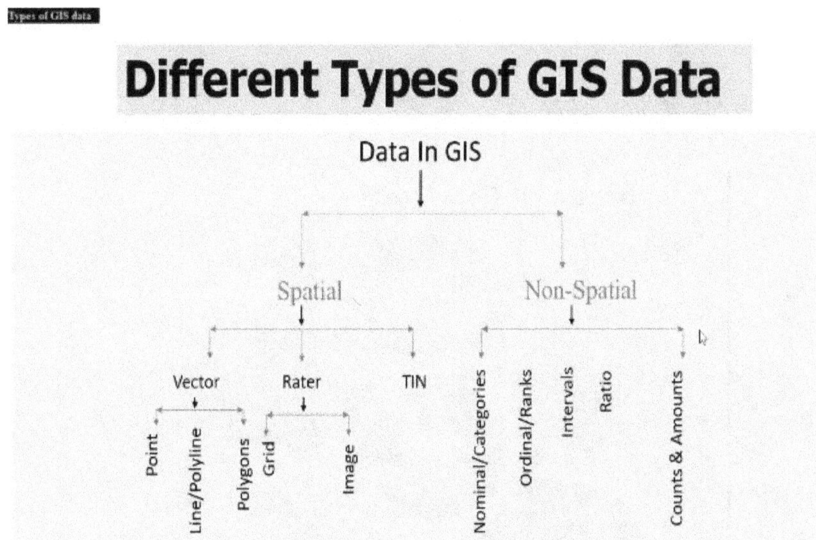

In contrast, raster data organizes information into a grid of cells, each holding a value representing a specific attribute. This format is well-suited for continuous phenomena, enabling the detailed representation of satellite imagery, elevation models, and land cover information. Raster data finds applications in remote sensing, terrain analysis, and climate modelling, offering insights into broad and continuous geographic patterns.

Data Sources

A large collection of data, that directly or indirectly references a specific geographical area or location, collected in different ways from many sources is called geospatial data. Data such as census, satellite images, weather, mobile phones, photographs, social media, and others can all be included as sources. Drawing features on a map is enabled by accurate information found in geospatial data such as latitude and longitude. It can be used in GIS data analysis and mapping.

1. Satellite Imagery:

Satellite imagery is a primary source of geospatial data, capturing high-resolution images of the Earth's surface from orbiting satellites. These images offer valuable insights into land cover, changes over time, and can be used for applications such as environmental monitoring, agriculture, and urban planning. Organizations like NASA and commercial satellite providers contribute to the wealth of satellite imagery available.

Figure 3. Sources of Spatial Data

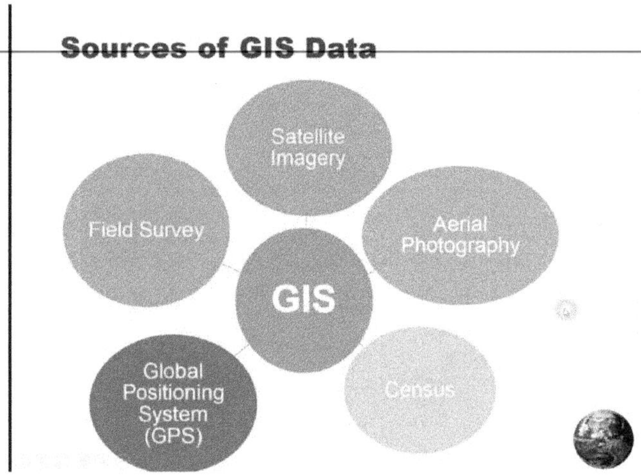

2. Aerial Photography:

Aerial photography involves capturing images from aircraft flying at various altitudes. These high-resolution images provide detailed views of the Earth's surface, contributing to applications like urban planning, cartography, and disaster response. Aerial surveys are often conducted for specific projects or areas of interest.

3. Global Positioning System (GPS):

GPS is a fundamental technology for collecting precise location data. By triangulating signals from satellites, GPS receivers determine accurate geographic coordinates. This data source is widely used in navigation, mapping, and location-based services. GPS data is generated by devices ranging from smartphones to specialized GPS receivers.

4. OpenStreetMap (OSM):

OpenStreetMap is a collaborative mapping project where volunteers contribute geospatial data to create a free, editable map of the world. OSM includes information about roads, buildings, land use, and more. The data is accessible for various applications and is particularly valuable in regions where authoritative mapping data may be limited.

5. Sensor Networks:

Sensor networks, such as weather stations, environmental monitoring stations, and IoT devices, generate real-time geospatial data. These sources contribute to weather forecasting, climate studies, and the monitoring of air and water quality. Sensor data is crucial for understanding dynamic environmental conditions.

6. Geospatial Databases:

Geospatial databases house a variety of structured geospatial data. These databases can be managed by organizations, research institutions, or governments and may include information about infrastructure, demographics, and natural resources. They serve as centralized repositories for spatial data.

Common Formats

Geospatial data can be stored in many formats. Choosing right format for geospatial data has significant impact in geospatial data processing. Selecting the right format make sures that data can be read, shared, analysed accurately and efficiently across various platforms.

Geojson

It supports various geometric types like Point, Line String, Polygon, MultiPoint, Multiline String, and Mult Polygon. it also allows feature collections, which is a combination of these geometry types along with its associated properties. Many popular mapping libraries and platforms like Leaflet, map box and open Layers supports Godson making it easy to integrate with various mapping tools. To save a Godson file with the file extension ".godson" in Python. It is a simple and effective way to represent complex geographic information systems.

Shapefile

An Esri vector data storage format called a shapefile is used to store the coordinates, characteristics, and form of geographical features. It has a single feature class and is kept as a collection of connected files. Shapefile is primarily made up of multiple files. There are three primary files, a .dB file for attribute data, .shx file for indexing, and .she file containing the geographic data.Numerous GIS platforms and applications use Shapefiles.

A shapefile can only have two gigabytes of storage for each of its component files. As a result, the only files that are likely to be very large are.dbf and.shp files, which have a limit of 2 GB each.

KML

Keyhole Markup Language (KML) is an XML notation is used to express geographic annotation and visualization in two-dimension and three-dimensional Earth browsers and maps. With the help of KML files, users may add unique map overlays to Google Earth that include a variety of features like points, lines, and polygons. This makes it possible to see the data in a more individualized and detailed way.

Applications And Significance of Geospatial Data

Geospatial data plays a pivotal role in a multitude of applications across various industries, shaping decision-making processes, improving resource management, and enhancing our understanding of the world. The significance of geospatial data lies in its ability to provide spatial context to information, enabling informed actions based on geographic relationships. Here, we explore some key applications and the broader significance of geospatial data:

1. Urban Planning and Development:

Application: Geospatial data assists urban planners in designing sustainable and efficient cities. It helps analyse infrastructure needs, plan transportation networks, allocate land use, and optimize urban spaces.

Significance: By understanding the spatial relationships between buildings, roads, and public spaces, urban planners can create cities that are more resilient, environmentally friendly, and conducive to a high quality of life.

Figure 4. Applications of spatial data

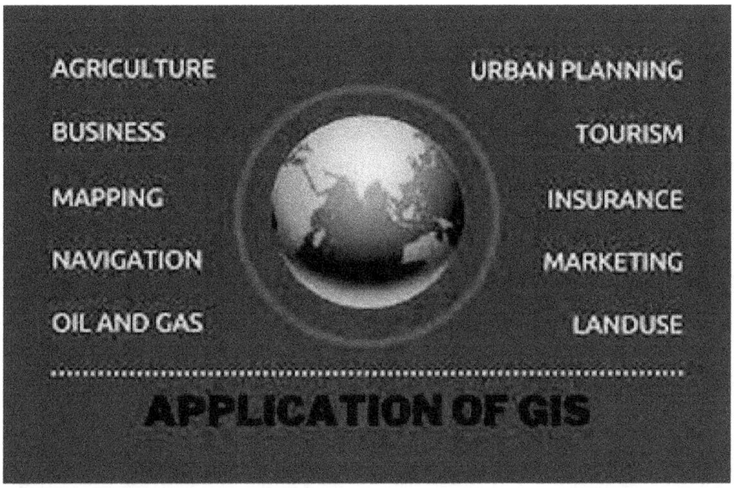

2. Environmental Monitoring and Conservation:

Application: Geospatial data is essential for monitoring environmental changes, tracking deforestation, analysing land cover, and assessing the impact of climate change on ecosystems

Significance: It enables scientists and conservationists to identify areas at risk, implement conservation measures, and track the effectiveness of environmental protection initiatives.

3. Disaster Management and Response:

Application: Geospatial data aids in disaster risk assessment, early warning systems, and post-disaster response by mapping affected areas, assessing damage, and planning rescue operations.

Significance: Timely access to accurate spatial information helps emergency responders and decision-makers allocate resources effectively, minimizing the impact of natural disasters on communities.

4. Precision Agriculture:

Application: Geospatial data is used in precision agriculture for optimizing crop management, monitoring soil health, and implementing efficient irrigation and fertilization practices.

Significance: Farmers can make data-driven decisions, enhancing crop yields, reducing resource use, and minimizing environmental impact through targeted and efficient farming practices.

5. Navigation and Location-Based Services:

Application: Geospatial data is the backbone of navigation systems and location-based services, facilitating accurate mapping, route planning, and real-time location information.

Significance: It enables seamless navigation for individuals, logistics optimization for businesses, and the development of location-aware applications, transforming the way people and goods move.

6. Health Planning and Epidemiology:

Application: Geospatial data is used to analyse the spread of diseases, identify high-risk areas, and plan healthcare resources. It plays a crucial role in epidemiological studies.

Significance: By understanding the spatial distribution of health-related data, public health officials can respond more effectively to disease outbreaks, allocate resources strategically, and implement targeted interventions.

BACKGROUND STUDY

In the realm of geospatial analysis, a comprehensive background study serves as the bedrock for understanding the complexities and nuances of working with spatial data. This section aims to provide an overview of key concepts, methodologies, and challenges associated with geospatial data analysis.

Data Preprocessing Techniques

Data preprocessing is a critical phase in the geospatial analysis pipeline, involving various techniques to enhance the quality and usability of spatial data.

Cleaning, Transforming, Normalizing, And Handling Missing Values

Cleaning: Involves identifying and rectifying errors, outliers, and inconsistencies in geospatial datasets, ensuring data integrity and reliability.

Transforming: Converts raw data into a suitable format for analysis, often involving coordinate system transformations or data aggregation.

Normalizing: Adjusts data scales to a common reference, facilitating fair comparisons and preventing biases in spatial analyses.

Handling Missing Values: Addresses the challenge of incomplete data by employing methods like imputation or interpolation to estimate missing values based on existing information.

Preparing Geographic Data For Analysis

Geographic Data Formatting: Focuses on organizing and structuring geospatial data, ensuring it adheres to specific standards and formats such as Godson, Shapefile, or KML.

Geocoding: Involves converting location descriptions into geographic coordinates, allowing for spatial analysis based on textual information.

Topological Data Structuring: Maintains spatial relationships and connectivity information, facilitating accurate analyses that rely on geometric and topological relationships.

Addressing Typical Problems In Geospatial Analysis

Scale and Resolution Issues: Considers challenges related to the scale and resolution of geospatial data, ensuring that the chosen level of detail aligns with the objectives of the analysis.

Data Integration Challenges: Involves addressing issues related to integrating diverse datasets from different sources, ensuring consistency and coherence in the combined spatial information.

Temporal Aspects: Incorporates temporal considerations in geospatial analysis, recognizing the dynamic nature of many spatial phenomena over time.

Spatial Analysis Methods And Algorithms

Spatial analysis is the core of deriving meaningful insights from geospatial data, involving various methods and algorithms tailored to different analytical objectives.

Point Pattern Analysis: Spatial Point Pattern Analysis: Examines the distribution and clustering of points in geographic space, providing insights into the underlying processes or phenomena.

Density-Based Approaches: Identify areas of high or low point concentrations, helping to uncover spatial patterns or anomalies.

Spatial Autocorrelation: Global and Local Spatial Autocorrelation: Assesses the degree of similarity or dissimilarity between spatial units, identifying clusters or spatial trends in the data.

Moran's I and Geary's C Indices: Quantify spatial autocorrelation, offering statistical measures to evaluate the spatial distribution of attributes.

Geospatial Modelling: Spatial Regression: Incorporates spatial dependencies into regression models, allowing for a more realistic representation of spatial relationships.

Agent-Based Modelling: Simulates individual agents in a geographic environment, providing a dynamic framework for understanding spatial phenomena.

BASICS OF PYTHON PROGRAMMING FOR GEOSPATIAL ANALYSIS

Understanding Python's role in geospatial analysis requires a grasp of fundamental programming concepts. This section unfolds across two key dimensions: an exploration of Python's syntax and an introduction to essential data structures.

Overview Of Python Programming

Python's strength lies in its syntax, defining the language's signature style. In this section, we delve into the fundamental aspects of Python's syntax, covering variable assignments, control structures, and essential programming constructs. This understanding of basic syntax forms the cornerstone of effective geospatial programming, ensuring clarity and precision in the manipulation of spatial data.

Basic Syntax

Python's syntax is designed for readability and simplicity. Understanding the basic syntax is crucial for writing clear and effective code in geospatial analysis and other programming tasks.

Variables and Data Types:

```
name = "john"
age = 25
```

```
height = 1.75
is_adult = true
```

In Python, variables can be created without explicitly declaring their data type. The interpreter dynamically assigns data types based on the assigned value.

Indentation and Code Blocks:

```
if age > 18:
print(name + " is an adult.")
else:
print(name + " is a minor.")
```

Indentation is a fundamental aspect of Python syntax. Instead of using braces { } or keywords like end to denote code blocks, Python uses indentation.

Conditional Statements:

```
if age > 21:
    print(name + " can legally drink.")
elif age == 21:
    print(name + " just reached the legal drinking age.")
else:
    print(name + " is underage.")
```

Python uses if, elif (else if), and else for conditional statements. The colon (:) indicates the start of a block, and indentation determines the scope.

Loops:

```
# Functions
def greet_person(name):
    print("Hello, " + name + "!")
# Call the function
greet_person("Alice")
```

Python supports for and while loops. Again, indentation is crucial for defining the loop body.

Functions:

```
coordinates = [4.5, 3.2]
city_names = ["New York", "Paris", "Tokyo"]
mixed_data = [1, "apple", True, 3.14]
```

Functions are defined using the def keyword, and parameters are specified in parentheses. The function body is indented.

Data Structures

Lists:

```
# Tuples
point = (4.5, 3.2)
color_rgb = (255, 0, 0)
```

A list is a versatile and mutable data structure that can hold elements of different data types.

Tuples:

```
# Dictionaries
attributes = {"name": "City Park", "type": "Public"}
```

Tuples are like lists but are immutable, meaning their elements cannot be changed after creation.

Sets:

```
# Strings
name = "Alice"
greeting = "Hello, " + name + "!"
```

Sets are unordered collections of unique elements, useful for tasks like eliminating duplicate values.

Dictionaries:

```
import geopandas as gpd
# GeoDataFrame
gdf = gpd.read_file("path/to/shapefile.shp")
# Accessing geometry column
geometry_series = gdf.geometry
```

Dictionaries are collections of key-value pairs, providing a way to associate data with descriptive labels.

Strings:

```
from osgeo import gdal
# Open a raster dataset
dataset = gdal.Open("path/to/raster.tif")
# Access raster properties
print("Raster Size:", dataset.RasterXSize, "x", dataset.
RasterYSize)
```

While not a traditional data structure, strings are fundamental for handling text data, and they support various operations.

These data structures are building blocks for more complex data manipulation in geospatial analysis. For instance, in GeoPandas, the GeoDataFrame is a specialized data structure that combines the tabular structure of a Pandas DataFrame with spatial information represented by a geometry column.

Key Libraries and Modules for Geospatial Analysis in Python

In the vast landscape of geospatial analysis, Python is empowered by a suite of key libraries and modules that elevate its capabilities. This section unravels the essence of these tools, offering insights into their functionalities and applications within geospatial data analysis.

Introduction to GDAL Library

Geospatial Data Abstraction Library (GDAL) is a free and open-source library for working with raster and vector geospatial data. It provides a unified interface for reading and writing geospatial data in a variety of formats, as well as performing various geospatial operations such as reprojection, resampling, and format conversion.

Figure 5. GDAL

- ***Data format support***: GDAL supports a wide range of geospatial data formats, including popular ones such as GeoTIFF, Shapefile, JPEG, and PNG.
- ***Data transformation***: GDAL allows users to perform various geospatial data transformations, such as reprojection, resampling, and format conversion. This is essential for ensuring data compatibility and consistency in geospatial analysis.
- ***Data access***: GDAL provides functions for reading and writing geospatial data, making it a valuable tool for data extraction, manipulation, and export. This is especially useful for working with remote sensing data, satellite imagery, and more.
- ***Geometric operations***: GDAL provides functionality for performing geometric operations on geospatial data, such as clipping and mosaicking. This allows users to extract specific regions of interest from datasets or combine multiple datasets into one.

Understanding Geopandas Functionality

Figure 6. Geopandas

GeoPandas

GeoPandas is a Python library that extends the Pandas Data Frame to provide support for geospatial data. It provides a few features for working with geospatial data, including:

- ***Data structures***: GeoPandas introduces two new data structures, GeoSeries and Geodata Frame, which are extensions of pandas Series and Data Frame. Geometries and GeoDataFrame objects can store geometric data, such as points, lines, and polygons, alongside attribute data.
- ***Data I/O***: GeoPandas provides convenient methods for reading and writing geospatial data in a variety of formats, including Shapefile, GeoJSON, and GeoPackage.
- ***Data exploration and analysis***: GeoPandas provides methods for exploring and analyzing geospatial data, such as head(), tail(), sample(), and plot().

It also supports a wide range of geometric operations, such as buffering, dissolving, intersection, union, and difference.

- ***Spatial indexing*:** GeoPandas can leverage spatial indexing to improve the performance of spatial queries on large datasets.
- ***Spatial joins*:** GeoPandas supports spatial joins between GeoDataFrames, which are useful for aggregating attribute data from one dataset to another.

GeoPandas is a powerful tool for geospatial data analysis in Python. It is used by a wide range of users, including GIS professionals, researchers, and developers.

Applications of GeoPandas

GeoPandas is used in a wide range of applications, including:

- ***Geographic Information Systems (GIS)*:** GeoPandas is a popular tool for developing and using GIS applications in Python.
- ***Geospatial research*:** GeoPandas is used by researchers in a variety of fields, such as geography, urban planning, and environmental science.
- ***Web mapping*:** GeoPandas can be used to develop interactive web maps and other geospatial visualization applications.
- ***Business intelligence*:** GeoPandas can be used to analyse geospatial data for business intelligence purposes, such as identifying market trends and assessing risk.

Exploring Fiona's Capabilities

Fiona, a geospatial data processing library, specializes in the seamless transfer of data to and from GIS formats such as GeoPackage and Shapefile. It simplifies real-world data handling with support for diverse GIS formats, compressed and in-memory virtual file systems, and storage locations including hard drives and cloud storage. Fiona, comprising Python modules and a command line interface, prioritizes code readability and productivity.

The library depends on GDAL but differs in its approach. Fiona's inclusion of plugins linking to libgdal adds installation complexity, especially when compared to binary distributions available on the Python Package Index. Users seeking more optional format drivers may find Anaconda and conda-forge to be advantageous installation methods.

Installation from source is possible using pip with the --no-binary option, allowing users to specify GDAL settings through the GDAL_CONFIG environment variable. Fiona 2.0 requires Python 3.7 or later and GDAL 3.2 or later. (Fiorini et al., 2016)

Some unique features of Fiona:

- ***Specialization in GIS Formats***: Fiona excels in handling GIS formats, specifically GeoPackage and Shapefile. Its specialization makes it a go-to tool for tasks related to geospatial data interchange.
- ***Clean and Readable Code***: Fiona is designed with a focus on simplicity and readability, making it accessible to users with varying levels of expertise. Its clean code structure enhances the overall user experience.
- ***Pythonic Interface and Integration***: With a Pythonic interface, Fiona seamlessly integrates into Python workflows. This integration extends to working well with GeoPandas, providing a cohesive environment for geospatial data analysis within the Python ecosystem.
- ***Flexibility in Installation***: Fiona offers flexibility in installation methods, allowing users to choose between binary distributions for ease of installation or opt for source installation for more control over settings and compatibility.
- ***Real-World Data Handling***: Fiona stands out in its ability to handle real-world data effectively. It supports multi-layered GIS formats, operates with compressed and in-memory virtual file systems, and accommodates data stored in various locations, including hard drives and cloud storage.

BASIC GEOSPATIAL OPERATIONS IN PYTHON

Geospatial operations in Python provide essential tools for spatial data manipulation and analysis. This section explores fundamental techniques, including coordinate transformations, spatial indexing, geocoding, and raster manipulation. By understanding these basic geospatial operations, practitioners can harness Python's capabilities for efficient geographical data processing and visualization. The discussion encompasses practical applications and code examples, showcasing the versatility of Python in handling spatial data.

Coordinate Reference Systems (CRS)

Figure 7. (Coordinate Reference Systems)

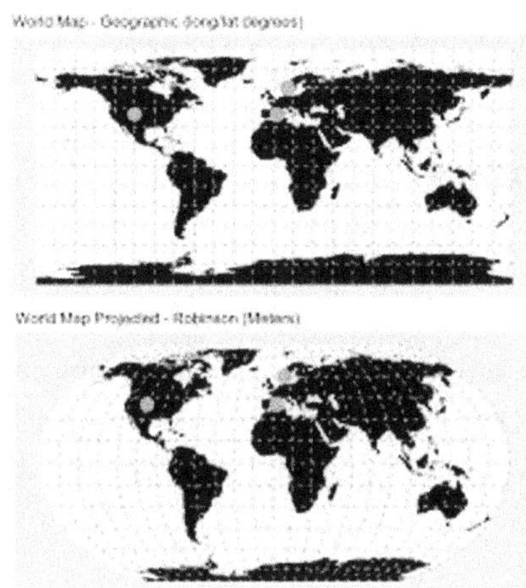

Coordinate Reference Systems (CRS) are essential for accurately representing and analyzing geospatial data on the Earth's surface.Explanation of how CRS defines the coordinate system used to represent geographic features, including latitude, longitude, and elevation.Importance of CRS in ensuring compatibility and consistency when working with geospatial datasets.

Common CRS Systems:

- Overview of commonly used CRS systems such as WGS84 (EPSG:4326), Universal Transverse Mercator (UTM), and State Plane Coordinate System (SPCS).
- Discussion of the characteristics and applications of each CRS system, including their advantages and limitations.

CRS Transformation:

- Explanation of CRS transformation, which involves converting coordinates from one CRS to another.
- Introduction to Python libraries like pyproj or geopandas for performing CRS transformations.
- Practical examples demonstrating how to transform coordinates between different CRS to ensure spatial data alignment.

Handling CRS in Geospatial Data:

- Guidance on how to assign, change, and check CRS information for geospatial datasets using Python.
- Explanation of methods for setting CRS metadata within geospatial data files (e.g., Shapefiles, GeoJSON) and GeoDataFrames.

Geometric Operations

Point, Line, and Polygon Geometries:

- Definition and illustration of basic geometric shapes used in geospatial data representation, including points, lines, and polygons.
- Explanation of how these geometries is defined by their coordinates and topology.

Figure 8. (Geometric Operations)

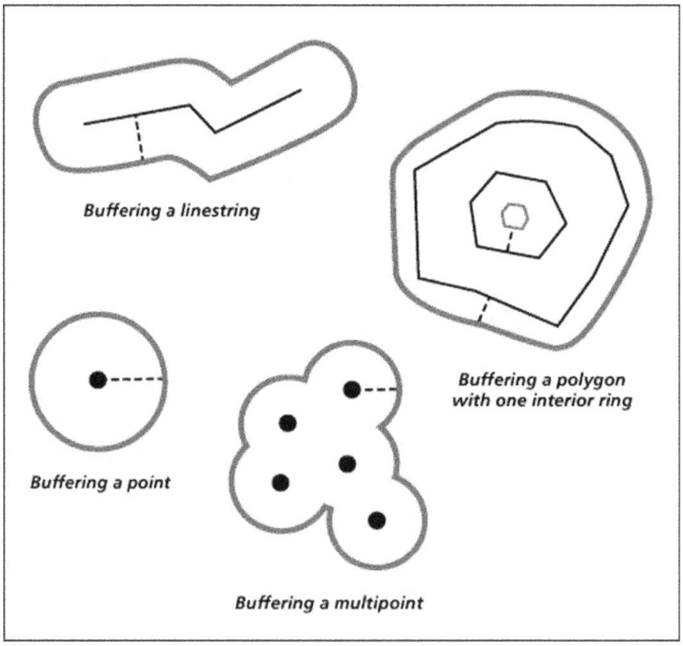

Geometric Operations with Shapely:

- Introduction to the Shapely library, which provides efficient and flexible tools for performing geometric operations in Python.
- Overview of common geometric operations supported by Shapely, such as buffering, intersection, union, difference, and simplification.
- Practical examples demonstrating how to use Shapely to manipulate and analyse geometric objects. (Huang et al., 2010)

Geometric Manipulations:

- Explanation of various geometric manipulations that can be performed using Shapely, including centroid calculation, convex hull construction, and boundary determination.
- Demonstration of how these manipulations can be applied to solve real-world geospatial problems.

Visualization of Geometric Operations:

- Guidance on visualizing geometric operations using Matplotlib or other plotting libraries.
- Examples of creating visualizations to illustrate the results of geometric operations, aiding in data interpretation and analysis.

Attribute Joins And Spatial Queries

Joining Geospatial Data:

- Explanation of attribute joins, which involve combining geospatial datasets based on common attribute values.
- Overview of different types of joins, including inner joins, outer joins, and left/right joins.
- Practical examples demonstrating how to perform attribute joins using Python libraries like GeoPandas.

Spatial Queries:

- Introduction to spatial queries, which involve retrieving geospatial data based on their spatial relationships.
- Explanation of common spatial query operations, such as finding points within polygons, identifying intersecting geometries, and determining nearest neighbours.

- Illustrative examples showcasing how spatial queries can be used to extract meaningful insights from geospatial datasets.

Performing Spatial Queries in Python:

- Step-by-step instructions on executing spatial queries using Python libraries like GeoPandas or SQLite/Spatialite.
- Discussion of the syntax and methods for performing spatial queries, including spatial indexing for improved query performance.

Optimizing Spatial Queries:

- Tips and best practices for optimizing spatial queries to enhance efficiency and reduce computational overhead.
- Guidance on implementing spatial indexing techniques to accelerate query processing and improve overall performance.
- Examples demonstrating the impact of optimization strategies on query execution time and resource utilization.

GEOSPATIAL DATA VISUALIZATION IN PYTHON

Geospatial data visualization is a crucial aspect of understanding and interpreting spatial patterns. Python, with its rich ecosystem of libraries, offers powerful tools for creating compelling and informative visualizations of geospatial data. In this section, we will explore various techniques and libraries for geospatial data visualization in Python.

Python Visualization For Geospatial Data

Geospatial data, also known as spatial data, refers to information that contains a geographic component, typically represented by latitude and longitude coordinates. This type of data has a wide range of applications in various fields, including urban planning, environmental research, agriculture, transportation, and more. Visualizing geospatial data effectively allows for insights into spatial patterns, trends, and relationships.

Common Types Of Geospatial Data Visualizations

There are many different types of geospatial data visualizations, each with its own strengths and weaknesses. Some of the most common types of geospatial data visualizations include:

- ***Choropleth maps:*** Choropleth maps use colour variations to represent data associated with different geographic regions. They are often used to show population density, income distribution, or other quantitative data.
- ***Point symbol maps:*** Point symbol maps use symbols to represent individual data points. They are often used to show the locations of businesses, crime incidents, or other point-based data.
- ***Line maps:*** Line maps use lines to connect data points. They are often used to show transportation routes, migration patterns, or other linear data.
- ***Heatmaps:*** Heatmaps use colour gradients to represent the intensity of data over an area. They are often used to show environmental data, such as air pollution levels or temperature distribution.

Benefits Of Using Python For Geospatial Data Visualization

There are several benefits to using Python for geospatial data visualization:

- ***Powerful and versatile language:*** Python is a powerful and versatile programming language that can be used to create a wide variety of visualizations, from simple maps to complex animations.
- ***Rich ecosystem of libraries:*** Python has a large and active community of developers who have created a wide variety of libraries for geospatial data visualization. This means that there are a variety of tools available to choose from, depending on your specific needs.
- ***Maturity and stability:*** Python is a mature and stable language that is well-supported by the developer community. This means that you can be confident that the libraries you use will be reliable and up-to-date.

Matplotlib For Geospatial Data Visualization

Matplotlib, a versatile plotting library for Python, can be employed to create a variety of visualizations, including maps. Despite not being specifically designed for geospatial data, Matplotlib's adaptability and flexibility make it a popular choice for generating basic maps and simple geospatial visualizations.

```
import geopandas as gpd
import matplotlib.pyplot as plt
data = gpd.read_file('data.csv')
fig, ax = plt.subplots()
data.plot(ax=ax)
ax.set_title('Geospatial Data Map')
ax.set_xlabel('Longitude')
ax.set_ylabel('Latitude')
plt.show()
```

- ***Plotting points, lines, and polygons:*** Matplotlib can plot individual data points, connecting lines, and polygons representing geographic areas. This enables the visualization of various geospatial data types, including point distribution, road networks, and administrative boundaries.
- ***Customizing map appearance:*** Matplotlib offers a range of options for customizing the appearance of maps, including colour palettes, symbol styles, and line properties. This allows users to create visually appealing and informative maps tailored to their specific needs.
- ***Integration with Geoplanids:*** Matplotlib can be used in conjunction with Geoplanids, a Python library that extends Pandas to support geospatial data. This integration allows for seamless manipulation and visualization of geospatial data within the Matplotlib framework.

Enhancing Data Representation With Matplotlib

Matplotlib, a versatile plotting library for Python, goes beyond basic data visualization and offers a range of features to enhance the representation of data, making it more informative, visually appealing, and tailored to specific needs.

Customizing Plot Appearance

```
import numpy as np
import matplotlib.pyplot as plt
import seaborn as sns
x = np.linspace(0, 10, 100)
y1 = np.sin(x)
y2 = np.cos(x)
colors = sns.color_palette("husl", 2)
plt.figure(figsize=(8, 6))
plt.plot(x, y1, label='Sin Function', color=colors[0],
```

Matplotlib provides a variety of colour palettes to customize the appearance of plots. These palettes range from simple colour gradients to perceptually uniform colour schemes, ensuring that colour variations effectively represent data values.

Output: A histogram with a smooth colour gradient from blue to red, representing the distribution of data values.

Figure 9. (Customizing colour in map)

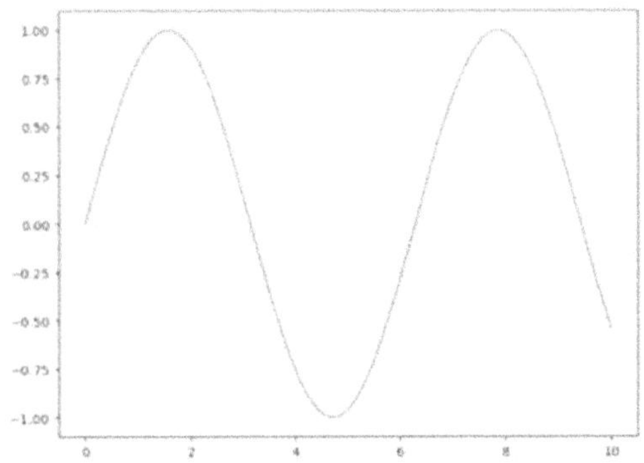

Line Properties

```
import numpy as np
import matplotlib.pyplot as plt
x = np.linspace(0, 10, 100)
y1 = np.sin(x)
y2 = np.cos(x)
plt.figure(figsize=(8, 6))
plt.plot(x, y1, label='Sin Function', linewidth=2,
linestyle='--', color='red')
```

Matplotlib provides various options for customizing line properties, including line width, style, and colour. This allows for differentiating between multiple data series or emphasizing specific trends.

Output: A line plot with two data series, each represented by a distinct line style (solid and dashed) and colour (blue and green).

Figure 10. (Customizing lines in map)

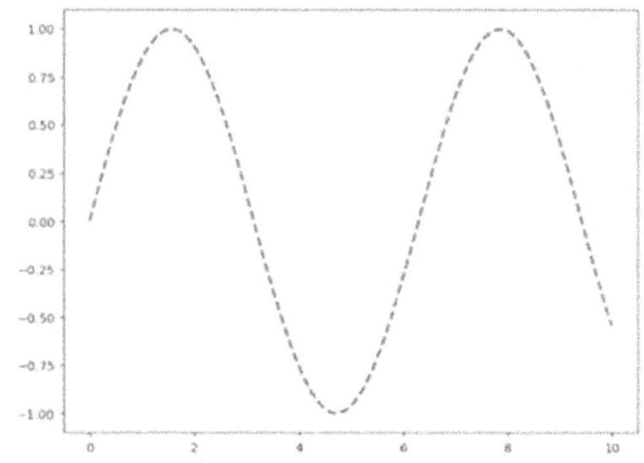

Symbol Styles

```
import numpy as np
import matplotlib.pyplot as plt
x = np.linspace(0, 10, 20)
y = np.sin(x)
plt.figure(figsize=(8, 6))
plt.scatter(x, y, label='Circle Marker', marker='o',
color='blue', s=50)
plt.scatter(x, -y, label='Square Marker', marker='s',
color='green', s=50)
plt.scatter(x, 2*y, label='Triangle Marker', marker='^',
color='orange', s=50))
```

Matplotlib offers a variety of symbol styles for plotting individual data points. These symbols can range from simple markers (circles, squares, triangles) to more intricate shapes, allowing for effective representation of different data types.

Output: A scatter plot with data points represented by red circles (marker='o'), each with a slightly transparent appearance (alpha=0.7).

Figure 11. (Customizing markers in map)

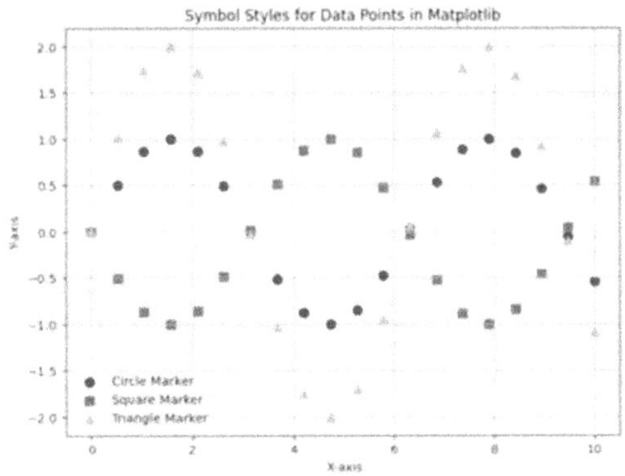

Creating Informative Geospatial Maps

Geospatial maps are a powerful tool for visualizing and understanding spatial data. They are used in a variety of fields, including urban planning, environmental research, agriculture, transportation, and more.

Creating informative geospatial maps requires careful consideration of several factors, including:

- ***Data selection:*** Choose data that is relevant to the purpose of the map and that provides insights into the spatial patterns or relationships of interest.
- ***Map projection:*** Select a map projection that is appropriate for the area being mapped and that minimizes distortions.
- ***Map elements***: Use symbols, lines, and polygons to represent different types of geospatial data. Choose symbols that are clear, easy to distinguish, and well-suited to the data being represented.
- ***Colour schemes:*** Use colour palettes that are perceptually uniform and that effectively communicate the data values.
- ***Labels:*** Add clear and concise labels for geographic features, data values, and legends.
- ***Balance and simplicity:*** Aim for a map that is both informative and easy to understand. Avoid overwhelming the viewer with too much information or complex colour schemes.
- ***Tools and libraries***: Utilize Python's rich ecosystem of geospatial data visualization libraries, such as Matplotlib, Folium, GeoPandas, and Geoplot, to simplify map creation.

By following these guidelines, you can create geospatial maps that effectively communicate insights and inform decisions.

Utilizing Matplotlib For Basic Geospatial Maps

Matplotlib, a versatile plotting library for Python, offers basic geospatial visualization capabilities, enabling the creation of simple maps and straightforward visualizations of geospatial data. While it is not as specialized as dedicated geospatial plotting libraries like Folium or Geoplot, Matplotlib provides a user-friendly and accessible tool for beginners and those who need to incorporate basic maps into their data analysis workflow.

Essential Steps for Creating Basic Geospatial Maps with Matplotlib:

- ***Data Preparation:*** Ensure the geospatial data is in a structured format, such as a CSV file containing latitude and longitude coordinates.
- ***Importing Libraries:*** Import the necessary libraries, including matplotlib. pyplot for plotting and geopandas for handling geospatial data.
- ***Data Loading:*** Load the geospatial data into a Pandas DataFrame or GeoDataFrame using the appropriate read_csv or read_file function.
- ***Map Creation:*** Create a figure and an axis using plt.subplots() and specify the projection type if necessary.
- ***Data Plotting:*** Plot the geospatial data using the appropriate Matplotlib functions, such as plot() for point data or pcolor() for raster data.

Customization: Customize the map appearance using Matplotlib's extensive customization options, including color palettes, symbol styles, line properties, labels, and titles. (Koperski et al., 2019)

Adding Annotations And Context To Maps With Matplotlib

Matplotlib's annotation capabilities extend the visual communication power of geospatial maps, allowing you to add labels, comments, explanations, and other visual elements to enhance the clarity, context, and interpretability of the data.

1. Text Annotations for Highlighting Key Points

Text annotations provide a direct way to add labels, comments, or descriptions to specific locations on the map, drawing attention to important features or patterns. Efficiently analyze geospatial data by summarizing key locations, identifying critical spatial details, and highlighting essential information. This concise approach streamlines decision-making processes, emphasizing key insights in the vast landscape of geospatial information for effective understanding and strategic applications.

Figure 12. (Text Annotations in map)

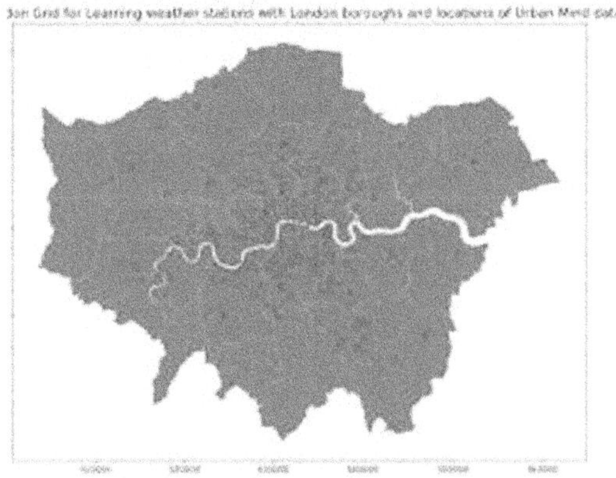

2. Colorbars for Interpreting Color-Mapped Plots

Colorbars are essential for interpreting color-mapped plots, such as heatmaps or choropleth maps, where color variations represent data values. They provide a visual guide to the color scale, enabling viewers to associate color variations with specific data ranges.This code demonstrates adding a colorbar to a choropleth map. The colorbar provides a visual representation of the color scale, enabling viewers to interpret the variations in data values represented by different colors.

Figure 13. (Colorbar in map)

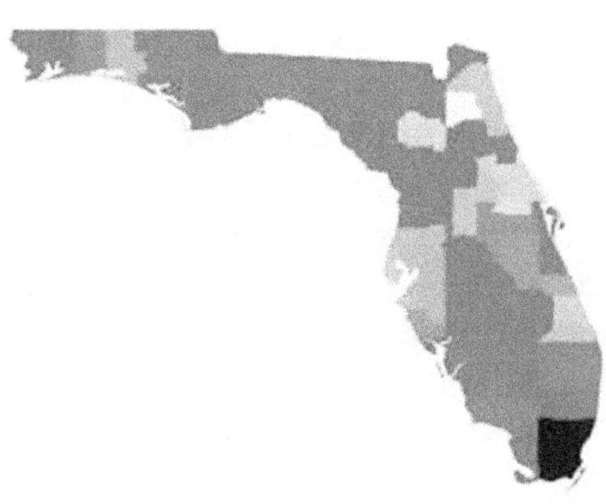

3. Legends for Differentiating Data Series

Legends are crucial for distinguishing between multiple data series in line plots, scatter plots, or other visualizations that combine different data elements. They provide a key to identify the meaning of each line, symbol, or color.

Figure 14. (Legends in map)

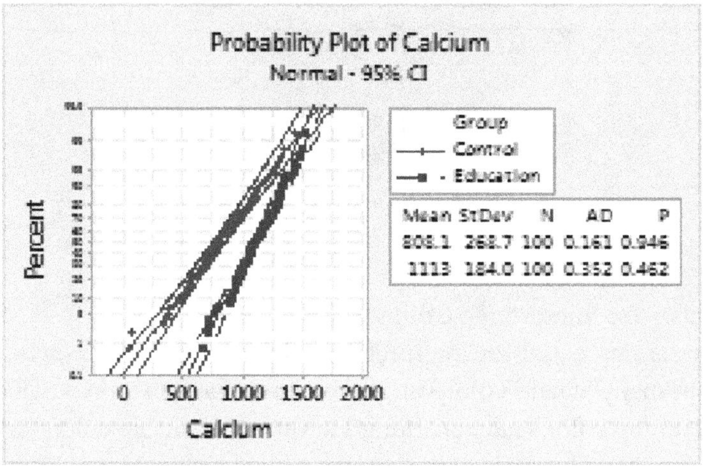

CASE STUDIES AND EXAMPLES

Introducing diverse case studies and examples, this section illustrates real-world applications of the discussed concepts. Through practical scenarios, we demonstrate the effectiveness and versatility of the presented methodologies.

Analyzing Point Of Interest (POI) Data

- Point of Interest (POI) data, encompassing locations like businesses, landmarks, and attractions, serves as vital information for various sectors, from urban planning to marketing.
- Diverse sources, including business directories, social media check-ins, and governmental databases, contribute to the richness of POI datasets.

Data Collection and Preprocessing:

- Explaining the process of acquiring POI data, which involves accessing APIs like Google Places API or OpenStreetMap, or leveraging existing datasets.

- Preprocessing steps encompass data cleaning, including handling missing values, removing duplicates, and standardizing categorical variables for consistency.

Exploratory Data Analysis (EDA):

- Techniques such as histograms, bar charts, and spatial visualizations are utilized to explore POI data attributes like categories, ratings, and geographical distributions.
- Insights gained during EDA aid in understanding the distribution patterns of different types of POIs and identifying potential anomalies or biases.

Spatial Analysis:

- Spatial analysis techniques, including proximity analysis, clustering, and hotspot detection, are applied to reveal spatial patterns and relationships within POI datasets.
- Examples involve identifying clusters of similar businesses, analysing the distribution of amenities in a city, and detecting areas with high tourist activity.

Case Study Example:

- A case study focusing on analysing restaurant POI data in a metropolitan area, exploring factors such as proximity to tourist attractions, competition between cuisines, and correlations with neighbourhood demographics.
- Utilizing spatial analysis techniques to identify optimal locations for new restaurant establishments based on factors like foot traffic, nearby amenities, and competitor density.

Recapitulating key insights derived from the analysis, including trends in POI distribution, correlations between POI attributes, and actionable recommendations for stakeholders. Highlighting the significance of POI analysis in informing strategic decisions in urban planning, business development, and tourism promotion.

Spatial Analysis Of Environmental Data

Environmental data encompasses a wide range of geospatial information, from climate variables and land cover to biodiversity indices and pollution levels. Understanding spatial patterns and trends in environmental data is crucial for ecosystem management, climate modelling, and natural resource conservation.

Data Collection and Preprocessing:

- Discussing methodologies for collecting environmental data through satellite remote sensing, ground-based sensors, and citizen science initiatives.
- Preprocessing steps involve data fusion, quality control, and spatial interpolation to address issues like missing data, sensor errors, and spatial heterogeneity.

Exploratory Data Analysis (EDA):

- Exploratory techniques, including time series analysis, spatial autocorrelation, and correlation matrices, help uncover temporal and spatial patterns in environmental variables.
- Visualizations like heatmaps, scatter plots, and spatial overlays facilitate the identification of trends, anomalies, and potential environmental hotspots.

Spatial Analysis:

- Spatial analysis methods, such as terrain analysis, habitat suitability modelling, and species distribution modelling, enable the assessment of environmental dynamics and ecosystem health.
- Examples include analysing changes in land cover over time, modelling the impact of climate change on species distributions, and identifying areas at risk of environmental degradation.

Case Study Example:

- A case study focusing on analysing climate data to assess the impact of urbanization on local microclimates, investigating phenomena such as urban heat islands and precipitation patterns.
- Utilizing spatial analysis techniques to quantify changes in land cover, evaluate biodiversity loss, and predict future environmental scenarios under different development scenarios.

Summarizing key findings from the analysis, including insights into environmental trends, vulnerabilities, and potential mitigation strategies. Emphasizing the importance of integrating spatial analysis into environmental management practices to promote sustainability, resilience, and ecosystem health.

Geospatial Visualization Of Social Media Data

Social media data offers valuable insights into human behaviour, social interactions, and cultural trends, with geolocation data providing spatial context to user-generated content. Geospatial visualization techniques enable the representation of social media data on maps, facilitating the exploration of spatial patterns and relationships.

Data Collection and Preprocessing:

- Describing methods for collecting geotagged social media data from platforms like Twitter, Instagram, and Flickr, utilizing APIs or third-party tools.
- Preprocessing steps involve geocoding, sentiment analysis, and topic modelling to extract spatially relevant information and identify trends in social media conversations.

Exploratory Data Analysis (EDA)

- Exploratory techniques, including sentiment analysis, network analysis, and temporal analysis, provide insights into user behaviours, community structures, and trending topics.
- Visualization methods like network graphs, word clouds, and sentiment maps aid in the exploration of social media data and the identification of spatial trends.

Geospatial Visualization:

- Geospatial visualization methods, such as point maps, cluster maps, and trajectory maps, enable the visualization of spatial patterns in social media activity.
- Examples include mapping the spatial distribution of tweets during events, identifying clusters of social media activity in urban areas, and visualizing the movement patterns of users over time.

Case Study Example:

- A case study focusing on analysing social media conversations during a major event, such as a music festival or sporting event, to understand attendee experiences and event impacts.
- Utilizing geospatial visualization techniques to map the spatial distribution of social media posts, identify popular event venues, and analyse sentiment trends across different locations.

Recapitulating key insights derived from the geospatial visualization of social media data, including spatial trends, user behaviours, and event dynamics.

Discussing the implications of social media analysis for marketing, event planning, and urban studies, and highlighting opportunities for future research and applications. (Pundt, 2010)

FUTURE TRENDS AND APPLICATIONS IN GEOSPATIAL ANALYSIS

As technology continues to evolve, the field of geospatial analysis is poised for transformative advancements. Emerging trends and developments suggest a future where geospatial data becomes more accessible, analytical techniques become more sophisticated, and applications become increasingly diverse.

Analysis Of Emerging Trends And Developments

The realm of geospatial analysis is undergoing dynamic transformations driven by emerging trends and developments. These advancements are poised to shape the future landscape of spatial data analytics, unlocking new possibilities and applications.

Exploring Cloud-Based Geospatial Data Analysis

In the ever-expanding domain of geospatial analysis, a profound transformation is underway, marked by the ascendancy of cloud-based methodologies. This paradigm shift signifies a fundamental change in the approach to harnessing the power of spatial data.

In this visionary landscape, spatial data transcends the limitations of traditional infrastructure, finding a new abode in the ethereal realms of cloud computing. The cloud emerges as an expansive horizon where scalability takes centre stage. Liberated from the constraints of fixed server capacities, geospatial analyses dynamically adapt to varying workloads and datasets of all sizes, offering unprecedented flexibility.

Collaboration takes on a new dimension in this cloud-driven environment. Geospatial analysts, regardless of their physical locations, converge in a collaborative ecosystem where insights are shared seamlessly. The cloud becomes a global meeting ground, dissolving physical boundaries and making collaboration the cornerstone of geospatial endeavours.

Accessibility becomes omnipresent in this cloud-driven landscape. Geospatial tools and datasets are available anytime, anywhere, liberating analysts from the constraints of physical proximity and fostering a globally distributed approach to spatial analysis.

The tangible benefits of this cloud-based approach are manifold. Operating on a pay-as-you-go model, cloud-based solutions optimize costs and align financial investments precisely with computational needs. The cloud becomes a beacon of cost-efficiency, enabling real-time insights where spatial data synchronously aligns with ongoing events. Swift decision-making and timely responses are the hallmarks of this agile and transformative era in geospatial data analysis, where the cloud unfolds new possibilities and reshapes the landscape of spatial understanding.

Impact Of Artificial Intelligence In Geospatial Analysis

In the evolving landscape of geospatial analysis, a profound influence is emerging through the integration of Artificial Intelligence (AI). This transformative synergy not only enhances analytical capabilities but also opens new avenues for understanding and interpreting spatial data.

Within this canvas of automated insight, AI introduces a paradigm shift by automating processes and infusing a level of intelligence that transcends traditional spatial interpretation boundaries. AI algorithms take centre stage in automated feature extraction, delicately unravelling nuances of land cover classification, object detection, and spatial entity identification without manual intervention.

The impact of AI extends to pattern recognition mastery, providing a deeper understanding of intricate spatial patterns. Analysts are equipped with assistants that not only observe but interpret underlying patterns, revealing insights that may elude the human eye. AI further elevates geospatial analysis to predictive modelling, forecasting changes in land use, environmental conditions, and dynamic spatial phenomena.

The infusion of AI into geospatial analysis represents a profound transformation, illuminating advantages that redefine the landscape of spatial understanding. AI algorithms, particularly neural networks, enhance accuracy in predictions, classifications, and interpretation of complex spatial relationships. This era of AI-infused geospatial analysis prioritizes operational efficiency, automating repetitive tasks and allowing analysts to focus on higher-level interpretation and decision-making. The canvas of spatial analysis is adorned with strategic insights, painted with the precision that reflects the learning and adaptability of AI.

Exploring Potential Applications

In the forward-looking domain of geospatial analysis, the exploration of potential applications unveils a horizon brimming with transformative possibilities. Within this realm, two distinct chapters emerge, each narrating a unique story of how geospatial analysis can shape critical aspects of our world. (Wang et al., 2017)

Geospatial Analysis In Healthcare And Epidemiology

Imagine a realm where geospatial analysis becomes a sentinel, guarding the frontiers of healthcare and epidemiology. In this chapter, the application of spatial insights becomes a crucial instrument in understanding and addressing health-related challenges.

Precision Health Mapping: Geospatial analysis, armed with the power of intricate spatial data, can create precision health maps. These maps do not merely depict geographical landscapes but serve as dynamic canvases illustrating the distribution of diseases, healthcare resources, and the movement of populations. The analyst, equipped with spatial tools, becomes a cartographer of health, charting courses for optimized healthcare delivery.

Epidemiological Surveillance: In the world of epidemiology, geospatial analysis becomes a watchtower. It monitors the spread of diseases, tracks the movement of pathogens, and identifies hotspots with a level of precision that transcends conventional surveillance methods. The spatial analyst is akin to a detective, deciphering the intricate clues embedded in the geography of health.

Healthcare Resource Allocation: Spatial insights guide the allocation of healthcare resources with surgical precision. From identifying optimal locations for healthcare facilities to predicting the demand for medical services in specific regions, geospatial analysis becomes the compass that directs healthcare strategies. The spatial analyst is a strategist, orchestrating the symphony of healthcare resources.

Sustainable Development And Geospatial Data Applications

As the narrative unfolds, another chapter emerges, one that intertwines geospatial analysis with the fabric of sustainable development. Here, spatial insights become the guiding star, illuminating pathways toward a harmonious coexistence with our planet.

Environmental Monitoring: Geospatial analysis becomes the custodian of our environment. It monitors deforestation, tracks changes in land cover, and assesses the health of ecosystems. The spatial analyst transforms into a guardian, preserving the delicate balance between human activities and the natural world.

Urban Planning for Sustainability: In the realm of urban planning, geospatial analysis becomes a tool for crafting sustainable cities. It assesses infrastructure needs, optimizes transportation networks, and envisions green spaces. The spatial analyst takes on the role of an architect, designing urban landscapes that harmonize with the environment.

Climate Change Mitigation: Geospatial analysis becomes a compass in the fight against climate change. It assesses the impact of rising temperatures, monitors sea-level changes, and identifies regions vulnerable to climate-related events. The spatial analyst becomes a strategist, navigating the complex terrain of climate resilience (Thomas et al., 2014).

CONCLUSION

In conclusion, this exploration of geospatial data and Python programming has provided a comprehensive understanding of the symbiotic relationship between technology and geography. Beginning with the fundamentals of geospatial data types, formats, and sources, the guide laid a strong foundation for navigating the intricate landscape of spatial analysis. The significance of Python in this context was underscored, demonstrating its prowess in handling diverse data types, and facilitating efficient geospatial analysis.

Key Python libraries such as GDAL, Geoplanids, and Fiona were dissected, showcasing their integral roles in processing, analysing, and visualizing geospatial data. Through practical examples, the guide illuminated the potential of Python scripts to automate tasks, manipulate data, and generate meaningful maps. Matplotlib's contribution to geospatial data visualization added a layer of insight, enhancing the communicative power of spatial information. The guide ventured into spatial queries, statistical analyses, and real-world applications, offering a glimpse into the versatility of geospatial analysis across domains.

REFERENCES

Balla, D. (2017). *Possibilities of spatial data visualization with web technologies for cognitive interpretation.* Research Gate.

Capata, A. (2016). *AIS Data Visualization for Maritime Spatial Planning (MSP)*. Research Gate.

Fernando, J. (2018). The Spatial-Perceptual Design Space: A New Comprehension for Data Visualization. Research Gate.

Huang, B. (2010). *An integration of GIS, virtual reality and the Internet for visualization, analysis and exploration of spatial data.* Research Gate.

Koperski, K2019). Spatial Data Mining. Progress and Challenges Survey paper.

Kwon, O. (2021). *Fostering Spatial Visualization Ability through Web-Based Virtual-Reality Program and Paper-Based Program*. Research Gate.

Pundt, H. (2010). *Visualization of spatial data for field-based GIS*. Research Gate.

Rogowitz, B. (2021). *Data visualization: the end of the rainbow*. Research Gate.

Shaito, M. (2021). *Map Visualization using Spatial and Spatio-Temporal Data: Application to COVID-19 Data.* Research Gate.

Thomas, B. (2014). Spatial augmented reality — A tool for 3D data visualization. Reseearch Gate.

Chapter 2
Python Programming for Geospatial Applications:
Web Mapping, Interactive Visualization, and Beyond

Deepthi Das
Christ University, India

Arul Kumar Natarajan
https://orcid.org/0000-0002-9728-477X
Samarkand International University of Technology, Uzbekistan

ABSTRACT

Geospatial solutions represent a pivotal toolset for analyzing, interpreting, and visualizing spatial data across diverse domains, facilitating informed decision-making and fostering innovation. This book chapter provides a comprehensive overview of geospatial solutions, emphasizing their critical role in addressing spatially explicit challenges and driving efficiency, productivity, and innovation across various sectors. Furthermore, it explores the integration of Python programming in geospatial applications, highlighting its versatility and extensive ecosystem of libraries and tools tailored for spatial data analysis and visualization. The fundamentals of web mapping are discussed in depth, elucidating spatial representation, technologies, and tools commonly employed in web mapping applications. Also, the chapter explores Python's role in retrieving geospatial data with Python, visualization methods, and interactive web mapping.

DOI: 10.4018/979-8-3693-1754-9.ch002

INTRODUCTION TO GEOSPATIAL SOLUTIONS

Geospatial solutions, a powerful tool that harnesses spatial data and technology, are not just theoretical concepts. They are practical solutions that help us analyze, interpret, and visualize information about our world. From urban planning to disaster response, these solutions are at the heart of various domains, providing insights into spatial relationships. This, in turn, enables informed decision-making, enhances efficiency, and boosts productivity.

Importance of Geospatial Solutions

Kamraju (2023) explains that geospatial technology education and training require significant resources. These resources include faculty positions, equipment acquisitions, space, and technical support. This underscores the commitment needed to advance proficiency in geospatial science and technology. Geospatial solutions result from integrating geographical data with analytical and visualization techniques. They are designed to address challenges specific to a particular location, known as 'spatially explicit challenges.' These solutions encompass various applications, including geographic information systems (GIS), remote sensing, cartography, and spatial analysis.

- Informed Decision-making: Geospatial solutions provide valuable insights into spatial relationships, facilitating better decision-making across sectors such as urban planning, environmental management, natural resource conservation, disaster response, and public health.
- Efficiency and Productivity: By leveraging geospatial data and tools, organizations can optimize resource allocation, streamline operations, improve workflow efficiency, and reduce costs.
- Innovation: Geospatial technologies drive innovation by enabling the development of advanced applications such as location-based services, precision agriculture, smart cities, autonomous vehicles, and spatially enabled IoT devices.
- Addressing Global Challenges: Geospatial solutions are crucial in addressing global challenges such as climate change, biodiversity conservation, sustainable development, infrastructure planning, and emergency management by providing spatially explicit information for analysis and decision support.

PYTHON PROGRAMMING IN GEOSPATIAL APPLICATIONS

Python has emerged as a leading programming language for geospatial applications due to its versatility, simplicity, and extensive ecosystem of libraries and tools tailored for spatial data analysis, visualization, and geoprocessing. Sládek (2023) examines the feasibility of employing Python packages, including pandas, elevation, geopy, suncalc, and gdal, for rapid geographical analyses of terrain, mainly focusing on ad hoc weather stations and terrain features such as elevation ratios, slope, exposure, hill-shading, and visibility to cardinal directions. Rajamani and Iyer (2023) highlight the extensive utilization of Python in ecological research, emphasizing its simplicity, active user community, and diverse libraries such as NumPy, pandas, matplotlib, Seaborn, ArcPy, GDAL/OGR, scikit-learn, TensorFlow, NumPy, and SciPy, which collectively support tasks ranging from data analysis to machine learning and scientific computing in the field (Kaswan et al., 2023) (Kumar et al., 2023).

Versatility: Python's versatility is underscored by its user-friendly syntax, which is accessible to beginners yet equally attractive to seasoned developers. Its cross-platform compatibility further enhances its appeal, enabling the development of geospatial applications that operate seamlessly across various operating systems (Gupta and Bagchi, 2024a).

Extensive Ecosystem: Python boasts a rich ecosystem of geospatial libraries and frameworks, including GDAL/OGR, Fiona, Shapely, GeoPandas, Pyproj, Cartopy, Folium, Bokeh, and Dash. These libraries provide robust functionalities for reading, writing, processing, analyzing, and visualizing geospatial data in various formats, such as shapefiles, GeoTIFFs, GeoJSON, and web services. Wu (2023) presents a comprehensive web-based teaching system constructed using front-end Vue/React and back-end Python with Django, incorporating user and course management modules, ORM technology for database access, and features such as rich text editors and PDF reporting tools to facilitate tasks within the online education process, exemplifying technical innovation and practicality in educational reform (Mohan et al., 2023).

Integration With GIS Platforms

Python seamlessly integrates with popular GIS platforms such as QGIS and ArcGIS, allowing users to expand the functionalities of these software packages through scripting and automation. Its scripting capabilities empower users to automate repetitive tasks, create custom tools, and optimize geoprocessing workflows within GIS environments. Through Python, users can enhance efficiency and productivity by streamlining processes and automating routine operations, making GIS analysis and management more effective and accessible. Wang and Trauth (2020) propose a methodology utilizing GIS-based Python scripts to calculate water surface profiles on a landscape, facilitating wetlands decision-making by examining locations suitable for developing mitigation wetlands, thereby addressing concerns regarding wetland loss due to land-use changes. Chen, Judge, and Hulse (2022) introduce PyLUSAT: Python for Land Use Suitability Analysis Tools, an open-source toolkit designed to address limitations of desktop GIS software in conducting land use suitability analysis, offering improved efficiency, cross-platform compatibility, and high extensibility for enhanced modeling workflows.

Python benefits from a vibrant community of geospatial developers, educators, and enthusiasts who contribute to open-source projects, share knowledge through forums and tutorials and collaborate on innovative solutions. The Python geospatial community actively maintains and updates libraries develops plugins and extensions, and supports users through online forums, mailing lists, and conferences. This collaborative environment fosters continuous improvement and innovation within the Python geospatial ecosystem, ensuring users can access the latest tools and resources for their geospatial projects (Upadhye et al., 2023).

FUNDAMENTALS OF WEB MAPPING

Understanding Web Mapping Principles

The foundational principles underpin web mapping, elucidating the core concepts necessary for effectively utilizing spatial data in online environments.

Spatial Representation: This section delves into the methods utilized for representing and structuring spatial data in web mapping applications, encompassing discussions on coordinate systems, projections, and spatial indexing techniques. The different web mapping techniques are given in Figure 1. Gupta and Bagchi (2024b) emphasize the crucial role of data visualization in facilitating the comprehension and analysis of large datasets, highlighting its ability to provide organized and pictorial representations that aid in understanding patterns, correlations, and trends, thereby enhancing data interpretation and model selection.

- Folium is a Python library that allows the creation of Leaflet.js maps directly in the Jupyter Notebook environment or as standalone HTML files. It provides a simple and intuitive interface for adding markers, polygons, heatmaps, and other spatial data overlays to maps, making it ideal for web mapping applications.
- Geopandas extends the capabilities of Pandas, a popular data manipulation library in Python, to handle spatial data. It enables users to read, manipulate, and visualize geospatial datasets effortlessly. Developers can create dynamic

maps with spatial analysis capabilities by integrating Geopandas with web mapping libraries like Folium.

- Pydeck is a Python library for creating 3D visualizations of large-scale geospatial datasets. It leverages WebGL and Deck.gl to render interactive maps with advanced features like point clouds, 3D terrain, and heat maps. Pydeck can be integrated with web frameworks like Flask or Django to build sophisticated web mapping applications.

Figure 1. Web mapping techniques

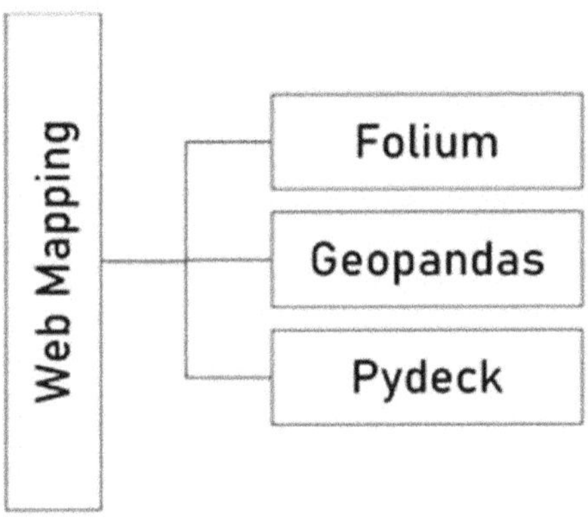

Technologies and Tools for Web Mapping

This section provides an overview of the technologies and tools commonly employed in developing web mapping applications, highlighting their functionalities and suitability for different use cases.

1. Client-side vs. Server-side Mapping
2. Mapping APIs and Libraries
3. Geospatial Data Services

Client-side versus Server-side Mapping: The distinction between client-side and server-side mapping approaches is fundamental to web mapping. Client-side mapping involves rendering map data directly within the user's browser, offering real-time interaction and dynamic visualization capabilities. Conversely, server-side mapping entails processing map data on the server before transmitting it to the client's browser, allowing for greater control over data manipulation and security (Farkas, 2020) (Vishwa Kiran et al., 2023).

Mapping Application Programming Interfaces (APIs) and Libraries: They play a pivotal role in developing web mapping applications. They provide developers access to various functionalities for incorporating maps, geospatial data, and interactive features into their projects (Figure 2).

Figure 2. APIs and libraries

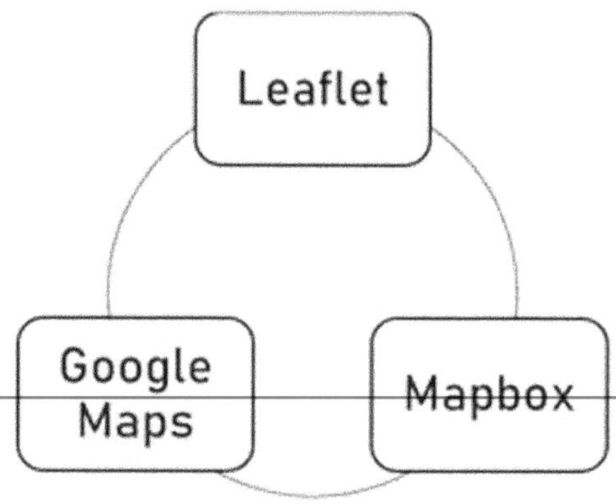

These APIs and libraries, such as Leaflet, Mapbox, and Google Maps API, offer diverse features, integrations, and customization options, empowering developers to create rich and interactive mapping experiences tailored to their needs. Airlangga and Rachman (2022) discuss the deployment of location mapping results based on an application programming interface (API), emphasizing the process of displaying user mapping data from the GIS web to personal websites and mobile GIS applications through API-based tokens. The study outlines the procedure for obtaining and utilizing Mapbox tokens to embed mapping data into personal websites, facilitating seamless integration and accessibility of mapping results across various platforms.

PYTHON'S ROLE IN WEB MAPPING

Python is crucial in web mapping, offering versatile tools and frameworks that empower developers to create sophisticated mapping applications. This discussion delves into Python's contributions to web mapping, exploring its integration with web frameworks, geospatial libraries, and web technologies to facilitate the development of interactive and dynamic mapping solutions. Python's role in web mapping is given in Figure 3. Kumar and Roy (2023) present a technique for data collection known as web scraping with Python, emphasizing its significance in the preparation stage of data essential for artificial intelligence applications. This technique involves extracting data from the internet using automated systems and utilizing open-source tools for its functionality.

The authors discuss how Python, with its extensive library support, implements web scraping to collect data from the internet.

- *Python Web Frameworks:* Python web frameworks like Django and Flask are robust foundations for constructing web mapping applications, enabling the integration of mapping functionalities into web applications. These frameworks are exemplified through real-world implementations.
- *Geospatial Libraries and Tools:* Python boasts an array of geospatial libraries and tools facilitating spatial data manipulation, analysis, and visualization. Libraries such as Folium, Bokeh, and Dash are utilized to create interactive web maps within the Python environment. Folium is an essential tool in Python's arsenal. Folium emerges as a versatile library enabling the creation of interactive leaflet maps directly within Jupyter notebooks. This discussion elucidates Folium's streamlined approach to crafting web maps with Python, offering a user-friendly interface for seamlessly incorporating layers, markers, and interactive elements.

Bokeh empowers users to generate interactive plots, dashboards, and applications quickly. This exploration delves into Bokeh's capacity for constructing interactive geospatial visualizations in Python, harnessing its support for interactivity, widgets, and server-based deployment. Dash facilitates the development of analytical web applications. This demonstration underscores Dash's utility in crafting interactive geospatial applications, leveraging Python's data manipulation capabilities to foster web-based interactivity and visualization.

- *Integration with Web Technologies:* Python's compatibility with web technologies, including HTML, CSS, and JavaScript, ensures seamless integration of mapping functionalities into web applications. Python enables dynamic web content generation and harnesses client-side scripting for enhanced interactivity.

Figure 3. Python's role in web mapping

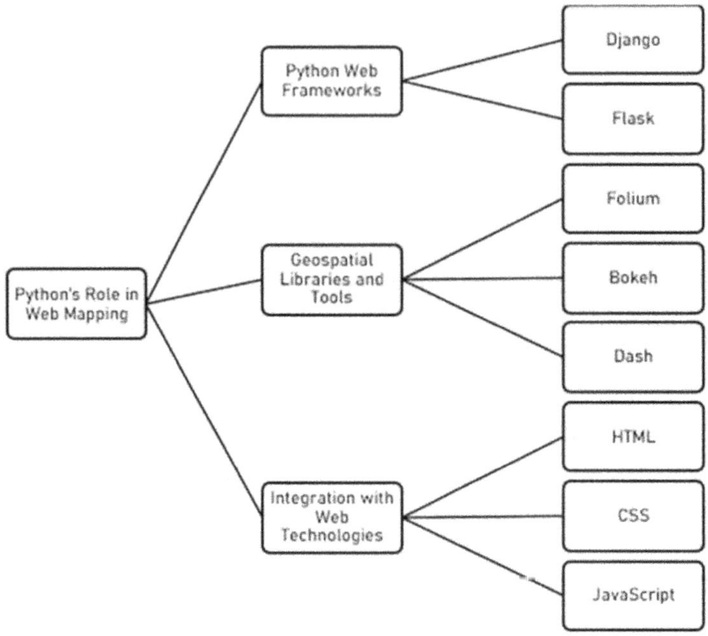

RETRIEVING GEOSPATIAL DATA WITH PYTHON

This section delves into the methods and techniques for retrieving geospatial data using Python, leveraging various data sources and APIs to acquire spatial information for web mapping applications.

- Diverse sources of geospatial data, including public repositories, governmental agencies, and commercial providers, are explored, along with an analysis of standard formats used for storing and distributing spatial data, such as shapefiles, GeoJSON, and Web Map Services (WMS) (Lanucara et al., 2024).
- Python offers a plethora of libraries and tools tailored for accessing geospatial data from diverse sources. Popular libraries like GeoPandas, OWSLib, and Requests are highlighted, showcasing their functionalities in retrieving data from APIs, web services, and local repositories. Joosten (2021) conducted a study focusing on detecting criminal activity in rural regions, explicitly examining the utility of satellite imagery for drug detection.

Python code for Retrieving Geospatial Data

```
import geopandas as gpd# Load the sample dataset
"naturalearth_lowres"world = gpd.read_file(gpd.datasets.get_
path('naturalearth_lowres'))# Display the first few rows of the
datasetprint(world.head())
```

The provided code snippet demonstrates the process of retrieving geospatial data using the Python library GeoPandas. Initially, the sample dataset "naturalearth_lowres" is loaded into memory. This dataset, curated within GeoPandas, encompasses geospatial information representing countries of the world at a low resolution. The dataset is accessed through the gpd.datasets.get_path() function, which retrieves the path to the sample dataset. Upon loading the dataset into a GeoDataFrame named "world," the script proceeds to display the first few rows of the dataset. This action provides researchers with a preliminary insight into the structure and content of the geospatial data, facilitating further exploration and analysis within their research endeavors.

VISUALIZATION TECHNIQUES

The Visualization Techniques section encompasses various methodologies employed in Python for visualizing geospatial data. Choropleth Maps and Scatter Plots are prominent techniques demonstrated using the GeoPandas library, providing insights into population density distribution and city locations worldwide. The Choropleth Map visually represents population density across countries by shading each country according to its density level. At the same time, the Scatter Plot illustrates city locations using individual markers on the map. Both visualizations employ Matplotlib for plotting and GeoPandas for geospatial data handling, enabling informative representations of spatial patterns.

Additionally, the section explores other visualization techniques, such as Heatmaps, Line Plots, Kernel Density Estimation (KDE) Maps, 3D Visualization, and Interactive Web Maps, each offering unique perspectives on spatial data analysis and interpretation. These techniques leverage various libraries and tools in Python, facilitating a comprehensive understanding of geospatial phenomena.

Bekker (2023) explored the effectiveness of general propensity choropleth maps in visualizing databases of protest incidents, mainly focusing on the South African context. Through the analysis of a large-scale protest event database provided by the South African Police Services, the study delineated protest occurrences per capita across the country's 234 municipalities. The findings highlighted a notably high proportion of rural protests and municipalities hosting major industries and provincial government seats exhibiting the highest propensity for protest.

Geetha et al. (2023) investigated the application of machine learning techniques in geospatial data exploration, focusing on accommodation decisions for students in various locations. Utilizing the K-means algorithm, commonly employed in geospatial analysis to identify spatial patterns or hotspots, the study organized data points into clusters based on their similarities. The research highlights the potential benefits of such analysis in urban planning, transportation analysis, epidemiological research, and other related fields. Examining and visualizing patterns includes Scatter plots, trends, and relationships within geolocational data, which is referred to as exploratory analysis of geospatial data, demonstrating its relevance and utility in decision-making processes.

The following are the various visualization techniques used in the Python.

- Choropleth Maps
- Scatter Plots
- Kernel Density Estimation (KDE) Maps
- Line Plots
- 3D Visualization
- Interactive Web Maps

Figure 4. Various types of visualization techniques

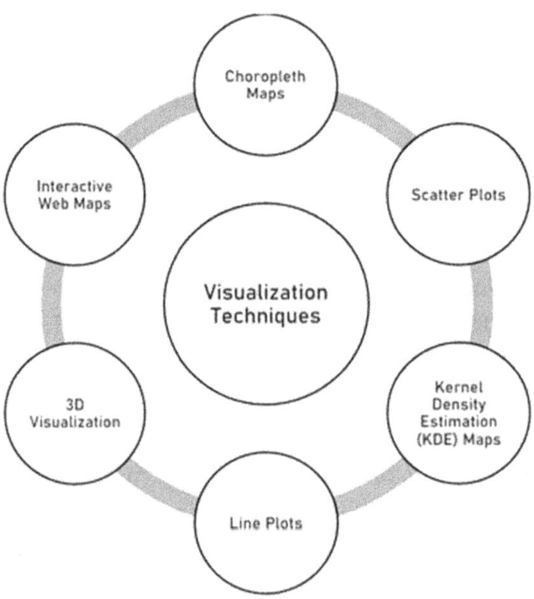

Choropleth Map and Scatter Plot

The provided Python code utilizes the GeoPandas library to visualize geospatial data from the "naturalearth_lowres" dataset.

- *Choropleth Map:* The first part of the code generates a choropleth map, representing the population density of countries worldwide. Each country is shaded according to its population density, with darker shades indicating higher densities. The resulting plot visually represents population distribution on a global scale (Figure 5).
- *Scatter Plot:* In the second part of the code, a scatter plot is created to display the locations of cities worldwide (Figure 6). Each city's latitude and longitude coordinates are plotted as individual points on the map, represented by red markers. This visualization offers insights into the spatial distribution of urban centers across different regions.

Both visualizations leverage Matplotlib for plotting and GeoPandas for handling geospatial data, providing informative representations of spatial patterns and distributions in the "naturalearth_lowres" dataset.

Figure 5. Choropleth map of population density

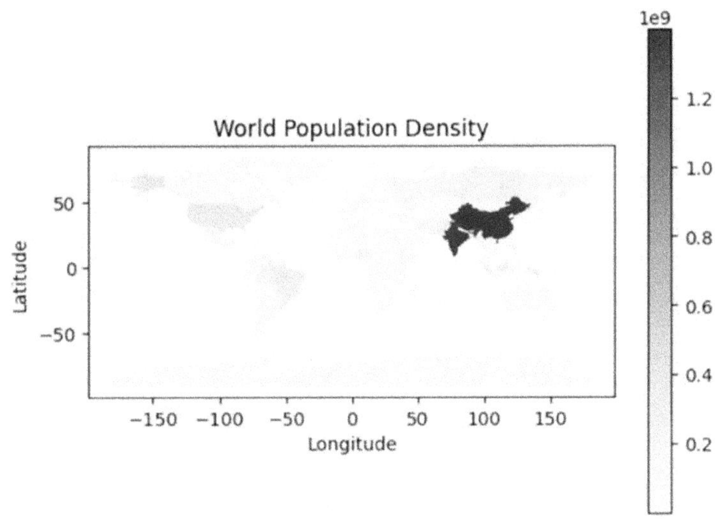

Python Code for Choropleth Map

```
import geopandas as gpdimport matplotlib.pyplot as plt# Load
the "naturalearth_lowres" datasetworld = gpd.read_file(gpd.
datasets.get_path('naturalearth_lowres'))# Create a choropleth
map of the population densityworld.plot(column='pop_est',
cmap='OrRd', legend=True)plt.title('World Population Density')
plt.xlabel('Longitude')plt.ylabel('Latitude')plt.show()
```

Python Code to Create a Scatter Plot of Cities

```
# Load the "naturalearth_cities" datasetcities = gpd.read_
file(gpd.datasets.get_path('naturalearth_cities'))# Create a
scatter plot of citiescities.plot(ax=world.plot(figsize=(10,
6)), marker='o', color='red', markersize=5)plt.title('Cities
Around the World')plt.xlabel('Longitude')plt.ylabel('Latitude')
plt.show()
```

Figure 6. Scatter plot of cities

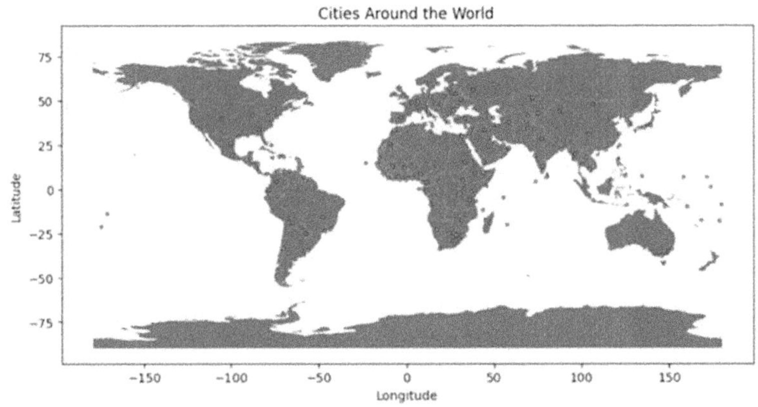

Heatmap of Population Density

Heatmaps visualize spatial data by representing the density or intensity of a phenomenon using color gradients. They are particularly effective for highlighting high or low-concentration areas, such as population density, crime rates, or environmental pollution. The provided Python code utilizes the GeoPandas library to generate a heatmap illustrating the population density across countries worldwide. First, the "naturalearth_lowres" dataset, containing geographical data, is loaded into the world variable. Subsequently, the plot() function is employed to create the heatmap, with the column parameter specifying the variable ('pop_est') to represent population density.

Bao et al. (2023) investigated the population distribution mapping in Beijing using Baidu heat map data, addressing the scarcity of high spatiotemporal resolution data. They proposed a spatial downscaling framework integrating ambient population, nighttime light, and building volume data, enabling hourly mapping at 100-meter resolution. The study validated the accuracy of their approach using mobile signaling data and highlighted spatiotemporal variation characteristics in Beijing's population distribution.

The heatmap in Figure 7 uses the 'YlOrRd' colormap and quantile classification scheme to distinguish different population density levels visually. Additionally, a legend is included to aid interpretation, positioned on the right side of the plot. Lastly, the plot is adorned with a title ('World Population Density Heatmap') and labeled axes denoting latitude and longitude.

Python Code for Heatmap of Population Density

```
import geopandas as gpdimport matplotlib.pyplot as plt# Load
the "naturalearth_lowres" datasetworld = gpd.read_file(gpd.
datasets.get_path('naturalearth_lowres'))# Plot a heatmap of
population densityworld.plot(column='pop_est', cmap='YlOrRd',
scheme='quantiles', legend=True, legend_kwds={'bbox_to_anchor':
(1, 0.5)})plt.title('World Population Density Heatmap')plt.
xlabel('Longitude')plt.ylabel('Latitude')plt.show()
```

Figure 7. Heatmap of population density

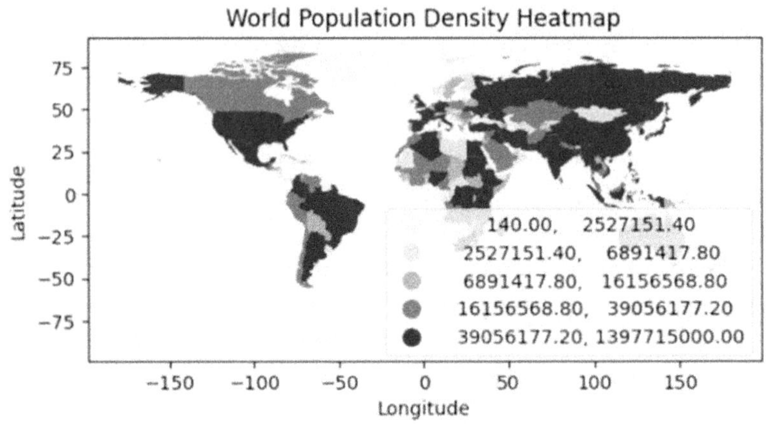

Line Plots

Line plots (Figure 8) display spatial data as lines or polylines on a map, representing linear features such as roads, rivers, or transportation networks. They provide insights into linear features' spatial distribution and connectivity across the landscape. Sharif et al. (2022) investigated the interaction experiences of screen-reader users with online geospatial data visualizations, aiming to address the challenges faced by this user group in accessing information from such visualizations. Through a user study involving 12 screen-reader users, they identified the information sought and questions asked by participants to extract data from online geospatial visualizations. Additionally, the researchers extended the functionalities of VoxLens, an open-source multi-modal solution, to enhance accessibility for screen-reader users interacting with geospatial data visualizations.

Lemenkova (2020) conducted a study aiming to perform modeling and spatial analysis of marine geological data in the Mariana Trench using a combination of QGIS and Python programming. The research involved digitizing cross-section profiles in QGIS and further data processing in Python to analyze the trench's geomorphology, subjected to various phenomena such as bathymetry, geodesy, gravimetry, tectonics plates, and geological settings. The study utilized statistical analysis methods implemented in Python's Seaborn library, including multiple facet grids, area charts, regression analysis, letter-value plots, and hexagonal and kernel density estimation, to understand the spatial variation of the trench's geomorphology. The findings demonstrate the effectiveness of Python in geographic data analysis, providing Python codes for reproducibility.

Python Code for Line Plots

```
import geopandas as gpdimport matplotlib.pyplot as plt# Load
the "naturalearth_lowres" datasetworld = gpd.read_file(gpd.
datasets.get_path('naturalearth_lowres'))# Create a line plot
to display linear featuresworld.boundary.plot(color='blue',
linewidth=1)# Customize the plotplt.title('World Boundary Line
Plot')plt.xlabel('Longitude')plt.ylabel('Latitude')# Display
the plotplt.show()
```

The provided Python code utilizes the GeoPandas library to generate a line plot illustrating the boundaries of countries worldwide. Initially, the "naturalearth_lowres" dataset, containing geographical data, is loaded into the 'world' variable. Subsequently, the boundary.plot() function is employed to create the line plot, specifying the color of the lines as blue and setting the linewidth to 1. The plot is customized by adding a title ('World Boundary Line Plot') and labeled axes denoting longitude and latitude. Lastly, the plot is displayed using the show() function from the Matplotlib library.

Figure 8. Line plot to display linear features

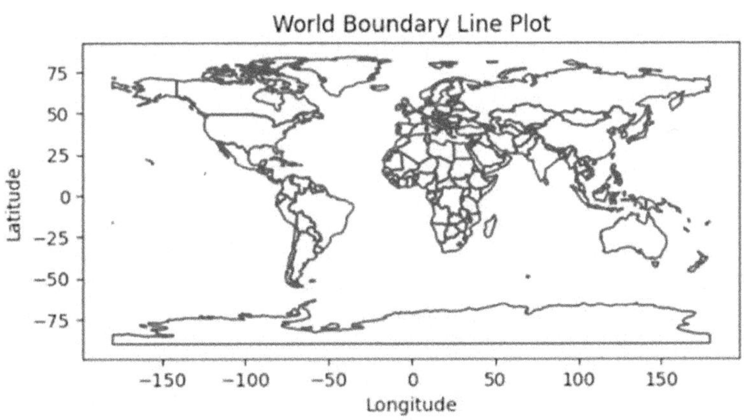

Kernel Density Estimation (KDE) Maps

KDE maps estimate the spatial distribution of point data by smoothing the density of point observations across the study area. They help identify clusters or hotspots of activity within spatial datasets. Srikanth and Srikanth (2020) conducted a case study aiming to identify and rank accident hotspots in Des Moines City, Polk County, United States, using Kernel Density Estimation (KDE) and Hotspot Analysis (Getis-Ord Gi*) methods. The study utilized five years of crash data from 2008 to 2012 to visually represent hotspots with KDE and statistically analyze their significance with Hotspot Analysis. The methodology demonstrated in the research could be applied to any study area with sufficient accident data to prioritize actions for reducing traffic accidents.

Shuo et al. (2021) investigated the identification of grassland fire point density using kernel density estimation (KDE) based on spatial distribution characteristics. The study emphasized the importance of understanding the risk of grassland fires and proposed a method to convert historical fire point records into continuous spatial distribution data using KDE. By considering the spatial distribution characteristics of the points, the study determined the bandwidth of KDE using Ripley's K function method, resulting in kernel density surfaces that accurately represent the risk of grassland fire occurrence. The research findings highlight the effectiveness of Ripley's K function method in selecting bandwidth for mapping and analyzing the risk of grassland fires, mainly when dealing with dependent or inaccurate point variables.

Python Code for KDE Maps

```
import geopandas as gpdimport seaborn as snsimport matplotlib.
pyplot as plt# Load the "naturalearth_lowres" datasetworld =
gpd.read_file(gpd.datasets.get_path('naturalearth_lowres'))#
Extract latitude and longitude coordinatesx = world.
centroid.x.valuesy = world.centroid.y.values# Create a KDE
plot to estimate spatial distributionsns.kdeplot(x=x, y=y,
cmap='YlOrRd', fill=True)# Customize the plotplt.title('Kernel
Density Estimation (KDE) Map')plt.xlabel('Longitude')plt.
ylabel('Latitude')# Display the plotplt.show()
```

The provided code utilizes the GeoPandas library to load the "naturalearth_lowres" dataset, representing a simplified world map with country boundaries (Figure 9).

- It then extracts the longitude and latitude coordinates of the centroids of each country polygon. Using Seaborn's kdeplot() function, a Kernel Density Estimation (KDE) plot is created to estimate the spatial distribution of the centroid coordinates.

- The colormap 'YlOrRd' is applied to the KDE plot, with filled contours representing areas of higher density. Finally, the plot is customized with a title, xlabel, and ylabel and displayed using Matplotlib.

Figure 9. KDE plot to estimate the spatial distribution

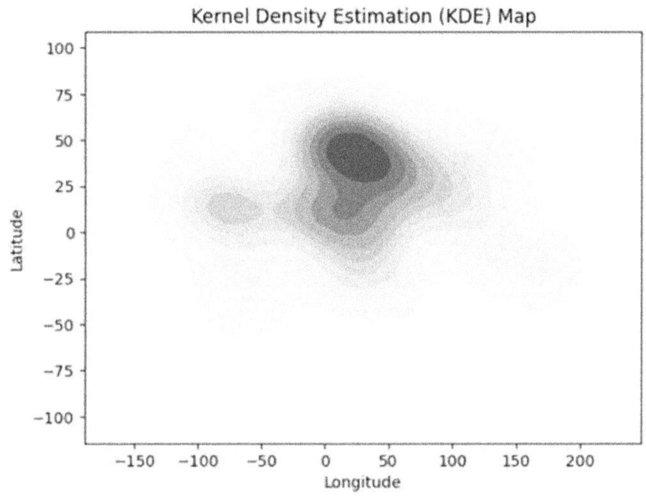

3D Visualization

3D visualization techniques allow geospatial data representation in three-dimensional space, providing a more immersive and realistic view of the data. This can be particularly useful for visualizing terrain, elevation, or urban landscapes. Cho, Kim, and Choi (2020) designed and implemented a mobile device service system to utilize the integrated underground geospatial information map in underground exploration fields. The system visualizes tiled maps, 3D terrain information, underground structures, facilities, and ground information based on the user's current position, providing real-time reflection of field exploration results. Additionally, the system transfers and visualizes integrated underground geospatial information maps in glTF format to accommodate wireless network constraints and device characteristics of mobile devices, thereby enhancing accident prevention and maintaining data accuracy.

Python Code for 3D Visualization

```
import geopandas as gpdimport matplotlib.pyplot as pltfrom
mpl_toolkits.mplot3d import Axes3D# Load the "naturalearth_
lowres" datasetworld = gpd.read_file(gpd.datasets.
get_path('naturalearth_lowres'))# Create a 3D plotfig
= plt.figure(figsize=(8, 5))ax = fig.add_subplot(111,
projection='3d')# Extract latitude, longitude, and
elevation coordinatesx = world.centroid.x.valuesy = world.
centroid.y.valuesz = [0] * len(x) # Assuming all countries
have the same elevation for simplicity# Plot the world map
in 3Dax.scatter(x, y, z, c='blue', marker='o')# Customize
the plotax.set_title('3D Visualization of World Map')
ax.set_xlabel('Longitude')ax.set_ylabel('Latitude')ax.set_
zlabel('Elevation')# Show the plotplt.show()
```

The provided code utilizes the GeoPandas library to load the "naturalearth_lowres" dataset, which contains geospatial information about world countries. It then creates a 3D plot using Matplotlib's Axes3D, setting each country's centroid's longitude and latitude coordinates as the x and y axes, respectively. The z-axis represents the elevation, which is assumed to be zero for all countries in this example. The resulting plot in Figure 10 visualizes the world map in three-dimensional space, allowing viewers to observe the spatial distribution of countries. Finally, the plot is customized with a title and axis labels before being displayed using Matplotlib.

Figure 10. KDE plot to estimate the spatial distribution

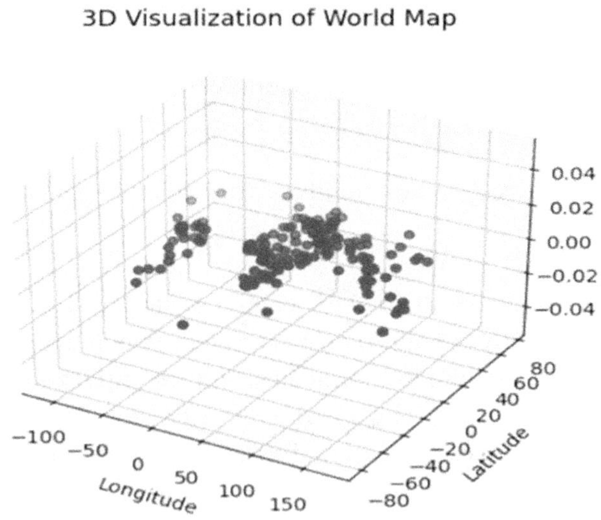

Interactive Web Maps

Interactive web maps leverage web technologies such as HTML, CSS, and JavaScript to create dynamic and interactive visualizations of geospatial data. They enable users to explore and interact with spatial data in real time, providing a rich and engaging user experience. Walker et al. (2020) introduced an interactive data visualization framework, the Interactive Catchment Explorer (ICE), designed for exploring geospatial environmental datasets and model predictions. The framework offers a highly interactive user experience, enabling users to discover spatial patterns, evaluate variable relationships, and identify specific locations using multivariate criteria. Through case studies covering various regions in North America, the authors demonstrated the application of the ICE framework. They highlighted its broader impacts, including facilitating stakeholder discussion and collaboration and supporting the iterative data collection, analysis, and decision-making process.

Python Code for Interactive Web Maps

```
import foliumimport geopandas as gpd# Load the "naturalearth_
lowres" datasetworld = gpd.read_file(gpd.datasets.get_
path('naturalearth_lowres'))# Create a Folium map centered
around the mean latitude and longitude of the datasetm
```

```
= Folium.Map(location=[world.centroid.y.mean(), world.
centroid.x.mean()], zoom_start=2, control_scale=True)#
Add the GeoDataFrame as a GeoJSON layer to the mapfolium.
GeoJson(world).add_to(m)# Display the interactive mapm.
save('interactive_map.html')
```

The provided code utilizes the Folium library in Python to generate an interactive web map.

- Initially, the "naturalearth_lowres" dataset is loaded using the GeoPandas library. Subsequently, a Folium map is created, centered around the dataset's mean latitude and longitude coordinates, with an initial zoom level of 2.
- The dataset, represented as a GeoJSON layer, is then added to the map. Finally, the interactive map is saved as an HTML file named "interactive_map.html" for display in a web browser (Figure 11).

Figure 11. Interactive web maps

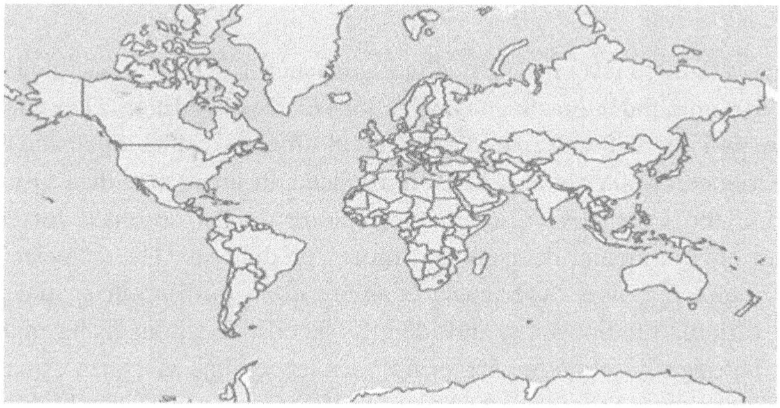

CONCLUSION

In conclusion, this book chapter underscores the indispensable nature of geospatial solutions in modern data analysis, offering unparalleled insights into spatial relationships and enabling informed decision-making across various domains. Integrating Python programming in geospatial applications has significantly enhanced the development of sophisticated mapping tools, leveraging its versatility and extensive library support. Moreover, the fundamentals of web mapping provide a robust foundation for creating interactive and dynamic mapping applications, facilitating spatial data visualization and analysis. As the demand for spatial data analysis continues to grow, combining geospatial solutions and Python programming presents immense opportunities for innovation and advancement in spatial data science. Researchers, practitioners, and decision-makers can unlock new possibilities for effectively understanding and addressing spatial challenges by leveraging these tools and techniques.

REFERENCES

Airlangga, P., & Rachman, A. A. (2022). Deployment of Location Mapping Results Based on Application Programming Interface. *NEWTON: Networking and Information Technology*, *1*(3), 111–118. doi:10.32764/newton.v1i3.1922

Bao, W., Gong, A., Zhang, T., Zhao, Y., Li, B., & Chen, S. (2023). Mapping population distribution with high spatiotemporal resolution in Beijing using baidu heat map data. *Remote Sensing (Basel)*, *15*(2), 458. doi:10.3390/rs15020458

Bekker, M. (2023). Depends on how you count them: The value of general propensity choropleth maps for visualising databases of protest incidents. *Journal of Maps*, *19*(1), 2064778. doi:10.1080/17445647.2022.2064778

Chen, C., Judge, J., & Hulse, D. (2022). PyLUSAT: An open-source Python toolkit for GIS-based land use suitability analysis. *Environmental Modelling & Software*, *151*, 105362. doi:10.1016/j.envsoft.2022.105362

Cho, S. K., Kim, Y. T., & Choi, J. Y. (2020). Design and implementation of mobile 3D visualization service system on the integrated underground geospatial information map. *Journal of The Korea Society of Computer and Information*, *25*(12), 173–186.

Farkas, G. (2020). Possibilities of using raster data in client-side web maps. *Transactions in GIS*, *24*(1), 72–84. doi:10.1111/tgis.12588

Geetha, K., Srivani, A., Gunasekaran, S., Ananthi, S., & Sangeetha, S. (2023, September). Geospatial Data Exploration Using Machine Learning. In *2023 4th International Conference on Smart Electronics and Communication (ICOSEC)* (pp. 1485-1489). IEEE. 10.1109/ICOSEC58147.2023.10275920

Gupta and Bagchi. (2024a). Python Language Basics. In *Essentials of Python for Artificial Intelligence and Machine Learning* (pp. 87–126). Springer Nature Switzerland. doi:10.1007/978-3-031-43725-0_3

Gupta and Bagchi. (2024b). Data Visualization with Python. In *Essentials of Python for Artificial Intelligence and Machine Learning* (pp. 237–282). Springer Nature Switzerland. doi:10.1007/978-3-031-43725-0_7

Joosten, T. (2021). *Detecting criminal activity in rural areas: A study into the application of satellite imagery in drugs detection.* [Master Thesis, Eindhoven University of Technology].

Kamraju, M. (2023). Overview of Geospatial Science and Technology. In *Temporal and Spatial Environmental Impact of the COVID-19 Pandemic* (pp. 1–6). Springer Nature Singapore. doi:10.1007/978-981-99-1934-5_1

Kaswan, K. S., Dhatterwal, J. S., & Balamurugan, B. (2023). *Python for Beginners.* Chapman and Hall/CRC. doi:10.1201/9781003202035

Kumar, S., & Roy, U. B. (2023). A technique of data collection: web scraping with Python. In *Statistical Modeling in Machine Learning* (pp. 23–36). Academic Press. doi:10.1016/B978-0-323-91776-6.00011-7

Kumar, T. S., Jothilakshmi, S., James, B. C., Prakash, M., Arulkumar, N., & Rekha, C. (2023). HHO-based vector quantization technique for biomedical image compression in cloud computing. *International Journal of Image and Graphics, 23*(03), 2240008. doi:10.1142/S0219467822400083

Lanucara, S., Praticò, S., Pioggia, G., Di Fazio, S., & Modica, G. (2024). Web-Based Spatial Decision Support System for Precision Agriculture: A tool for delineating dynamic management unit zones (MUZs). *Smart Agricultural Technology, 8*, 100444. doi:10.1016/j.atech.2024.100444

Lemenkova, P. (2020). Python libraries matplotlib, seaborn and pandas for visualization geo-spatial datasets generated by QGIS. *Analele stiintifice ale Universitatii" Alexandru Ioan Cuza" din Iasi-seria Geografie, 64*(1), 13-32.

Mohan, P., Neelakandan, S., Mardani, A., Maurya, S., Arulkumar, N., & Thangaraj, K. (2023). Eagle strategy arithmetic optimisation algorithm with optimal deep convolutional forest based fintech application for hyper-automation. *Enterprise Information Systems, 17*(10), 2188123. doi:10.1080/17517575.2023.2188123

Rajamani, S. K., & Iyer, R. S. (2023). Use of Python Modules in Ecological Research. In *Perspectives on the Transition Toward Green and Climate Neutral Economies in Asia* (pp. 182–206). IGI Global. doi:10.4018/978-1-6684-8613-9.ch011

Sharif, A., Zhang, A. M., Shih, A., Wobbrock, J. O., & Reinecke, K. (2022, October). Understanding and improving information extraction from online geospatial data visualizations for screen-reader users. In *Proceedings of the 24th International ACM SIGACCESS Conference on Computers and Accessibility* (pp. 1-5). 10.1145/3517428.3550363

Shuo, Z., Jingyu, Z., Zhengxiang, Z., & Jianjun, Z. (2021). Identifying the density of grassland fire points with kernel density estimation based on spatial distribution characteristics. *Open Geosciences*, *13*(1), 796–806. doi:10.1515/geo-2020-0265

Sládek, D. (2023, May). Applicability of Python-based geospatial analysis for operational assessment of weather station deployment. In *2023 International Conference on Military Technologies (ICMT)* (pp. 1-8). IEEE. 10.1109/ICMT58149.2023.10171287

Srikanth, L., & Srikanth, I. (2020, January). A case study on kernel density estimation and hotspot analysis methods in traffic safety management. In *2020 international conference on communication systems & networks (COMSNETS)* (pp. 99-104). IEEE.

Upadhye, S., Neelakandan, S., Thangaraj, K., Babu, D. V., Arulkumar, N., & Qureshi, K. (2023). Modeling of Real Time Traffic Flow Monitoring System Using Deep Learning and Unmanned Aerial Vehicles. J. *Mobile Multimedia*, *19*(2), 477–496.

Vishwa Kiran, S., Kaur, I., Thangaraj, K., Saveetha, V., Kingsy Grace, R., & Arulkumar, N. (2023). Machine learning with data science-enabled lung cancer diagnosis and classification using computed tomography images. *International Journal of Image and Graphics*, *23*(03), 2240002. doi:10.1142/S0219467822400022

Walker, J. D., Letcher, B. H., Rodgers, K. D., Muhlfeld, C. C., & D'Angelo, V. S. (2020). An interactive data visualization framework for exploring geospatial environmental datasets and model predictions. *Water (Basel)*, *12*(10), 2928. doi:10.3390/w12102928

Wang, Z., & Trauth, K. M. (2020). Development of GIS-based Python scripts to calculate a water surface profile on a landscape for wetlands decision-making. *Journal of Hydroinformatics*, *22*(3), 628–640. doi:10.2166/hydro.2020.167

Wu, N. (2023, November). *Construction and Application of Financial Management Network Teaching System in Colleges and Universities under Python Language*. In 4th International Conference on Modern Education and Information Management, ICMEIM 2023, Wuhan, China. 10.4108/eai.8-9-2023.2340077

Chapter 3
Python-Powered Remote Sensing Data

Aamir Raza
https://orcid.org/0009-0001-1867-2660
University of Agriculture, Faisalabad, Pakistan

Sheraz Maqbool
University of Agriculture, Faisalabad, Pakistan

Muhammad Safdar
https://orcid.org/0009-0006-1779-6967
University of Agriculture, Faisalabad, Pakistan

Hasnain Ali
University of Agriculture, Faisalabad, Pakistan

Ikram Ullah
https://orcid.org/0000-0001-5386-7917
Yangzhou University, China

Ali Akbar
Liaoning University, Shenyang, China

Avery Williams
University of North Texas, USA

Mohammed Saleh Al Ansari
https://orcid.org/0000-0001-9425-0294
University of Bahrain, Bahrain

Mubashir Ahmed
University of Agriculture, Faisalabad, Pakistan

Awn Abbas
University of Agriculture, Faisalabad, Pakistan

Abdul Malik
University of Agriculture, Faisalabad, Pakistan

ABSTRACT

Remote sensing is a crucial technique in environmental and spatial investigations, and Python is a popular programming language for analyzing this data. This chapter provides a comprehensive guide to using Python for remote sensing data analysis, covering various data types, attributes, and practical implementations. It introduces

DOI: 10.4018/979-8-3693-1754-9.ch003

Python and its data processing libraries, discusses preprocessing operations like data conversion and import, geometric rectification, and radiometric correction, and covers image enhancement techniques like edge detection, contrast enhancement, and filtering. It also covers image analysis techniques like band mathematics, indices, classification, and segmentation. The chapter also covers exporting data and generating visualization maps and charts. Python's application in remote sensing data analysis is illustrated through case studies.

INTRODUCTION TO REMOTE SENSING DATA

Remote Sensing

Remote sensing is the practice of collecting data about an object or occurrence without physically interacting with it, primarily used to gather knowledge about Earth and celestial bodies, particularly planets (Campbell and Wynne, 2011). This data is acquired through sensors installed on aircraft or satellites, which capture electromagnetic radiation from the Earth's surface. The data is then transformed into visual representations or other forms of data outputs, which can be used to identify and delineate Earth's terrain characteristics. Remote sensing data has diverse applications, including mapping different land cover types, identifying pests and diseases, assessing crop yields, and tracking agricultural progress Mirzakarimova (2023). Figure 1 shows how different land cover types, such as forests, grasslands, urban areas, and water bodies, may be mapped using remote sensing data. It has also proven effective in the surveillance of natural disasters like earthquakes, wildfires, and floods, as well as monitoring deforestation, water quality, and air quality.

Figure 1. A Satellite Image Showing Different Land Cover Type

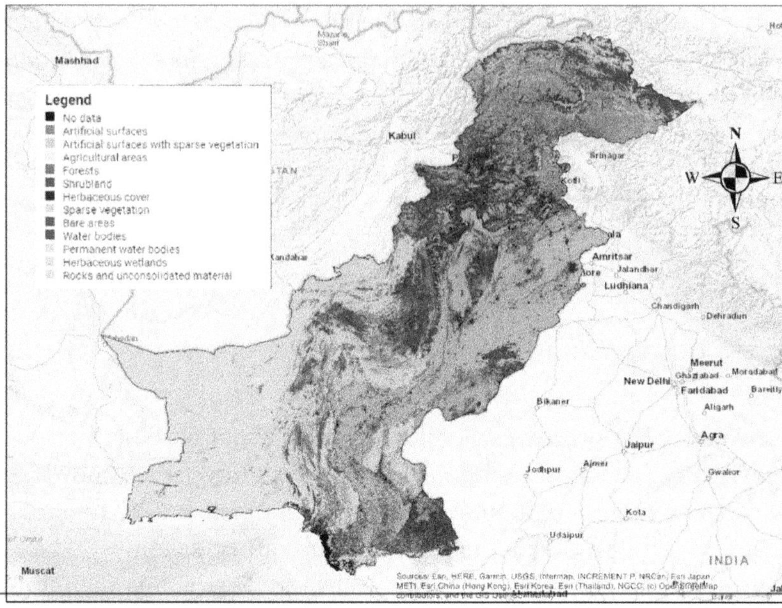

Remote sensing methodologies have revolutionized Earth's mapping, enabling the exploration of unexplored geological features and resources. These methods have also improved weather forecasting accuracy. Remote sensing data includes aerial photography, satellite images, LiDAR (Light Detection and Ranging) data, thermal infrared imagery, and hyperspectral data. These data sources provide valuable insights into Earth's surface, enabling the identification of previously unexplored geological features. Other methods used for data analysis include thermal imaging, radar, sonar, and aerial photography. The use of remote sensing technology has significantly enhanced the accuracy of weather forecasting and the overall understanding of Earth's environment.

Types of Remote Sensing Data

According to Jia et al. (2021), Data from remote sensing may be divided into two main categories: passive and active. As seen in Figure 2, passive remote sensing instruments are made to collect electromagnetic radiation that is emitted or reflected from the surface of the Earth. Most passive remote sensing data comes from sunlight. Thermal and optical systems are the two main categories of passive remote sensing systems.

- Devices for optical remote sensing are made to detect electromagnetic radiation in the visible and near-infrared ranges. To create Earth surface pictures, optical remote sensing data is used. This makes it easier to identify and map different types of land cover, such as forests, grasslands, cities, and water bodies.
- The purpose of thermal remote sensing equipment is to detect thermal infrared radiation. Temperature measurements on the Earth's surface are made using thermal remote sensing data. Applications for thermal remote sensing data are numerous and include mapping water currents, tracking volcanic activity, and detecting forest fires.

Active remote sensing systems employ the emission of electromagnetic radiation and subsequently assess the reflection or backscatter originating from the Earth's surface, as depicted in Figure 2. Radar and lidar are widely recognized as the predominant active remote sensing techniques.

Figure 2. Types of Remote Sensing Data

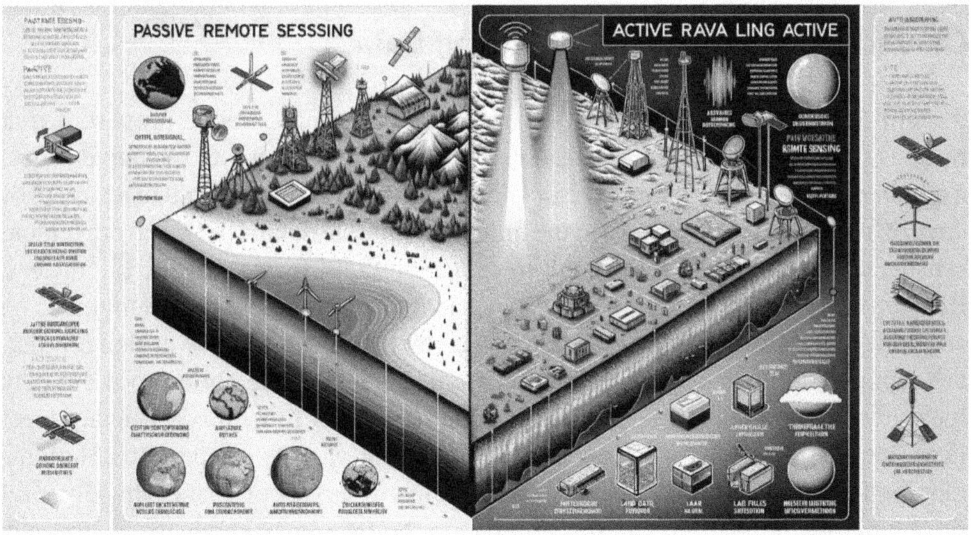

- Radar systems employ the emission of microwave radiation followed by the measurement of the backscatter originating from the Earth's surface. Radar data possesses the capability to provide visual representations of the Earth's surface, even under circumstances characterized by diminished illumination or cloud cover. Radar data finds utility across a diverse range of applications, encompassing the mapping of sea ice, the monitoring of weather patterns, and the detection of airplanes.

- Lidar systems employ laser pulses to emit and subsequently gauge the duration it takes for the pulses to retrace their path back to the sensor. Lidar data may be used to produce high-resolution images of the Earth's surface as well as precisely measure the elevation of different geographical features. Lidar data is useful for many different purposes, such as making topographic maps, tracking deforestation, and mapping floodplains.

Remote sensing data is a valuable resource for mapping the Earth's surface, including areas difficult to access on the ground (Navalgund, *et al.*, 2007). It consists of LIDAR, active radar, and passive optical data. These kinds of data are utilized in many different applications, such as lidar, radar, thermal, and satellite imagery. The advantages and disadvantages of various data kinds vary depending on the user's particular requirements. While LIDAR data offers high-resolution photos and precise elevation measurements, active radar data takes pictures of the Earth's surface at night and under cloud cover. The user's particular demands determine which data to choose.

Characteristics of Remote Sensing Data

According to Liu (2015), remote sensing data has several unique properties that make it extremely useful for a wide range of applications. Remote sensing data has a number of noteworthy characteristics. The degree of clarity or detail in a picture or dataset, particularly with regard to the tiniest perceptible details or Data from remote sensing may be collected at different spatial resolutions, from meters to kilometers. Resolutions, ranging from a solitary wavelength to numerous wavelengths. The spectral resolution of a remote sensing image refers to the extent of wavelengths that are captured by the sensor. Temporal resolution refers to the ability to accurately measure and represent changes or events that occur over time. It is a crucial aspect in various Remote sensing data can be acquired at various temporal resolutions, ranging from short intervals of seconds to longer intervals of years. The temporal resolution of a remote sensing image refers to the frequency at which the image is acquired. The synoptic view refers to a comprehensive perspective or overview of a certain subject or topic. Remote sensing data offers a comprehensive and panoramic perspective of the Earth's surface. Consequently, it is possible to capture and observe extensive regions of the Earth's surface simultaneously. The ability to penetrate certain forms of remote sensing data, such as radar and lidar, has the capability to permeate cloud cover and smoke. Remote sensing data can be acquired in situations when the collection of other forms of data, such as optical data, is not feasible.

The utilization of high-resolution satellite imagery has facilitated the process of mapping and monitoring temporal alterations in the Earth's surface. Spectral remote sensing data has been employed for the purpose of identifying and cartographically representing various categories of flora, minerals, and other substances. Temporal remote sensing data has been employed for the purpose of monitoring agricultural growth, tracking storm movements, and detecting alterations in land usage. Maps of floodplains, ice sheets, and other topographical features that are difficult or impossible to map using optical methods have been created using synthetic aperture radar (SAR) images. Lidar data has been used to correctly determine the elevation of various geographical components and to create high-resolution cartographic depictions of the Earth's surface.

APPLICATIONS OF REMOTE SENSING

Remote sensing, a method of acquiring information about the Earth's surface without direct contact, has become an indispensable tool in numerous sectors, as detailed in Segarra, et al. (2020) and expanded upon by Sishodia, et al. (2020). The applications of remote sensing span across diverse areas, each benefiting from the unique capabilities of this technology.The many sectors that can profit from remote sensing are shown in Figure 3. Remote sensing, a technology for acquiring information about the Earth's surface without physical contact, has revolutionized our understanding and management of the planet. This note delves into the various applications of remote sensing, highlighting its significance across different sectors. Remote sensing is a non-invasive, highly effective tool for gathering data about the Earth's surface. Its diverse applications play a critical role in various fields, from enhancing agricultural productivity to aiding in disaster response and environmental conservation. As technology advances, the scope of remote sensing continues to expand, offering new possibilities for sustainable management and protection of our planet's resources.

Figure 3. Applications of Remote Sensing

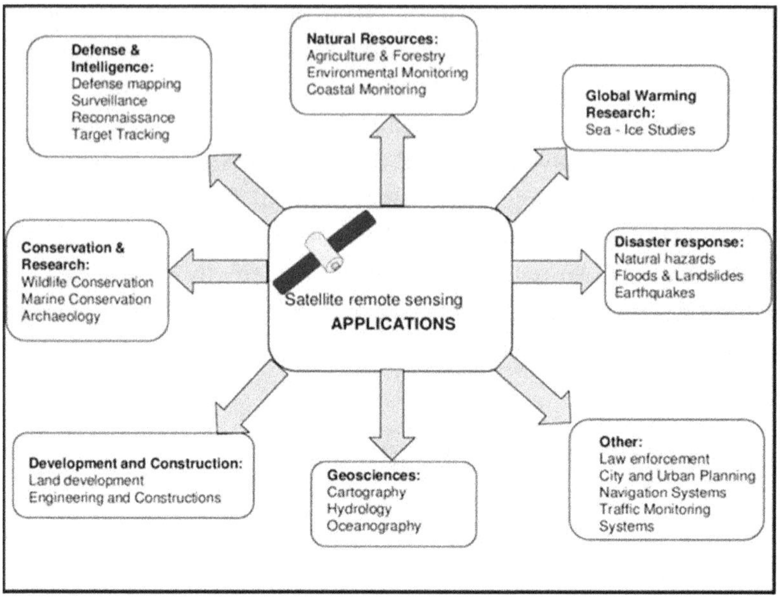

Land Cover Mapping

Land cover mapping involves the identification and categorization of different types of land surfaces within a specific area (Sreelakshmi and Chandra 2023). This process uses remote sensing data to map various land categories, such as forests, grasslands, urban areas, and bodies of water. These maps are invaluable for urban planning, environmental management, and disaster response. They provide critical data for government policies, urban development, and conservation efforts.

Agriculture

Remote sensing has transformed agricultural practices. It enables the detection of pests and diseases, assessment of crop yields, and monitoring of agricultural progress (Faisal *et al.,* 2023). By leveraging remote sensing data, precision agriculture has become more efficient. Farmers can now make informed decisions about pest control, fertilizer application, and irrigation, leading to increased productivity and reduced environmental impact.

Disaster Management

Remote sensing is crucial for disaster management, including preparation, coordination, and execution of strategies to mitigate natural or man-made disasters (Sreelakshmi and Chandra 2024). It provides real-time data on disasters like earthquakes, wildfires, and floods, which is essential for planning rescue and relief operations. This data helps in assessing the extent of damage and streamlining recovery efforts.

Environmental Monitoring

Systematic collection and analysis of environmental data are key aspects of environmental monitoring, remote sensing plays a vital role in this field (Talal *et al.,* 2024). Authorities use remote sensing data to monitor air quality, water quality, and deforestation. This data is critical for enforcing environmental legislation and protecting natural habitats.

Additional Applications

In oceanography, remote sensing aids in the study of ocean currents and marine ecosystems. In hydrology, it's used for water resource surveillance, and in geology, for mapping geological features. The applications extend to climate change research, wildlife monitoring, and even archaeology, demonstrating the technology's versatility.

PYTHON FOR REMOTE SENSING DATA PROCESSING

Introduction to Python

Python is a widely used general-purpose programming language in many different disciplines, one of which being data processing for remote sensing (Helmus and Collis, 2016). Python boasts a sizable and vibrant development community in addition to being a reasonably simple language to learn. Python is also available to everyone as free and open-source software. Python is a good choice for processing data from remote sensing because of a number of its capabilities. The data processing capabilities are highly robust and efficient. Python possesses a multitude of modules and tools that facilitate the seamless handling, manipulation, and analysis of extensive datasets. Python possesses a multitude of libraries and tools that facilitate the execution of scientific computing tasks, encompassing image processing, statistical analysis, and machine learning. Python offers a diverse range of modules and tools that facilitate the creation of visual representations of data, encompassing many forms such as maps, charts, and graphs. Python is renowned for its high extensibility, which allows for the development of novel libraries and tools tailored to cater to the unique requirements of its users. Python is seen as a potent instrument for the processing of remote sensing data due to its ability to adapt to the continuous development of new sensors and data outputs.

Python is a powerful and versatile tool for processing remote sensing data, offering various libraries for reading and writing data in various formats like Geo TIFF, NetCDF, and HDF5 (Esmaili 2021). These libraries enable easy import and export of data for processing and analysis. They also perform image processing tasks like cropping, masking, band math, and filtering, preparing images for analysis and extracting meaningful information. Python also offers statistical analysis libraries to identify patterns and trends in data, drawing conclusions about the Earth's surface. Additionally, Python has libraries for developing and training machine learning models, which can classify images, detect changes in the Earth's surface, and predict future events. As a free and open-source software, Python has a sizable development community and is simple to learn.

Python Libraries for Remote Sensing

sentinelhub-py

sentinelhub-py is a Python library that provides easy access to Sentinel Satellite data and other Earth observation data sources. It simplifies the process of searching, downloading, and processing satellite imagery for various applications, including remote sensing and geospatial analysis. Installing the sentinelhub-py library for Python programming is illustrated in Figure 4.

Figure 4. Python Programming to Install sentinelhub-py

```
#To install
pip install sentinelhub
#To Import
from sentinelhub import SHConfig, BBox,
    CRS, DataCollection, SentinelHubRequest osgeo
```

GDAL (Geospatial Data Abstraction Library)

The GDAL library is a highly adaptable open-source tool used for the manipulation and management of geographical data. Installing the GDAL library for Python programming is illustrated in Figure 5. This software offers a comprehensive suite of tools that facilitate the interpretation, composition, and manipulation of diverse geographic data types.

Figure 5. Python Programming to Install GDAL Library

```
#To install
pip install GDAL
#To Import
from osgeo import GDAL
```

Rasterio

Rasterio is a software library that leverages the capabilities of GDAL to facilitate the manipulation and analysis of geospatial raster data. The Python library in question is designed to facilitate the reading and writing of geographic raster files, with a particular emphasis on user-friendliness. Installing the Rasterio for Python programming is illustrated in Figure 6.

Figure 6. Python Programming to Install Rasterio

```
#To install
pip install rasterio
#To Import
import rasterio
```

Xarray

Xarray is a library for working with scientific data, including remote sensing data. It provides a number of useful features for remote sensing data processing, such as the ability to easily create and manipulate multidimensional arrays, perform data analysis, and create visualizations. Installing the Xarray library for Python programming is illustrated in Figure 7.

Figure 7. Python Programming to Install Xarray

```
#To install Xarray
pip install xarray
#To import Xarray
import xarray as xr
```

Earth Engine

A cloud-based platform called Earth Engine is used to handle and examine data from remote sensing. In addition to a multitude of tools for processing and evaluating the data, it offers access to a vast archive of remote sensing data. Earth Engine is a good option for users who need to process large datasets or who need access to a wide range of remote sensing data. To use Earth Engine, you will need to create an account and obtain an API key. Once you have an API key, you can install the Earth Engine Python library and import it into your Python code as depicted in Figure 8.

Figure 8. Python Programming to Install Earth Engine

```
#To install the Earth Engine Python library
pip install earthengine-api
#To import the Earth Engine Python library
import ee
```

GeoPandas

GeoPandas is a library for working with geospatial vector data. It builds on top of the Pandas library to provide a number of features for working with geospatial data. Reading and writing geospatial vector data in a variety of formats, such as Shapefile, GeoJSON, and GeoPackage. Performing geospatial analysis tasks, such as spatial joins, buffer analysis, and network analysis. Installing the Geopandas library for Python programming is illustrated in Figure 9. GeoPandas is a popular library for remote sensing data processing, as it can be used to work with the vector data that is often associated with remote sensing data, such as land cover maps and administrative boundaries.

Figure 9. Python Programming to Install Geopandas

```
#To install GeoPandas
pip install geopandas
#To import GeoPandas
import geopandas as gpd
```

Scikit-Image

A library for image processing is called Scikit-Image. It offers several helpful capabilities for processing data from remote sensing, including the ability to read and write pictures from remote sensing in many formats, including GeoTIFF, PNG, and JPEG. Carrying out image processing operations include filtering, band math, cropping, and masking. Because it can be used for a variety of image processing tasks on remote sensing data, Scikit-image is a popular library for processing data from remote sensing. The process of installing the Scikit-image library for Python programming is demonstrated in Figure 10.

Figure 10. Python Programming to Install Scikit-image

```
#To install Scikit-image
pip install scikit-image
#To import Scikit-image:
import skimage
```

NumPy

NumPy is a scientific computing package. It offers several tools for working with numerical data, including statistical analysis, multidimensional array creation and manipulation, and mathematical operations on arrays. Because it can be used to do a wide range of numerical operations on remote sensing data, including band math, statistical analysis, and machine learning, NumPy is a well-liked library for processing data from remote sensing systems. To install and import the NumPy library follow the script as shown in Figure 11.

Figure 11. Python Programming to Install NumPy

```
#To install NumPy
pip install numpy
#To import NumPy
import numpy as np
```

Matplotlib

A data visualization library is called Matplotlib. It offers several tools for making maps, graphs, and charts. Because it can be used to generate a wide range of remote sensing data visualizations, including time series plots, change detection maps, and land cover maps, Matplotlib is a well-liked toolkit for processing remote sensing data. The process of installing the Matplotlib library for Python programming is demonstrated in Figure 12.

Figure 12. Python Programming to Install Matplotlib

```
#To install Matplotlib
pip install matplotlib
#To import Matplotlib
import matplotlib.pyplot as plt
```

Basic Python Concepts for Remote Sensing Data Processing

Python is a highly adaptable programming language that demonstrates exceptional proficiency in managing and examining data obtained via remote sensing. Python facilitates the handling of extensive geographic datasets because to its user-friendly learning curve, extensive development community, and diverse range of modules and utilities. The programming language accommodates a range of data types, including integers, floats, texts, and lists, which enables it to effectively express many forms of information in remote sensing data. Variables are employed for performing computations and operations, whereas conditional expressions govern the execution of code. Loops enhance the efficiency of iterating through code blocks, whereas functions encapsulate code to enable its reuse. Python is extensively utilized in the scientific world, boasting a plethora of scientific libraries at its disposal. Python is a powerful tool for studying and manipulating data, allowing for the handling, manipulation, and analysis of extensive geographic datasets. It has the capability to generate a diverse array of applications, spanning from basic scripts to complex machine learning models. Python is a costless and open-source software with a lively development community and is straightforward to acquire.

PREPROCESSING REMOTE SENSING DATA

In order to ensure that the collected data is reliable, consistent, and prepared for additional analysis, preprocessing is essential to the study of remote sensing data. This chapter explores the many ways that Python may be used to preprocess data from remote sensing applications, including data conversion and import, geometric correction, and radiometric correction.

Data Import and Conversion

Understanding Data Formats

Remote sensing data comes in various formats (e.g., raster, vector, spectral). The chapter will begin by discussing common file formats (such as GeoTIFF, NetCDF, etc.) and methods for importing and converting these formats into workable data structures using Python libraries like GDAL, Rasterio, and xarray.

Exploring Metadata

Discuss the significance of metadata in remote sensing data, including spatial and spectral information. Figure 13 demonstrate how Python libraries assist in extracting and utilizing metadata for analysis and processing. Python Implementation using Libraries like GDAL or Rasterio.

Figure 13. Installation of GDAL or Rasterio

```
import rasterio
# Open the input image
with rasterio.open('input_image.tif') as src:
    # Extract metadata, bands, and other relevant information
    metadata = src.meta
    bands = src.read()
# Perform data processing, format conversion, or reprojection if necessary
# Save the processed data in a consistent format
with rasterio.open('output_image.tif', 'w', **metadata) as dst:
    dst.write(bands)
```

Geometric Correction

Image Registration and Georeferencing

Explain the importance of geometric correction in aligning remote sensing images to a common coordinate system. Detail methods for image registration and georeferencing using Python, covering concepts like orthorectification, resampling, and coordinate system transformation.

Python Implementation

Provide Python code snippets using libraries like Geospatial Data Abstraction Library (GDAL) and OpenCV for implementing geometric correction techniques, showcasing step-by-step procedures and best practices in Figure 14.

Figure 14. Python For Geometric Correction Techniques

```
import gdal
# Open the raw image and extract georeferencing data
image = gdal.Open('raw_image.tif')
georef_data = image.GetGeoTransform()
# Apply geometric correction methods (e.g., resampling, warping) to rectify distortions
# Update or reproject the image for accurate spatial alignment
# Example: we'll assume you've applied correction and created a corrected_image
# Save the corrected image to a new file
corrected_image_path = 'corrected_image.tif'
corrected_image = gdal.GetDriverByName('GTiff').Create(corrected_image_path, image.RasterXSize, image.RasterYSize,
                                                        image.RasterCount, image.GetRasterBand(1).DataType)
corrected_image.SetGeoTransform(georef_data)
corrected_image.GetRasterBand(1).WriteArray(corrected_image_data)
corrected_image = None
```

Radiometric Correction

Calibration and Enhancements

Describe the necessity of radiometric correction in removing distortions caused by sensor characteristics and atmospheric conditions. Cover topics like sensor calibration, atmospheric corrections, and image enhancement techniques using Python libraries (e.g., Spectral Python, OpenCV).

Python Implementation

Illustrate how Python enables the implementation of radiometric correction techniques, including histogram matching, normalization, and transformation. Figure 15 could incorporate code samples that utilize appropriate Python libraries to perform radiometric correction procedures. Python enables the implementation of preprocessing procedures in the analysis of remote sensing data. The techniques of data import and conversion, geometric correction, and radiometric correction are employed to guarantee the reliability and quality of the data prior to any subsequent analysis. The extensive libraries and tools available in Python allow for effective preprocessing activities and enable the seamless preparation of remote sensing data for useful analysis and insights.

Figure 15. Python for Radiometric Calibration

```
import spectral
# Read the spectral image and perform radiometric calibration
img = spectral.open_image('raw_image.hdr')
# Implement radiometric correction techniques (e.g., sensor calibration, atmospheric correction)
# Normalize or adjust pixel values for consistency and accuracy in data analysis
```

IMAGE ENHANCEMENT

Remote sensing, a vital technology for capturing data from a distance, provides a wealth of information for various fields like environmental monitoring, agriculture, and urban planning. However, raw images often contain imperfections due to noise, limited contrast, or blurred details. Image enhancement techniques play a pivotal role in improving the quality and interpretability of these images. The Figure 16 provides an illustrative infographic demonstrating the concept of image enhancement in remote sensing, showing the transformation of a remote sensing image before and after the application of enhancement techniques. Image enhancement techniques, namely contrast enhancement, edge detection, and filtering, are essential in remote sensing data analysis. Python's versatile libraries such as OpenCV empower analysts and researchers to enhance and extract critical information from remote sensing imagery. Understanding and applying these techniques enables more accurate data interpretation, paving the way for informed decision-making in various fields relying on remote sensing technology.

Figure 16. Image Enhancement

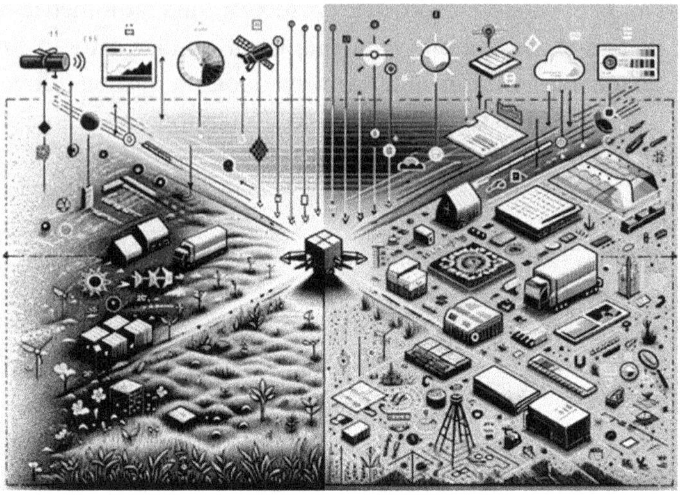

Contrast Enhancement

Contrast augmentation is crucial for enhancing the quality of remote sensing photographs. It enhances the perception of crucial elements in the image, simplifying the process for analysts to comprehend and extract useful data. This chapter will examine the significance of contrast enhancement and present a Python implementation utilizing OpenCV to accomplish this. Implementation of Python programming language. The OpenCV library is employed to augment the contrast of remote sensing photos, as illustrated in Figure 17. This code sample exemplifies a straightforward and efficient approach to accomplish this.

Figure 17. Python Programming to Apply Contrast Enhancement

```
import cv2
import numpy as np
# Load your remote sensing image
image=cv2.imread('remote_sensing_image.jpg',cv2.IMREAD_GRAYSCA    LE)
# Apply contrast enhancement using CLAHE (Contrast Limited Adaptive Histogram Equalization)
clahe = cv2.createCLAHE(clipLimit=2.0, tileGridSize=(8, 8))
enhanced_image = clahe.apply(image)
# Save the enhanced image
cv2.imwrite('enhanced_remote_sensing_image.jpg',enhanced_image)
```

contrast enhancement using CLAHE (Contrast Limited Adaptive Histogram Equalization) applied. CLAHE is an adaptive method that enhances contrast while preventing over-amplification of noise. You can adjust the clip Limit and tile Grid Size parameters to control the extent of enhancement. The enhanced image is saved using cv2.imwrite(). By implementing this code, you can significantly improve the visibility of critical features within your remote sensing images, making it easier for analysts to extract valuable insights.

Edge Detection Technique

A basic method of image processing called edge detection is used to locate boundaries inside an image. It is particularly helpful for remote sensing photos, which can provide a plethora of data regarding both natural and artificial characteristics. Through the extraction of edge data from remote sensing photos, analysts may learn more about a variety of uses. Roads, rivers, houses, and plants are just a few of the elements that may be extracted from remote sensing photos using edge detection. The creation of maps, the updating of land cover databases, and the tracking of environmental changes over time may all be done using this information. In remote sensing photographs, things like cars, planes, and ships may be found using edge detection. Numerous uses for this data exist, including border security, disaster response, and traffic monitoring. Remote sensing photos may be processed to assess the landscape using edge detection. Digital elevation models (DEMs), which may be utilized for a number of tasks including planning and building, hazard assessment, and watershed management, can be made using this information. Using the OpenCV library, the edge detection technique is implemented in the following Python code snippet to extract edge information from remote sensing images as shown in Figure 18.

Figure 18. Python Programming to Apply the Edge Detection Techniques

```
import cv2
import numpy as np
# Load the remote sensing image
image = cv2.imread("city_image.jpg", cv2.IMREAD_GRAYSCALE)
# Apply Canny edge detection
edges = cv2.Canny(image, 100, 200)
# Display the edges
cv2.imshow("Edges", edges)
cv2.waitKey(0)
```

The cv2.Canny() function is a popular edge detection algorithm that uses a Gaussian filter to blur the image and then applies two thresholds to identify pixels that are likely to be edges. The two thresholds are: Lower threshold this threshold controls the sensitivity of the edge detector. Lower values will result in more edges being detected but may also lead to more noise in the image. Upper threshold this threshold controls the strength of the edges that are detected. Higher values will result in stronger edges being detected but may also lead to some edges being missed.

The cv2.imshow() function displays the edges image. The cv2.waitKey(0) function waits for the user to press a key before closing the window.

Image Filtering

A wide range of image processing procedures are categorized as filtering techniques, and they are used to alter the look of images. There are several uses for filtering. Image noise in remote sensing applications is a prevalent issue that may be attributed to several sources, including air dispersion, sensor noise, and quantization errors. Filtering can be used to reduce noise in remote sensing images, making it easier to identify and interpret subtle features. Image sharpening is the process of enhancing the contrast between adjacent pixels in an image. This can make it easier to see small details in the image, such as edges and textures. Filtering can be used to sharpen remote sensing images, making them more informative for analysis. Smoothing is the process of reducing the overall contrast in an image. This can make it easier to see broad patterns and trends in the image. Filtering can be used to smooth remote sensing images, making them more suitable for certain types of analysis. Python implementation of Gaussian Blur: an illustrative example Figure 19 demonstrates the application of a filtering technique, specifically Gaussian blur, to reduce noise and enhance the legibility of the image.

Figure 19. Python Programming to Apply the Filtering Techniques

```
import cv2
import numpy as np
# Load the remote sensing image
image = cv2.imread("city_image.jpg")
# Apply Gaussian blur
blurred_image = cv2.GaussianBlur(image, (5, 5), 0)
# Display the blurred image
cv2.imshow("Blurred Image", blurred_image)
cv2.waitKey(0)spectral
```

The image is subjected to a Gaussian filter via the cv2.GaussianBlur () method. A Gaussian filter is a kind of low-pass filter that averages the values of nearby pixels to blur the image. The Gaussian filter's size is determined by the kernel size parameter. Greater blurring will be the outcome of larger kernel sizes.

IMAGE ANALYSIS

Image analysis is a fundamental aspect of remote sensing data processing, allowing researchers and analysts to extract valuable information from images captured by satellites, drones, and other sensors. This chapter focuses on how Python facilitates image analysis in remote sensing by exploring essential techniques such as band math, indices, classification, and segmentation. Image analysis techniques in remote sensing data using Python. The examples provided for band math, indices calculation, classification, and segmentation showcase Python's adaptability and efficiency in handling diverse image analysis tasks. Understanding and implementing these techniques allow for detailed extraction of valuable information from remote sensing imagery, facilitating informed decision-making in various domains relying on remote sensing data.

Band Math Calculations

Band math is a potent method that enables us to conduct mathematical calculations on various bands of multispectral or hyperspectral data. This procedure facilitates the formation of new composite bands, discerning distinct characteristics, or augmenting particular components within the images. This chapter will examine the significance of band math and present a Python application utilizing Sentinel Satellite imagery. In order to illustrate band mathematics using Sentinel Satellite imagery, we will utilize the Sentinel Hub Python package. This library streamlines the procedure of retrieving and modifying Sentinel Satellite data. If you have not yet done so, it will be necessary for you to install the library. The code in Figure 20 allows you to set up the Sentinel Hub by providing your client ID, client secret, and instance ID. Specify the region of interest, duration, and indicate the preferred spectral bands (specifically, the Red and NIR bands in this instance). Retrieve Sentinel-2 data and download the designated spectral bands. Execute spectral band calculations, such as the computation of the Normalized Difference Vegetation Index (NDVI), which is a widely used use of this method. Display the outcome using Matplotlib.

Indices Computation

Indices such as NDVI, NDBI, and SAVI are crucial derived metrics employed to delineate distinct attributes in remote sensing imaging. These indices allow us to evaluate many characteristics of the Earth's surface, such as the health of vegetation, the level of urban development, or the moisture content of soil. The code presented in Figure 21 is intended to compute the Normalized Difference Vegetation Index (NDVI) using satellite image data and store the resulting NDVI values in a separate image file. Additionally, it generates a colormap depiction of the NDVI data, depicted in Figure 22. The purpose of this code is to do image processing, compute the Normalized Difference Vegetation Index (NDVI), and store the output according to the provided requirements. Ensure that you have installed the required libraries and accurately specify the file locations for your data.

Figure 20. Python Programming you to set up the Sentinel Hub

```
import cv2
import numpy as np
# Load the remote sensing image
pip install sentinelhub-py
from sentinelhub import SHConfig, FisRequest, BBox, Geometry, CRS, WcsRequest
import numpy as np
import matplotlib.pyplot as plt
# Configuration for Sentinel Hub
config = SHConfig()
config.sh_client_id = 'your_client_id'
config.sh_client_secret = 'your_client_secret'
# Define the area of interest and time frame
bbox = BBox(bbox=[13.8222, 45.8524, 13.9958, 45.9194], crs=CRS.WGS84)
geometry = Geometry(geometry=bbox, crs=CRS.WGS84)
time_interval = ('2023-01-01', '2023-01-31')
# Request Sentinel-2 data
bands = ['B02', 'B03', 'B04']  # Specify the bands you want to use
data = WcsRequest(data_folder='sentinel_data',
                  layer='BANDS-S2-L2A',
                  bbox=bbox,
                  time=time_interval,
                  data_source=DataSource.SENTINEL2_L2A,
                  instance_id='your_instance_id',
                  custom_url_params={CustomUrlParam.SHOWLOGO: False},
                  config=config).get_data()
# Perform band math (e.g., NDVI calculation)
red_band = data[bands.index('B04')]
nir_band = data[bands.index('B08')]
ndvi = (nir_band - red_band) / (nir_band + red_band)
# Visualize the NDVI result
plt.imshow(ndvi, cmap='RdYlGn')
plt.colorbar()
plt.show()
```

Figure 21. Python Programming to Calculate the NDVI Index

```
import rasterio
import numpy as np
from xml.dom import minidom
import matplotlib.pyplot as plt
# Input image file and metadata XML file
image_file = "20161228_101647_0e26_3B_AnalyticMS.tif"
metadata_file = "20161218_101700_0e0d_3B_AnalyticMS_metadata.xml"
# Load red and NIR bands
with rasterio.open(image_file) as src:
    band_red = src.read(3)
    band_nir = src.read(4)
# Parse metadata to get reflectance coefficients
xmldoc = minidom.parse(metadata_file)
nodes = xmldoc.getElementsByTagName("ps:bandSpecificMetadata")
coeffs = {}
for node in nodes:
    bn = node.getElementsByTagName("ps:bandNumber")[0].firstChild.data
    if bn in ['3', '4']:
        i = int(bn)
        value = node.getElementsByTagName("ps:reflectanceCoefficient")[0].firstChild.data
        coeffs[i] = float(value)
# Multiply by corresponding coefficients
band_red = band_red * coeffs[3]
band_nir = band_nir * coeffs[4]
# Calculate NDVI
ndvi = (band_nir.astype(float) - band_red.astype(float)) / (band_nir + band_red)
# Output NDVI as GeoTIFF
with rasterio.open('ndvi.tif', 'w', **src.meta) as dst:
    dst.write(ndvi, 1)
# Output NDVI as colormap visualization
plt.imsave("ndvi_cmap.png", ndvi, cmap=plt.cm.summer)
```

Figure 22. Calculated NDVI Visualization

Image Classification

Classification is an essential method in remote sensing that allows for the grouping of pixels in an image into certain classes or categories. This procedure is essential for tasks such as mapping land cover, identifying features, and analyzing temporal changes. Figure 23 (A) and (B) programming examines the importance of classification and presents a Python implementation using Scikit-learn for a basic supervised classification technique, specifically Random Forest, for the classification of land cover, as depicted in Figure 24. Scikit-learn is utilized for land cover categorization using a Random Forest classifier. Make sure that you have Scikit-learn and any other required libraries installed.

Figure 23. Python Programming for Image Classification

```
import rasterio as rio
import geopandas as gpd
import pandas as pd
import numpy as np
from sklearn.model_selection import train_test_split
from sklearn.svm import SVC
from sklearn.metrics import classification_report, accuracy_score, confusion_matrix
import matplotlib.pyplot as plt
import seaborn as sns
# Input files
raster_loc = 'materials/rasters/s2image.tif'
points_loc = 'materials/shapefiles/samples.shp'
temp_point_loc = 'materials/temp/temp_y_points.shp'
lulc_name = ['Water', 'Dense Veg', 'Veg', 'Impervious']
# Read bands from input raster
with rio.open(raster_loc) as img:
    bands = img.count
print('Bands of input image:', bands)
# Create a list of feature names
features = ['band' + str(i) for i in range(1, bands + 1)]
print('Bands names:', features)
f_len = len(features)
# Read points from shapefile
points = gpd.read_file(points_loc)
# Add a new column 'id' with range of points
points['id'] = range(len(points))
# Save the new point file with 'id'
points.to_file(temp_point_loc)
# Convert GeoDataFrame to DataFrame and remove the 'geometry' column
points_df = pd.DataFrame(points.drop(columns='geometry'))
# Initialize an empty DataFrame to store sampled values
sampled = pd.DataFrame()
# Read input shapefile with Fiona and iterate over each feature
with fiona.open(temp_point_loc) as shp:
    for feature in shp:
        siteID = feature['properties']['id']
        coords = feature['geometry']['coordinates']
        # Read pixel values at the given coordinates using Rasterio
        with rio.open(raster_loc) as stack_src:
            values = [v for v in stack_src.sample([coords])]
        # Create a temporary DataFrame for the sampled values
        temp_df = pd.DataFrame(values, columns=features)
        temp_df['id'] = siteID
        sampled = sampled.append(temp_df, ignore_index=True)
# Split the data into features (X) and labels (Y)
X = sampled.iloc[:, :f_len].values
Y = sampled['labels'].values
# Split the data into training and testing sets
X_train, X_test, Y_train, Y_test = train_test_split(X, Y, test_size=0.30, stratify=Y)
print(f'X_train Shape: {X_train.shape}\nX_test Shape: {X_test.shape}\nY_train Shape:
      {Y_train.shape}\nY_test Shape: {Y_test.shape}')
# Initialize and train an SVM classifier
clf = SVC(kernel='rbf')
clf.fit(X_train, Y_train)
# Make predictions on the test data
Y_pred = clf.predict(X_test)
# Evaluate the classifier's performance
accuracy = accuracy_score(Y_test, Y_pred)
report = classification_report(Y_test, Y_pred)
confusion = confusion_matrix(Y_test, Y_pred)
print(f"Accuracy: {accuracy * 100:.2f}%")
print(report)
print('Confusion Matrix:\n', confusion)
```

Image Segmentation

Image segmentation is a significant technique in image processing that involves dividing an image into meaningful pieces or segments based on shared properties. This technique greatly facilitates the identification of certain regions, objects, or patterns within a picture. This chapter will examine the significance of picture segmentation and present a Python solution that utilizes Scikit-image. In order to segment images, Scikit-image is employed, as depicted in Figure 25. Scikit-image is a robust library for image processing. Make sure that you have Scikit-image and any other required libraries installed.

Figure 24. Python Programming to Perform the Image Classification

```
# Visualize the confusion matrix
cm_percent = confusion / np.sum(confusion)
plt.figure(figsize=(7, 7), facecolor='w', edgecolor='k')
sns.set(font_scale=1.5)
sns.heatmap(cm_percent, xticklabels=lulc_name, yticklabels=lulc_name, cmap="YlGn", annot=True, fmt='.2%',
            cbar=False, linewidths=2, linecolor='black')
plt.title('SVM')
plt.xlabel('Predicted')
plt.ylabel('Actual')
# Export the prediction results as a GeoTIFF
exp_name = 'materials/results/lulc_SVM.tif'
with rio.open(raster_loc) as img:
    img_arr = img.read()
    bands = img_arr.shape[0]
    height, width = img_arr.shape[1], img_arr.shape[2]
    crs = img.crs
    transform = img.transform
# Reshape the input data for prediction
img_n = np.moveaxis(img_arr, 0, -1)
img_n = img_n.reshape(-1, f_len)
# Perform predictions on the entire raster
pred_full = clf.predict(img_n)
img_reshape = pred_full.reshape(height, width)
# Export the results as a GeoTIFF
out_raster = rio.open(exp_name, 'w', driver='GTiff', height=height, width=width, count=1, dtype='uint8',
                      crs=crs, transform=transform, nodata=255)
out_raster.write(img_reshape, 1)
out_raster.close()
```

Figure 25. Predicted ML-Based LULC Results for SVM, RF, and DT

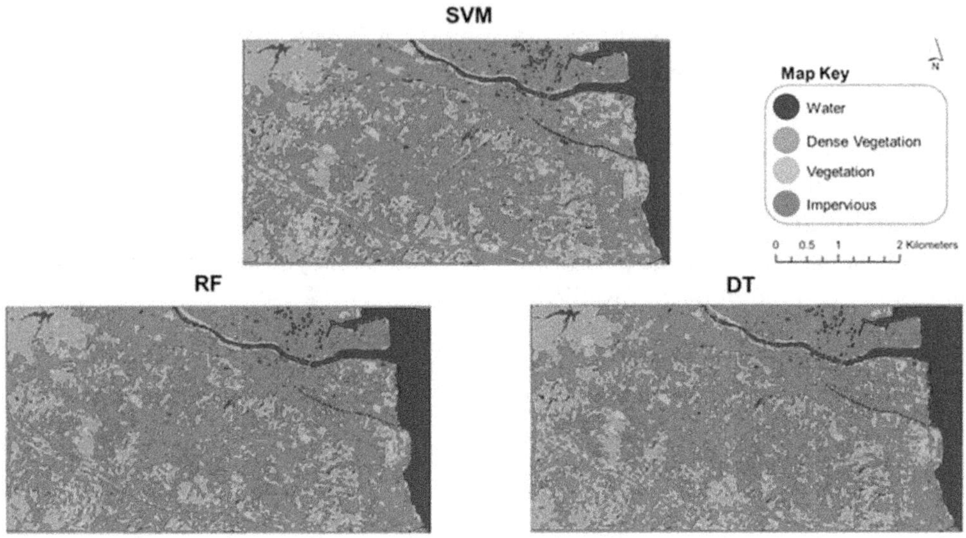

POSTPROCESSING AND VISUALIZATION

Postprocessing and visualization are crucial steps in remote sensing data analysis. They involve exporting processed data and creating informative maps and charts to communicate your findings effectively. Postprocessing and visualization are important steps in remote sensing data analysis. By exporting processed data and creating informative maps and charts, you can effectively communicate your findings to others. Python offers a variety of libraries and tools for postprocessing and visualization, making it a powerful platform for remote sensing data analysis. Postprocessing and visualization are crucial steps in remote sensing data analysis. They involve exporting processed data and creating informative maps and charts to communicate your findings effectively. This chapter explore the importance of postprocessing and visualization in remote sensing and provide Python implementations for exporting data, creating maps, and generating charts.

Exporting Data

Exporting data is crucial for disseminating your findings and engaging in collaborative efforts. Python provides a range of techniques for exporting data in many formats, including CSV, Geo TIFF, and shapefiles. Figure 26 illustrates the process of exporting data as a CSV file using the Pandas library.

Figure 26. Python Programming to Apply the Segmentation

```
import numpy as np
import matplotlib.pyplot as plt
from skimage import data
from skimage.color import rgb2gray
from skimage.filters import sobel
from skimage.segmentation import slic, mark_boundaries
# Load an example image (replace with your image)
image = data.coins()
# Convert the image to grayscale
image_gray = rgb2gray(image)
# Apply edge detection (e.g., Sobel filter)
edges = sobel(image_gray)
# Perform superpixel segmentation using SLIC
segments = slic(image, n_segments=100, compactness=10, sigma=1)
# Visualize the segmentation result
fig, ax = plt.subplots(1, 2, figsize=(10, 5))
ax[0].imshow(image, cmap='gray')
ax[1].imshow(mark_boundaries(image, segments, color=(1, 0,0)))
plt.show()
```

Creating Maps and Charts

Charts and maps are effective tools for visualizing data obtained through remote sensing and disseminating information to a wider audience. Python offers a variety of tools, such as Matplotlib, Folium, and Plotly, that facilitate the creation of maps and charts. Figure 27 depicts the process of utilizing Matplotlib to generate a fundamental scatter plot.

Figure 27. Python Programming to Export Data as CSV File

```
import pandas as pd
# Load the processed data
df = pd.read_csv('processed_data.csv')
# Export the data to a CSV file
df.to_csv('exported_data.csv', index=False)
```

More Advanced Examples

Another type of theme map is the choropleth map, which is a type of theme map that uses different colors or shades to show different values for a statistical variable across a number of different geographical regions. Figure 28 illustrates the program that can be used to build a choropleth map using the Folium package, which is in the Python programming language.

Figure 28. Python Programming to Create Maps and Charts

```
import matplotlib.pyplot as plt
# Create a scatter plot
plt.scatter(df['x'], df['y'])
# Set the title and labels
plt.title('Scatter Plot of Processed Data')
plt.xlabel('X')
plt.ylabel('Y')
# Show the plot
plt.show()
```

A time series chart is a type of chart that shows the variation of a variable over time. To create a time series chart in Python, you can use the Plotly library as depicted in Figure 29.

Figure 29. Python Programming to Create a Folium Map

```
import folium
# Create a Folium map
map = folium.Map()
# Add a choropleth layer
folium.Choropleth(
    geo_data=geojson_file,
    data=df,
    columns=['region', 'value'],
    fill_color='YlOrRd',
    fill_opacity=0.7,
    line_opacity=0.2,
    legend_name='Value'
).add_to(map)
# Show the map
map.show()
```

Figure 30. Python Programming for Time Series Chart

```
import plotly.express as px
# Create a time series chart
fig = px.line(df, x='date', y='value')
# Set the title and labels
fig.update_layout(title='Time Series Chart of Processed Data', xaxis_title='Date', yaxis_title='Value')
# Show the chart
fig.show()
```

CASE STUDIES

Land Cover Classification

Land cover classification refers to the systematic procedure of recognising and delineating several categories of land cover, encompassing diverse terrain types such as forests, grasslands, urban regions, and bodies of water (Reddy, *et al.*, 2015). The land cover classification of the World pictures was conducted using the Python programming language, as depicted in Figure 7. Python is widely utilised for land cover categorization due to its status as a freely available and open-source computer language, which is further enhanced by its extensive developer community. Python is equipped with various modules and tools that facilitate the processing of remote sensing data. A prevalent method involves the utilisation of machine learning algorithms, which can undergo training utilising annotated remote sensing photos in order to discern and classify distinct land cover categories. Python offers a range of machine learning tools, including scikit-learn, TensorFlow, and PyTorch, that facilitate the development and training of models for land cover classification.

Figure 31. Land Cover Classification of the World Images By Using Python

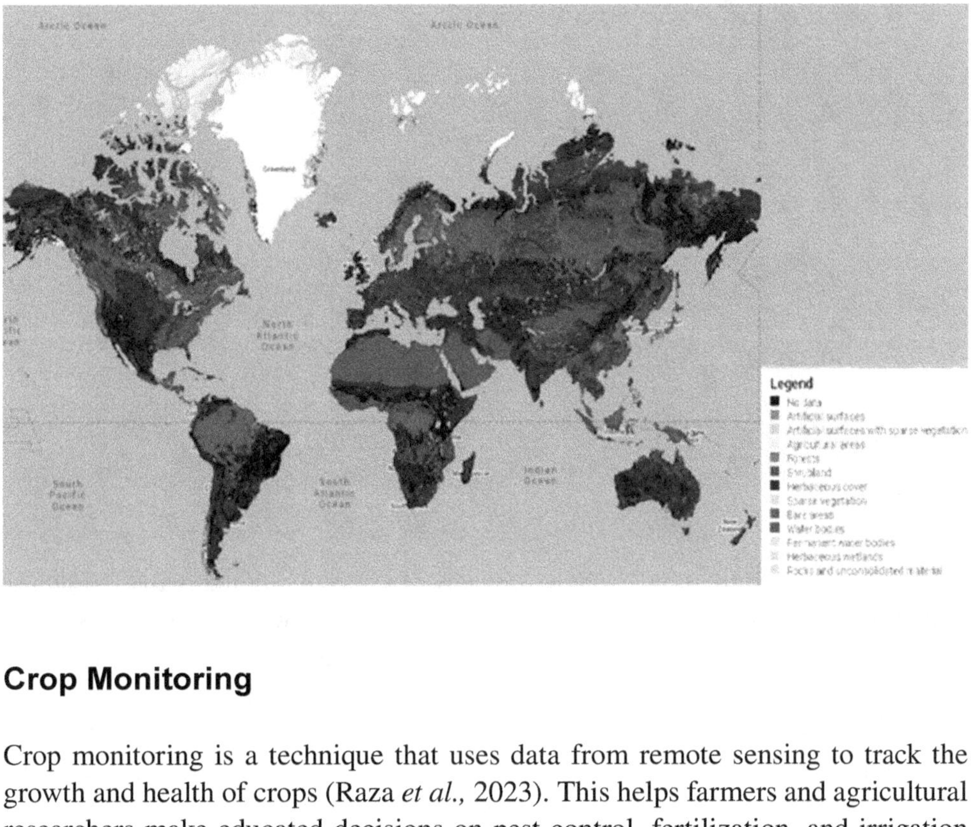

Crop Monitoring

Crop monitoring is a technique that uses data from remote sensing to track the growth and health of crops (Raza *et al.,* 2023). This helps farmers and agricultural researchers make educated decisions on pest control, fertilization, and irrigation (Barman, et al., 2022). Python is a freely available open-source programming language that is commonly used for agricultural monitoring. It contains an abundance of tools and modules that make data processing easy. Using a vegetation index—like the Normalized Difference Vegetation Index (NDVI), which measures the amount of greenness displayed by vegetation—is a commonly used method (Figure 8). A picture obtained from remote sensing is processed by using its red and near-infrared bands to calculate the Normalized Difference Vegetation Index (NDVI). Higher vegetation density is indicated by higher NDVI readings. For tracking crop health and growth, the Normalized Difference Vegetation Index (NDVI) is a useful instrument. Its application helps farmers to recognize regions of crop stress and then take the necessary actions, such applying fertilizer or irrigation, to alleviate the problems they see.

Figure 32. NDVI Based Crop Health Monitoring

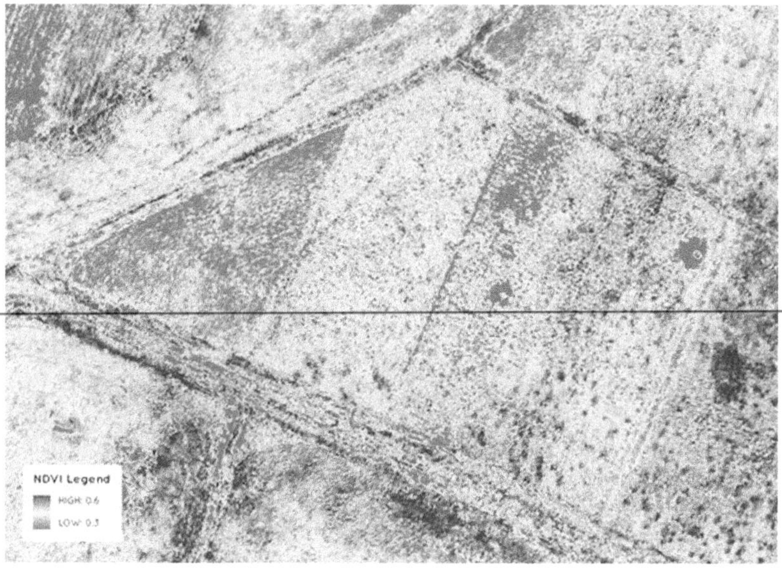

The African continent is home to a variety of crops, including maize, sorghum, and cassava. However, crop yields in Africa are often low due to factors such as drought, pests, and diseases. A team of researchers used Python to develop a crop monitoring application to help farmers in Africa improve their crop yields. The application used a random forest classifier to classify remote sensing images of crops into different health classes, such as healthy, stressed, and diseased. The application was used to monitor crops in Africa over a period of 5 years. The results showed that farmers who used the application were able to increase their crop yields by an average of 15%. This case study demonstrates the power of Python for crop monitoring. Applications for agricultural monitoring that assist farmers increase crop yields may be created with Python. Python is a strong and adaptable agricultural monitoring tool. It is free and open-source software, has a sizable and vibrant development community, and is simple to learn.

Disaster Response

Disaster response is the process of providing assistance to people who have been affected by a disaster. Disaster response can involve a variety of activities, such as search and rescue, medical assistance, and providing food and shelter. Python is a popular tool for disaster response. It is an open-source, free programming language with a sizable and vibrant development community. Additionally, Python offers many tools and frameworks that make data processing and analysis simple—a crucial skill for disaster response. Creating damage assessment tools is one way that Python is applied to disaster response. Damage assessment is the process of determining and evaluating the harm that a disaster has caused. Recovery activities can be planned and resourced using this knowledge.

CONCLUSION

This chapter examines how Python may be used to analyze data from remote sensing, emphasizing the platform's adaptability in gathering and processing earth-related data. This data can be processed, analyzed, and visualized well with Python, which makes it useful for a variety of tasks including classifying land cover and responding to disasters. The chapter introduces readers to various libraries and tools, such as GDAL, Rasterio, Xarray, Earth Engine, GeoPandas, Scikit-image, NumPy, and Matplotlib. Basic Python concepts are also covered to provide a solid foundation for those new to the language. The chapter covers preprocessing operations, such as data import and conversion, geometric correction, and radiometric correction, as well as image enhancement techniques like contrast enhancement, edge detection, and filtering. It then delves into image analysis, covering band mathematics, indices, classification, and segmentation.

Additionally covered are post-processing and visualization, which show how to export data and make educational maps and charts. The use of Python in remote sensing data processing for tasks including land cover categorization, agricultural monitoring, and disaster response is illustrated through case studies. In conclusion, there is no denying Python's ability to analyze remote sensing data effectively. It gives professionals and academics the means to fully realize the potential of remote sensing data for a range of uses. Python will remain a fundamental tool in the sector as technology develops and the value of remote sensing data increases, helping us to comprehend and safeguard our planet.

REFERENCES

Barman, D., Saha, R., Bhowmick, T., Bagui, A., & Dutta, G. (2022). Role of GIS, remote sensing and agro advisory in conservation agriculture. In *Conservation Agriculture and Climate Change* (pp. 233–248). CRC Press.

Campbell, J. B., & Wynne, R. H. (2011). *Introduction to remote sensing*. Guilford press.

Esmaili, R. B. (2021). *Earth Observation Using Python: A Practical Programming Guide* (Vol. 75). John Wiley & Sons. doi:10.1002/9781119606925

Faisal, G., Sreelakshmi, S., & Chandra, S. S. V. (2023, April). Crop yield prediction for smart agriculture with climatic parameters using random forest. In *International Conference on Advances in Computing and Data Sciences* (pp. 367-376). Cham: Springer Nature Switzerland. 10.1007/978-3-031-37940-6_30

Helmus, J. J., & Collis, S. M. (2016). The Python ARM Radar Toolkit (Py-ART), a library for working with weather radar data in the Python programming language. *Journal of Open Research Software, 4*(1), 4. doi:10.5334/jors.119

Ilie, G. A., Vasile, G., d'Urso, G., & Boldo, D. (2011). *Spaceborne SAR tomography: application in urban environment* [Doctoral dissertation, GIPSA-LAB].

Jia, J., Sun, H., Jiang, C., Karila, K., Karjalainen, M., Ahokas, E., Khoramshahi, E., Hu, P., Chen, C., Xue, T., Wang, T., Chen, Y., & Hyyppä, J. (2021). Review on active and passive remote sensing techniques for road extraction. *Remote Sensing (Basel), 13*(21), 4235. doi:10.3390/rs13214235

Liu, P. (2015). A survey of remote-sensing big data. *Frontiers in Environmental Science, 3,* 45.

Mirzakarimova, G. M. (2023). Remote sensing data: international experiences and applications. *Italy" actual problems of science and education in the face of modern challenges" 14*(1).

Navalgund, R. R., Jayaraman, V., & Roy, P. S. (2007). Remote sensing applications: An overview. *Current Science*, 1747-1766.

Raza, A., Shahid, M. A., Safdar, M., Tariq, M. A. R., Zaman, M., & Hassan, M. U. (2023, October). Exploring the Impact of Digital Farming on Agricultural Engineering Practices. In *Biology and Life Sciences Forum* (Vol. 27, No. 1, p. 10). MDPI. 10.3390/IECAG2023-15386

Reddy, C. S., Jha, C. S., Diwakar, P. G., & Dadhwal, V. K. (2015). Nationwide classification of forest types of India using remote sensing and GIS. *Environmental Monitoring and Assessment*, *187*(12), 1–30. doi:10.1007/s10661-015-4990-8 PMID:26615560

Segarra, J., Buchaillot, M. L., Araus, J. L., & Kefauver, S. C. (2020). Remote sensing for precision agriculture: Sentinel-2 improved features and applications. *Agronomy (Basel)*, *10*(5), 641. doi:10.3390/agronomy10050641

Sishodia, R. P., Ray, R. L., & Singh, S. K. (2020). Applications of remote sensing in precision agriculture: A review. *Remote Sensing (Basel)*, *12*(19), 3136. doi:10.3390/rs12193136

Sreelakshmi, S., & Chandra, S. V. (2024). Visual saliency-based landslide identification using super-resolution remote sensing data. *Results in Engineering*, *21*, 101656. doi:10.1016/j.rineng.2023.101656

Sreelakshmi, S., & Vinod Chandra, S. S. (2023, January). Landslide classification using deep convolutional neural network with synthetic minority oversampling technique. In *International Conference on Distributed Computing and Intelligent Technology* (pp. 240-252). Cham: Springer Nature Switzerland. 10.1007/978-3-031-24848-1_17

Talal, M., Raza, A., Safdar, M., Al Ansari, M. S., Ali, S. K., & Sattar, J. (2024). Optical Sensing for Precision Agriculture. In Agriculture and Aquaculture Applications of Biosensors and Bioelectronics (pp. 210-237). IGI Global. doi:10.4018/979-8-3693-2069-3.ch011

Chapter 4
Towards Spatial Data Visualization With Python:
Unveiling Geographic Patterns and Trends

Soressa Beyene Lemu
Bule Hora University, Ethiopia

Desalegn Aweke Wako
https://orcid.org/0000-0003-2803-7959
Bule Hora University, Ethiopia

Tesfaye Fufa
Space Science and Geospatial Institute, Ethiopia

Mohammad Gouse Galety
https://orcid.org/0000-0003-1666-2001
Samarkand International University of Technology, Uzbekistan

ABSTRACT

This chapter aims to explore the power of Python in spatial data visualization. Spatial data visualization is the process of representing spatial information visually, enabling one to explore and communicate patterns, distributions, and relationships within the data. An informative spatial data visualization with Python effectively represents and communicates spatial information using visual elements, enabling users to gain insights and make informed decisions related to geospatial data. Python provides many sets of libraries and tools for handling and visualizing geospatial data to enhance understanding, facilitate exploration, and present geographic patterns and relationships clearly and intuitively. The chapter demonstrates the capabilities of Python for spatial data visualization by showcasing various techniques, spatial data formats, and tools with Geopandas, Matplotlib, Plotly, and Folium libraries. Examples and code snippets are provided for the readers to gain solid knowledge about spatial data visualization using Python.

DOI: 10.4018/979-8-3693-1754-9.ch004

INTRODUCTION TO SPATIAL DATA VISUALIZATION

Spatial data visualization is the process of representing geographic information and patterns using visual elements such as maps, graphs, and charts. It is a powerful tool for analyzing and understanding complex spatial relationships and patterns that are difficult to discern through raw data alone. Python is an open-source programming language that has become increasingly popular for spatial data analysis and visualization. With its rich collection of libraries and tools, Python provides a powerful platform for working with spatial data. An introduction to spatial data visualization provides a foundational understanding of how geographical information can be effectively represented and communicated. It encompasses a variety of techniques and methods used to visually convey spatial patterns, relationships, and distributions. Spatial data visualization explains understanding Geographic data and understanding different types of geographical data such as points, lines, polygons, and rasters, as well as their associated attributes. This helps in understanding how to represent and visualize the data effectively.

There are several common types of spatial visualizations, including choropleth maps, symbol maps, heatmaps, flow maps, 3D visualizations, and interactive web maps, each serving different purposes and offering unique ways to explore and understand spatial data. The tools and libraries commonly used for spatial visualization, such as GeoPandas, Matplotlib, Folium, Plotly, and others in the Python ecosystem, provide a range of functionalities to create diverse and engaging visualizations. Techniques for visualizing spatial data using Python with cover topics such as data preparation, map creation, and spatial analysis, and provide examples of how these techniques can be applied to real-world problems.

Spatial data visualization enables us to explore, analyze, and communicate geographical patterns, trends, and insights, helping to make informed decisions in various domains such as urban planning, public health, and environmental studies, and loading the dataset format and this data used for GeoPandas data files (shapefile, GeoJSON) containing the attributes in the datasets. Spatial data visualization can create a map visualizing the form of a choropleth map showing on different applications.

Spatial analysis is performed of the spatial analysis to understand the spatial autocorrelation and explore the attributes that are distributed across the regions. Data visualization in the statistics and computer science literature is principally concerned with graphic tools. Spatial data visualization also needs to employ cartographic tools and different forms of map display referred to as cartographic visualization. The concerns of cartographic visualization include but extend beyond those of (Fairbairn et al., 2001). The spatial data visualization also extends to multi-media and visual reality (virtual landscape) representation. It also encompasses cartographic visualization, which involves the use of various map displays and cartographic techniques to represent spatial data effectively. Cartographic visualization goes beyond the concerns of exploratory spatial data analysis (ESDA) and includes considerations specific to map design, spatial relationships, and geographic context.

These modern technologies allow for immersive and interactive visualizations of spatial data, such as virtual landscapes and 3D models, which can provide a richer understanding of spatial phenomena and support more engaging and informative presentations of spatial information. Therefore, spatial data visualization encompasses a broad range of visual and interactive tools, including traditional cartographic displays, multimedia representations, and virtual reality applications, all aimed at effectively communicating spatial information and enabling users to explore patterns, relationships, and trends within spatial data. The introduction to spatial data visualization supports and serves as the stepping stone for delving into the diverse and powerful techniques used to visualize spatial information effectively. It sets the stage for building a strong foundation in spatial analysis and understanding the nuances of different visualization approaches to derive valuable insights from geospatial data.

SPATIAL DATA VISUALIZATION INTEGRATION WITH PYTHON

The popularity of spatial data visualization with Python has grown due to the abundance of powerful libraries and tools. A significant part of the history of spatial data visualization with Python is attributed to the widely used plotting library, Matplotlib. This library provides a diverse set of functions for creating different types of plots, including basic maps and spatial visualizations, making it foundational in the early development of spatial data visualization in Python. A more recent library that succeeded Basemap, Cartopy supplies an interface for crafting geospatial visualizations using Matplotlib(Coppock et al., 1994). It supports a broad spectrum of map projections and streamlines the process of plotting spatial data on maps. It also offers additional features such as the integration of web map tile services and shape file reading capabilities(Fairbairn et al., 2001).

GeoPandas is a potent library that extends the functionality of the panda's library to manage geospatial data(Lemenkova, 2020, pp. 20–31). It allows for the manipulation, analysis, and visualization of spatial data within a familiar tabular data structure. GeoPandas seamlessly integrates with other spatial libraries and provides features for generating maps and spatial visualizations.

The interactive data visualization with folium and Bokeh are libraries that facilitate the creation of interactive spatial visualizations(Lemenkova, 2020). Folium, built on the Leaflet JavaScript library, enables the development of interactive maps with diverse overlays and markers. Bokeh, a general-purpose interactive visualization library, includes support for spatial data, enabling the creation of interactive maps and geospatial visualizations.

Python's integration with Jupyter Notebooks has played a pivotal role in the advancement and widespread adoption of spatial data visualization(Lemenkova, 2020). Jupyter Notebooks offer an interactive environment for data exploration, analysis, and visualization, seamlessly integrating code, visualizations, and narrative explanations, making it an ideal platform for spatial data visualization workflows.

Spatial visualization integration with Python has seen significant advancements in recent years, driven by the growing demand for geospatial analysis, interactive mapping, and data-driven decision-making. The developments and trends in the background of spatial visualization integration with Python:

Geospatial Libraries and Tools

Python has a rich ecosystem of geospatial libraries and tools that facilitate spatial visualization. Libraries such as Geopandas, Shapely, Fiona, and Pyproj provide capabilities for working with geospatial data, performing spatial operations, and preparing data for visualization. Additionally, tools like GDAL (Geospatial Data Abstraction Library) and OGR (Simple Features Library) offer extensive support for reading and writing geospatial data formats.

Web Mapping and Interactive Visualization

The integration of Python with web mapping frameworks, such as Leaflet, Folium, and Ma box, has enabled the creation of interactive web-based spatial visualizations. These tools allow for the display of spatial data with interactive maps, markers, popups, and custom overlays, providing a dynamic and engaging user experience.

Data Visualization Libraries

Python's data visualization libraries, including Matplotlib, Seaborn, Plotly, and Bokeh, have expanded their support for spatial data visualization. These libraries offer features for creating static and interactive spatial visualizations, such as choropleth maps, scatter plots, and 3D visualizations, while also allowing for the integration of spatial data with other data types.

Spatial Analysis and Statistics

Python's spatial analysis and statistics libraries, such as PySAL (Python Spatial Analysis Library) and Geopandas, have enhanced support for visualizing spatial patterns, spatial autocorrelation, and spatial relationships. These libraries enable the creation of spatial visualizations for exploratory data analysis, spatial clustering, and spatial regression analysis.

Integration with Jupyter Notebooks

Python's integration with Jupyter Notebooks used frameworks, such as Voila and Panel, has facilitated the development of interactive spatial visualizations within a notebook environment. This integration allows for the creation of interactive dashboards that combine spatial visualizations with text, images, and other interactive elements.

Application Area of Spatial Data Visualization

Spatial data visualization is the process of representing spatial information visually, enabling us to explore and communicate patterns, distributions, and relationships within the data. It plays a crucial role in understanding complex spatial phenomena, such as population density, land use, climate patterns, and transportation networks. The complex spatial phenomena refer to intricate patterns, relationships, and interactions that exist within geographic. They encompass a wide range of phenomena, including population density, land use, climate patterns, and transportation networks.

Understanding and analyzing these complex spatial phenomena is crucial for a variety of disciplines, such as urban planning, environmental science, transportation management, and public health. Spatial data visualization is a powerful technique that involves representing and displaying geospatial information in a visual format. It leverages maps, charts, graphs, and other visual representations to communicate complex spatial patterns, relationships, and trends. This field plays a crucial role in various disciplines, including geography, environmental science, urban planning, and agriculture, among others. Spatial data visualization with Python has a wide range of application areas, spanning various industries and disciplines.

Main Objective of Spatial Data Visualization

The objective of spatial data visualization with Python can be demonstrated through several factors, including the depth and breadth of coverage, the focus on practical applications, the integration of cutting-edge tools, and techniques, and the clarity and accessibility of the content. The main objective of spatial data visualization with Python, as outlined in the provided factors, is to emphasize comprehensive coverage of various spatial data visualization libraries and tools in Python, practical applications of spatial data visualization in real-world scenarios, integration of advanced techniques for spatial data visualization, and ensuring accessibility and clarity in conveying complex concepts related to spatial data visualization. This approach aims to equip readers with the knowledge and skills to create compelling spatial visualizations using Python.

Spatial Data Visualization Techniques

The definitions of the spatial data visualization techniques that include a variety of methods and tools used to represent and explore spatial information. These techniques help in understanding patterns, relationships, and trends in spatial data. Visualization technique depends on the nature of the spatial data, the research question or objective, and the target audience. Often, a combination of techniques is employed to provide a comprehensive understanding of spatial patterns and relationships are.

Interactivity and Web Mapping

Techniques for creating interactive maps that enable users to explore and engage with spatial data, often employing technologies such as JavaScript and web mapping libraries. Understanding how spatial data is published and consumed through web map services, enabling access to spatial data over the internet.

Spatial Data Uncertainty and Visualization

The methods for representing and visualizing uncertainty associated with spatial data, critical for conveying the reliability and limitations of spatial information. The representing and communicating errors, inaccuracies, and variability in spatial data and measurements through appropriate visualization techniques. By having a solid understanding of these key concepts, practitioners can create more effective and informative visualizations of spatial data, contributing to better decision-making and insights across a wide range of domains.

Spatial Visualization Techniques

Spatial data visualization techniques aim to represent geospatial data in a visual format for analysis and communication(Lansley & Cheshire, 2016). Common techniques include choropleth maps, proportional symbol maps (using symbol size to represent data values), heat maps (using color intensity to represent data density), and spatial interpolation (estimating values for locations between measured data points). Spatial data visualization means geospatial data that representation of the graphical to gain insights, discovery patterns, and the effective communication of the information. The spatial data visualization is consists visual elements including maps, charts, graphs, and diagrams, to show the spatial relation, distributions, and attributes of geographic features and phenomena.

Spatial visualization techniques are essential for effectively communicating and analyzing spatial data. These techniques help in representing geographic information, patterns, and relationships in a visually informative manner. Here are several spatial visualization techniques commonly used in geographic information systems (GIS) and spatial analysis:

Choropleth Maps

Choropleth maps use color shading or patterns to represent different values of a variable across geographic areas, such as countries, states, or administrative units. This technique is effective for visualizing spatial patterns and variations in data, such as population density, income levels, or disease prevalence.

Heatmaps

Heatmaps visualize the density or intensity of point data or events across a geographic area. They use color gradients to represent the concentration or frequency of occurrences, making it easy to identify hotspots or clusters of activity with proportional symbol Maps, Isopleth Maps and Flow Maps.

Cartograms

Cartograms distort the geographic space to represent statistical data, such as population or economic indicators. The size of geographic areas is adjusted based on the variable being visualized, allowing for comparisons based on the chosen attribute. When working with spatial visualization techniques, it's important to consider the nature of the data, the audience, and the specific insights or patterns you want to communicate. Geographic information software (GIS), programming languages like Python and visualization tools such excel file formats like Command Separate Value(CSV) create and customize these spatial visualizations.

Spatial Data Visualization with Python Tools

Visualizing spatial data allows us to perceive and understand complex spatial patterns and relationships more effectively. Visual representations enable us to grasp information quickly, identify patterns, and gain insights that may not be apparent in raw data. Visualization enhances our cognitive abilities to comprehend and interpret spatial information. Spatial data visualization with Python, several powerful tools and libraries are available, each offering unique capabilities for handling, analyzing, and visualizing geographic and geometric data.

Python offers a variety of powerful tools and libraries for visualizing spatial data. These tools enable users to create compelling and informative visualizations of geographic information, spatial patterns, and relationships. Here are some popular Python libraries and tools for spatial data visualization:

Matplotlib: Matplotlib is a widely-used plotting library that provides a range of functions for creating static, interactive, and publication-quality spatial visualizations. Seaborn, built on top of Matplotlib, offers additional statistical visualization capabilities. Both libraries are used to create choropleth maps, scatter plots, line plots, and more.

Geopandas: Geopandas is an essential library for working with geospatial data in Python. It extends the capabilities of Pandas to include spatial operations and provides a high-level interface for working with geometric objects and spatial data structures. Geopandas integrates seamlessly with Matplotlib and allows for the creation of various spatial visualizations, including choropleth maps, point maps, and spatial overlays.

Folium: Folium is a Python library that allows for the creation of interactive leaflet maps directly within Jupyter notebooks. It leverages the Leaflet.js library and provides a simple and elegant way to visualize spatial data with interactive maps, markers, popups, and other interactive elements. Folium is particularly useful for creating web-based visualizations and dashboards.

Plotly: Plotly is a versatile graphing library that supports the creation of interactive, web-based visualizations, including maps, scatter plots, and 3D visualizations. Dash, built on top of Plotly, allows for the creation of interactive web applications for data visualization. Both Plotly and Dash are excellent choices for creating dynamic and interactive spatial visualizations.

PySAL (Python Spatial Analysis Library): PySAL is a comprehensive library for spatial data analysis and visualization. It offers tools for spatial autocorrelation analysis, spatial econometrics, and spatial statistics. PySAL includes visualization capabilities for spatial autocorrelation plots, Moran scatterplots, and spatial lag maps, among others. These Python libraries and tools for spatial data visualization offer a wide range of capabilities for visualizing spatial data, creating interactive maps, and communicating geographic information effectively. When working with spatial data visualization in Python, it's important to consider the specific requirements of the visualization task, the nature of the spatial data, and the intended audience.

Input Data for Spatial Visualization

Spatial data visualization is important to visualize the data properly on a spatial. The data format varies according to the requirement of spatial data visualization based on tools. Before, inputting data for visualization, the data format will be identified, and the spatial data to ensure it is in a suitable format for visualization(Lansley & Cheshire, 2016). In this input data spatial visualization, the guide goes through the essential steps are setup in the environments, installing the required libraries, importing required libraries, and applying tools used for data visualization with Python. Spatial visualization typically involves working with geospatial data, which includes information tied to geographic locations or areas. Here are some common types of geospatial data that can be used for spatial visualization:-

Geographic Information System (GIS) Data

GIS data includes various types of spatial data such as vector data (points, lines, polygons), raster data (elevation, satellite imagery), and geodatabases. This data can represent features such as roads, rivers, administrative boundaries, land parcels, and more. We are looking the data catalog from World Bank consists[1] Ethiopian GIS dataset for integrate to data visualization graphical and map representation

Geospatial Datasets

Geospatial datasets may include information about population density, land use, environmental factors, demographic data, economic indicators, and other attributes associated with specific geographic locations. We are looking for websites where obtain raster datasets related to weather and climate, there are several sources we explore website[2] offers access to a comprehensive collection of environmental data, including climate and weather-related datasets such as satellite imagery, climate models, and historical weather observations.

Install Python and Setup the Environment

Setting up the environment for spatial data visualization in Python involves installing and configuring the necessary libraries and tools. Here are the steps to set up an environment for environmental spatial data visualization using Python.

The Python official website [3]and download based on users' interest the oldest and the latest version of Python are exist on it. Follow the installation instructions on operating system and ccurrently, the latest version is python 3.11 within different points version as adaptable software and hardware evolutions. Python is flexible and adaptable programming language with different operating system like windows, Linux and mac operating systems. After download from website you must be copy the path of python installed or select add during installed the path of your python software installed under edit the system Environment Variables. The advantage of set directory of Python version installed on operating system to tackles error like pip is unrecognized displayed and effectively accessing and writing programming what we want based on library requires.

Install Python Package Manager (pip)

After, installed the python installed, the pip is a package manager for Python that allows you to easily install and manage Python packages. Ensure that pip is installed by running the following command in your command prompt or terminal. The pip used for install packages on terminal commands and examples like pip install jupyter, pip install numpy and etc.

Install Jupyter Notebook

Jupyter Notebook provides an interactive computational environment for Python. It allows you to create and share documents that combine code, text, and visualizations. The Jupyter Notebook for visualization on library(it work on any plattform and anacodada).

Look: To install Jupyter Notebook, run the following command: **pip install jupyter**

Remember: import is shortly call main package/s based on python version.If there are no import errors, it indicates that the installation was successful. By following these steps, we should have a functional environment for environmental spatial data visualization using Python. We can create maps, perform geospatial analysis, and visualize environmental data using the installed libraries.

Approaches to Spatial Data Visualization

Data visualization approaches and tasks amd the taxonomy proposed by (Shadare et al., 2016) for classifying work in data visualization divides the approachesto data visualization (in contrast to the individual tools themselves) into two areas:rendering and manipulation. Rendering refers to the decision as to what to show in a plot and in particular in deciding what type of plot to construct example techniques for displaying distributions (histograms, boxplots, and rankit (Shadare et al., 2016)lots) and time series (plots). In the case of multivariate data identify scatterplots (where cases are depicted as points), visualization methods traces (where cases are depicted as functions of a parameter as in the case of parallel coordinate plots) and scatterplots are the basic visual tool. There are several approaches to spatial data visualization, each with its own advantages and use cases. Some common approaches include:

Static Maps: This traditional approach involves creating static maps using tools such as Matplotlib, Seaborn, or Geopandas. Static maps are useful for visualizing spatial patterns and relationships at a single point in time. They are often used for presentation purposes.

Interactive Mapping: Interactive mapping tools, such as Folium and Plotly, allow for the creation of dynamic and interactive maps. These maps can be zoomed, panned, and clicked to explore spatial data in more detail. Interactive mapping is useful for exploring spatial data and engaging with end-users.

3D Visualization: For spatial data that has a vertical component, such as elevation or building heights, 3D visualization techniques can be employed. Tools like Plotly and Three.js can be used to create 3D visualizations of spatial data, providing a more immersive and detailed view of the data. An animation of data visualization can be used to visualize changes in spatial data over time. This approach is useful for visualizing temporal trends, dynamic processes, or the evolution of spatial patterns. Tools like Matplotlib and Plotly can be used to create animated visualizations of spatial data.

Spatial Statistical Analysis: Spatial data visualization can also involve the integration of spatial statistical analysis techniques, such as spatial autocorrelation or point pattern analysis. Visualizing the results of these analyses can help in understanding spatial patterns and relationships within the data.

Each of these approaches has its own strengths and can be chosen based on the specific requirements of the spatial data and the intended audience for the visualization. By considering these approaches, practitioners can effectively communicate spatial patterns and insights derived from the data.

Data format in Spatial Visualization

Spatial data can be represented using various data formats, each designed to capture and convey geographic and geometric information. Here are some common data formats used for spatial data visualization:

1. **Shapefile (.shp)** is a shapefile is a popular format for storing vector geospatial data. It consists of multiple files consists such as (.shp, .shx, .dbf) to represent different aspects of the spatial data such as geometry, attributes, and indices for vector data formats.
2. **GeoJSON:** A lightweight data interchange format for encoding geographic data structures, based on JavaScript Object Notation (JSON). It's commonly used for web-based mapping and visualization.

3. **Keyhole Markup Language(KML/KMZ):** is an XML-based format to store geographic annotations and visualizations. KMZ is a compressed version of KML, often used for sharing complex map features.
4. **Geography Markup Language(GML):** An XML-based encoding standard for the transport and storage of geographic information. It's particularly suitable for complex spatial data structures.
5. **Raster Data Formats:** is consist two basic file formats are Tagged Image File Format (TIFF) and image file format PNG, JPEG with World File.
6. **Tabular Data Formats**: is supported the Comma-Separated Values(CSV) While not strictly a spatial format, CSV files can be used to represent point-based spatial data with features like latitude, longitude and other attributes data linked to spatial features. These data formats each have their own advantages and uses, depending on the type and scale of the spatial data being visualized. Python libraries such as GeoPandas, Fiona, rasterio, and others support reading and writing of these formats, making them accessible for spatial data visualization and mapping libraries(Pundt, 2020). Matplotlib is a widely used 2D plotting library offering basic mapping functionalities and the foundation for other spatial visualization libraries.

Importing Necessary Libraries for Spatial Data Visualization

To start, import the necessary libraries for data preparation and spatial visualization. Common libraries include: pandas, geopandas, matplotlib, and other libraries specifically to needs, such as shapely for geometric operations and rasterio for working with raster data.

Check or Test the Installation

To ensure that your environment is set up correctly, open a Python interpreter or Jupyter Notebook and import the installed libraries as necessary libraries for Spatial data visualization.

Table 1. Installing packages and test the installation

Library	Installation Command	Import Statement
GeoPandas	pip install geopandas	import geopandas as gpd
Matplotlib	pip install matplotlib	import matplotlib.pyplot as plt
Descartes	pip install descartes	import descartes
Seaborn	pip install seaborn	import seaborn as sns
Plotly	pip install plotly	import plotly.express as px
Folium	pip install folium	import folium
Shapely	pip install shapely	from shapely.geometry import Point, Polygon
Basemap	pip install basemap	from mpl_toolkits.basemap import Basemap

This *Table 1* libraries provide various functionalities for spatial data visualization in Python. GeoPandas is particularly useful for working with geospatial data. Matplotlib, Descartes, and Seaborn offer versatile plotting capabilities. Plotly provides interactive and aesthetically pleasing visualizations(Lemenkova, 2020). Folium is focused on creating interactive maps. Shapely enables geometric operations, and Basemap is useful for creating maps and projections.

The library provides a rich set of functionalities for data manipulation, analysis, and visualization. For more detailed information and additional functions, refer to the panda's documentation on the internet[4]. Analyzing these complex spatial phenomena requires specialized tools and techniques, including geospatial data analysis, spatial statistics, spatial modeling, and data visualization. Python, with its geospatial libraries such as Geopandas,

Spatial Data Visualization With Python

There are several powerful tools and libraries available for spatial data visualization in Python. **Matplotlib: -** Matplotlib is a widely-used plotting library in Python that can be used for basic spatial data visualization. It provides a wide range of functionality for creating static, interactive, and animated visualizations.

Table 2. Spatial data visualization with matplotlib

```
import matplotlib.pyplot as plt
# Sample spatial data
x = [1, 2, 3, 4, 5]
y = [5, 4, 3, 2, 1]
# Create a scatter plot
plt.scatter(x, y)
plt.title('Basic Scatter Plot of Spatial Data')
plt.xlabel('X')
plt.ylabel('Y')
plt.show()
```

A basic scatter plot is a type of data visualization that is used to display the relationship between two numerical variables. In the context of spatial data visualization, a scatter plot can be used to show the relationship between two spatial variables, such as latitude and longitude.

Figure 1. Basic scatter plot of spatial data visualization

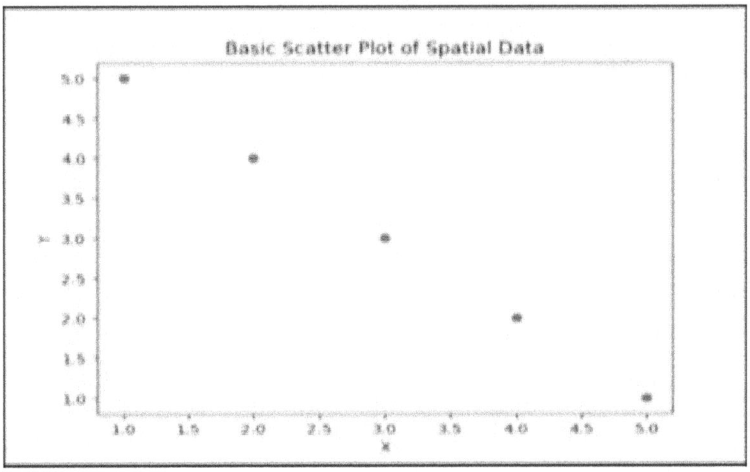

In *Figure 1* a basic scatter plot of spatial data visualization, each data point represents a specific location on a map. The x-axis typically represents one spatial variable (longitude) while the y-axis represents another spatial variable (latitude). Each data point is plotted at the intersection of its corresponding x and y values, allowing viewers to see the spatial distribution of the data.

Based on GeoPandas Library Spatial Data Visualization

GeoPandas is a powerful library that extends the functionality of the Pandas library to include spatial data types and operations. It provides a high-level interface for working with geometric objects and GeoDataFrames, enabling powerful spatial data visualization in Python. Here's an example of Python code using GeoPandas for spatial data visualization.

The steps of the visualization with geopandas code this code, we import the necessary libraries, including GeoPandas and Matplotlib. We then read a shapefile into a GeoDataFrame using the gpd.read_file() function. The shapefile contains the spatial data that we want to visualize. Next, we use the plot() method of the GeoDataFrame to create a basic plot of the spatial data. For example, we can set the title, x-label, and y-label of the plot using plot. title(), plot. xlabel(), and plt. ylabel() respectively. We create informative and visually appealing visualizations of spatial data in Python. Finally, we display the plot using the plot. show(). Data for the practical can be downloaded from the Introduction to Spatial Data Analysis and Visualisation in Python. In this activity View and explore the data, and descriptive statistics, Observe and compare the data using univariate plots.

To Import the Required Packages in Python:

- import geopandas as gpd
- import pandas as pd
- import pathlib
- import matplotlib.pyplot as plt
- import mapclassify as mc
- import numpy as np

We can also select which columns and rows we wish to print by entering the numerical ranges of the array within square brackets after the variable name (Ethipia_basin. head(8) where the comma separates the rows and columns. In this example we are selecting rows 0 to 9 and all columns to view the head 8 cases of a data frame.

Figure 2. Source of text load of Ethiopian rivers

```
Ethiopia_basin = gpd.read_file('eth_basins.shp')
Ethiopia_basin.head([illegible])
```

	BASIN_NAME	COUNT	SUM_AREA	SUM_SQKM	FIRST_BASI	BASIN_AREA	geometry
0	Abay	[illegible]	[illegible]e+11	[illegible]	[illegible]	Abay [illegible] SqKm	POLYGON ((23200.000 1046000.000, 23083.059 104...
1	Angereb	9	[illegible]e+10	[illegible]	[illegible]	Angereb [illegible] SqKm	POLYGON ((302400.000 1436000.000, 301980.469 1...
2	[illegible]	5	[illegible]e+09	[illegible]	[illegible]	[illegible] SqKm	POLYGON ((203450.951 1472391.251, 204219.781 1...
3	Awash	52	[illegible]e+11	[illegible]	[illegible]	Awash [illegible] SqKm	POLYGON ((563200.000 896400.000, 562800.000 89...
4	Baro	34	[illegible]e+10	[illegible]	[illegible]	Baro [illegible] SqKm	POLYGON ((150000.000 978400.000, 149984.828 97...
5	Bilate	12	[illegible]e+10	[illegible]	[illegible]	Bilate [illegible] SqKm	POLYGON ((347200.000 715600.000, 347200.000 71...
6	Danakil	17	[illegible]e+10	[illegible]	[illegible]	Danakil [illegible] SqKm	POLYGON ((552632.634 1612785.086, 552259.688 1...
7	Genale	[illegible]	[illegible]e+11	[illegible]	[illegible]	Genale [illegible] SqKm	POLYGON ((443328.500 693007.063, 443362.281 69...

geopandas.geodataframe.GeoDataFrame In this example, each vector data is represented as GeoDataFrame The geometry column is the most essential of the columns & it contains the x and y coordinate values of the vector data. For instance, we can look at the values using the .geometry function.

Figure 3. Geometry of polygon for Ethiopian rivers

```
print(Ethiopia_basin.geometry)
0      POLYGON ((23200.000 1046000.000, 23083.059 104...
1      POLYGON ((302400.000 1436000.000, 301980.469 1...
2      POLYGON ((203450.951 1472391.251, 204219.781 1...
3      POLYGON ((563200.000 896400.000, 562800.000 89...
4      POLYGON ((150000.000 978400.000, 149984.828 97...
5      POLYGON ((347200.000 715600.000, 347200.000 71...
6      POLYGON ((552632.634 1612785.086, 552259.688 1...
7      POLYGON ((443328.500 693007.063, 443362.281 69...
8      POLYGON ((659600.000 1396400.000, 659170.188 1...
9      POLYGON ((542535.082 1605247.134, 542817.563 1...
10     POLYGON ((241129.438 641633.688, 241122.500 64...
11     MULTIPOLYGON (((754000.000 1509600.000, 753667...
12     POLYGON ((280400.000 695200.000, 280994.188 69...
13     POLYGON ((640400.000 952800.000, 640386.125 95...
14     POLYGON ((581200.000 1548000.000, 581162.000 1...
15     POLYGON ((450562.531 935680.188, 451180.719 93...
16     MULTIPOLYGON (((423600.000 453200.000, 423048....
Name: geometry, dtype: geometry
```

The description of *Figure 3* above functions to extract the xy axis to coordinate values of a given polygon in this case, the first polygon the array([-161882.5625, 376388.125,1495282.875, 1645935.625]) and the shapefile ranges from 33.00541327 to 47.99567693 longitudes & 3.40510906 to 14.88539753 latitudes. This used to get a quick overview of the shape file by using the geometry values. For instance, to view the first data, you can use the code below.

Figure 4. Geometry values of Ethiopian sample basin

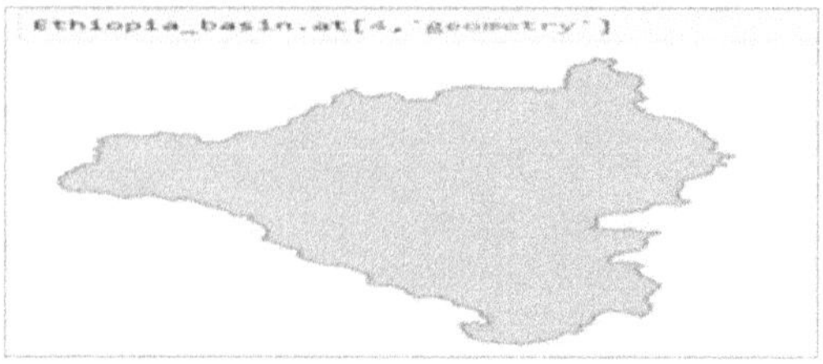

To see *Figure 4* shapefile plotting is to make a simple plot of the shapefile using the plot () function and colorize the map with the cmap function as follows.

Figure 5. Ethiopian map visualization

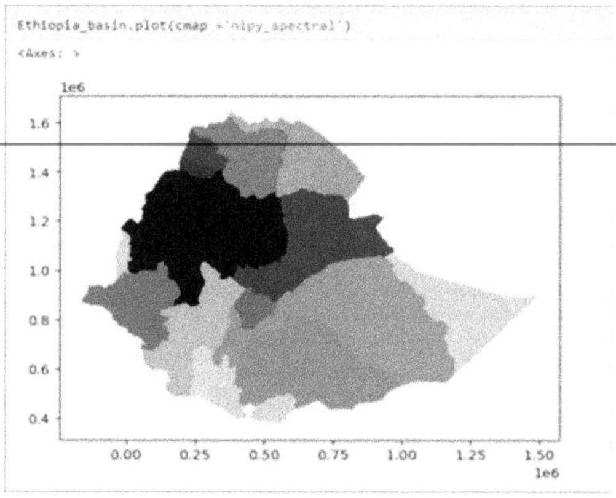

Begin by describing *Figure 5* and its content, using the matplotlib functions is a more convenient way to make plots. However, using the matplotlib functions is a more convenient way to make plots.

Figure 6. Most top level ethiopian rivers

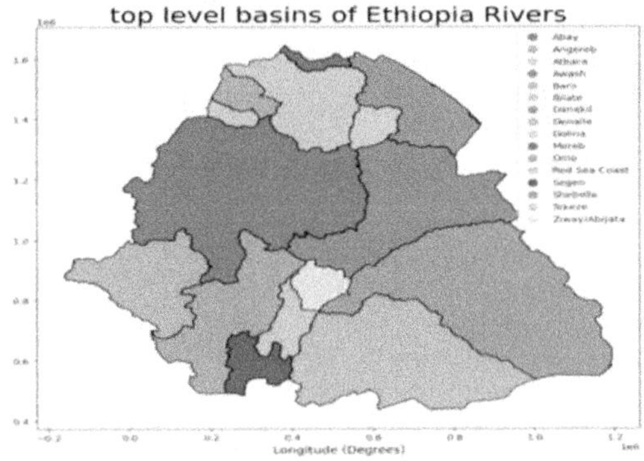

shapefile to clip the Ethiopia rivers to the area of interest and have a look at the loaded map.

Figure 7. Ethiopian river spatial data visualization

Figure 7 shows o clip of the Ethiopia Rivers shapefile to a specific area of interest, we can use GIS software to overlay the shapefile on the desired region and extract only the river data that falls within that area. This process helps to focus on the rivers within the specified boundary and exclude any extraneous data.

Visualizing 3D Spatial Data

Visualizing 3D spatial data involves creating visual representations of data in three-dimensional space. This can be done using various techniques and tools to effectively convey the spatial relationships and patterns within the data.

Table 3. Visualizing 3D spatial data snippet code

```
import plotly.graph_objects as go
# Sample 3D geospatial data (e.g., terrain elevation data)
# Assuming 'z' contains elevation data and 'x', 'y' contain coordinates
# Replace with actual 3D geospatial data
z = [[10, 10, 10, 10], [10, 20, 20, 10], [10, 20, 20, 10], [10, 10, 10, 10]]
x = [0, 1, 2, 3]
y = [0, 1, 2, 3]
# Create a 3D surface plot using Plotly
fig = go.Figure(data=[go.Surface(z=z, x=x, y=y)])
# Customize the 3D plot layout
fig.update_layout(scene=dict(aspectmode="data"))
# Show the 3D geospatial visualization
fig.show()
```

Figure 3, visualization of the above code as 3D data visualization is displayed in *Figure 6* with Python programming code. The left figure 10-20 indicates the form of the three-dimensional visualization data from sample geospatial data.

Figure 8. Visualize 3D spatial data using Python

Visualize 3D spatial data using Python with libraries such as Matplotlib, Plotly, and Mayavi interactive to visualization based from *Figure 8* . Depending on the specific requirements and nature of the data, different visualization techniques and libraries may be more suitable. Additionally, these examples can be extended and customized to accommodate various types of 3D spatial data and visualization needs.

Table 4. Using Matplotlib for 3D scatter plot

```
import matplotlib.pyplot as plt
from mpl_toolkits.mplot3d import Axes3D
import numpy as np
# Generate random 3D spatial data
x = np.random.rand(100)
y = np.random.rand(100)
z = np.random.rand(100)
# Create a 3D scatter plot
fig = plt.figure()
ax = fig.add_subplot(111, projection='3d')
ax.scatter(x, y, z)
# Set labels
ax.set_xlabel('X')
ax.set_ylabel('Y')
ax.set_zlabel('Z')
plt.show()
```

Based on the above Table 1. Visualization using Matplotlib for 3D Scatter Plot Visualization Spatial Data

Figure 9. Multi- Dimensional spatial data visualization

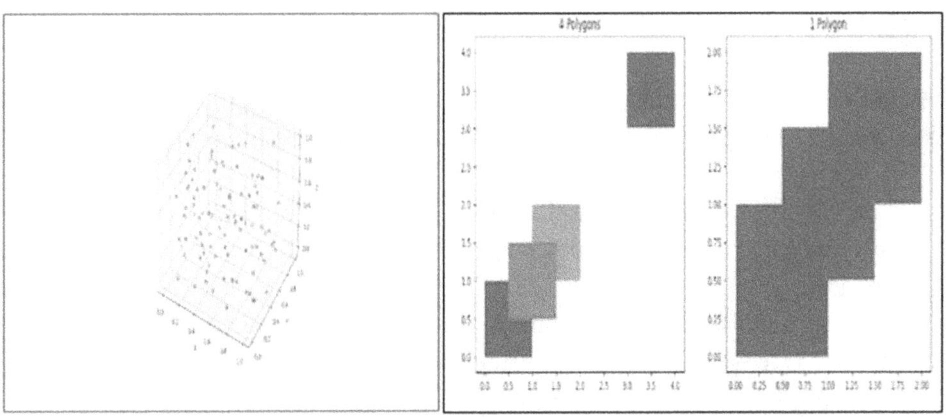

This *Table 9* shows the multi-dimensional spatial data visualization is the process of representing and displaying data that exists in multiple dimensions in a visual format. It involves the use of various techniques and tools to present complex data sets that have multiple attributes or dimensions in a way that is easy to understand and interpret.

- **Animating Spatial Temporal Data**

Animating spatial temporal data involves creating a visual representation of how data changes over time in a geographic location. Overall, animating spatial temporal data is a powerful way to visualize and understand the dynamics of data in a geographic context, and can be used in various fields such as urban planning, environmental monitoring, and transportation analysis. Animating spatial-temporal data in Python can be achieved using libraries such as Matplotlib, Plotly, and Mayavi.

Table 5. Animating spatial temporal data of snippet code

```
import plotly.express as px
import pandas as pd
data = { "latitude": [48.8583, 50.8503, 51.5074, -33.8688],
"longitude": [2.2945, 4.3528, -0.1276, 151.2093],
"time": [1, 2, 3, 4] # Replace with actual time data
}
df = pd.DataFrame(data)
fig = px.scatter_mapbox(df, lat="latitude", lon="longitude", hover_name="time", animation_frame="time",
zoom=3)
fig.update_layout(mapbox_style="open-street-map")
fig.update_layout(margin={"r":0,"t":0,"l":0,"b":0})
fig.show() # Show the animated map
```

Applying Folium for Spatial Data Visualization

A Python wrapper for Leaflet.js, Folium is used for creating interactive maps and visualizations that can be embedded within Jupyter notebooks and web applications.

- Start from importing the folium library. We create a map object using the folium.Map() function, specifying the center location and the initial zoom level.
- Next, we add markers to the map using the folium.Marker() function. Each marker is associated with a specific location and can have a popup message that appears when the marker is clicked.
- We also add a circle marker to the map using the folium.CircleMarker() function. This circle marker represents a specific location and can have a customizable radius, color, and fill color.
- Finally, we save the map as an HTML file using the m.save() method. The HTML file can be opened in a web browser, allowing you to interact with the map and view the markers and circle marker.

These libraries typically use JavaScript to create the map and add layers to it. Here are some popular web mapping libraries and how to use them in Python and to create a simple map with a marker using folium. Folium offers many more features and customization options, such as adding tiles from different map providers, overlaying GeoJSON data, and styling the map elements.By leveraging Folium's capabilities, we create interactive maps and visualizations that enhance the exploration and presentation of spatial data in Python.

Table 6. Folium data visualization snippet code

```
import folium
import pandas as pd
# Create a map centered on Finfinne
m = folium.Map(location=[9.02497, 38.74689], zoom_start=12)
# Add markers for some popular locations in Finfinnee
locations = pd.DataFrame({
'name': ['National Museum of Ethiopia', 'Holy Trinity Cathedral', 'Merkato'],
'latitude': [9.02742, 9.03171, 9.02497],
'longitude': [38.74069, 38.74166, 38.74689]
})
for i, row in locations.iterrows():
folium.Marker(
location=[row['latitude'], row['longitude']],
popup=row['name']
).add_to(m)
# Add a GeoJSON overlay for Finfinnee's boundaries
addis_geojson = 'path_to_addis_ababa_geojson_file.geojson'
folium.GeoJson(
addis_geojson,
name='geojson'
).add_to(m)
# Save the map to an HTML file
m.save('finfinnee_map.html')
```

Based on Add markers for some popular locations in Finfinnee python snippet code visualizations that enhance the exploration and presentation of spatial data in Python as Figure 10 below.

Figure 10. Ethiopian population in region number

This *Figure 10* is indicate the python snippet code visualization base on Table 6 with folium library.

Based on Plotly Spatial Data Visualization

Plotly's Python library provides interactive, web-based visualizations, including geospatial plots, choropleth maps, and 3D visualizations.Plotly is a Python library that offers interactive, web-based visualizations, including geospatial plots, choropleth maps, and 3D visualizations. It provides a high-level interface for creating interactive plots and supports a wide range of chart types. Here's an example of Python code using Plotly to create a choropleth map and create a DataFrame with the data for the choropleth map.

Table 7. Python code using Plotly spatial data

```
import plotly.express as px
# Create a DataFrame with the data for the choropleth map
data = {
'Region': ['Addis Ababa', 'Amhara', 'Oromia', 'Tigray', 'Southern Nations, Nationalities, and People\'s Region (SNNPR)'],
'Population': [3384569, 21436856, 34183493, 5708764, 20714756]
}
df = pd.DataFrame(data)
# Create the choropleth map
fig = px.choropleth(
df,
locations='Region',
locationmode='country names',
color='Population',
labels={'Population': 'Population'},
title='Ethiopia Population by Region'
)
# Show the map
fig.show()
```

From *Table 7* in this example, we import the necessary libraries, including plotly and express as px. We create a Data Frame df with the data we want to visualize on the choropleth map. Next, we use the px. Choropleth () function to create the choropleth map. We specify the DataFrame df, the column containing the locations (locations='Region'), the location mode (location mode='country names'), and the column containing the values to be represented by color (color='Population'). We also provide a label for the color legend using labels= {'Population': 'Population'} and set a title for the map using title='Ethiopia Population by Region'. Finally, we display the map using the fig. show () method. This code snippet demonstrates how to create a choropleth map to visualize the population of different regions in Ethiopia. Possibly modify the data and customize the map according to specific requirements. Plotly's flexibility and interactive features allow to creation engaging visualizations of geospatial data in Python, enabling effective exploration and communication of insights.

Figure 11. Ethiopian cities population visualization Python code using Plotly

Spatial Data Visualization on Web Integration Using Bokeh Library

Bokeh is a versatile library for creating interactive visualizations, capable of generating dynamic and interactive spatial plots. Bokeh is a powerful Python library for creating interactive visualizations that can generate dynamic and interactive spatial plots. It provides a flexible and high-level interface for building interactive plots, dashboards, and applications. Bokeh supports various types of spatial plots, such as scatter plots, line plots, heatmaps, and geospatial plots.

Table 8. Python code for creating interactive data visualizations

```
import libpysal as lp
import geopandas as gpd
spatial_data = gpd.read_file('spatial_data.shp')
# Calculate spatial autocorrelation
W=lp.weights.Queen.from_dataframe(spatial_data)
moran = lp.Moran(spatial_data['attribute'], w)
lp.Moran_Local(spatial_data['attribute'], w)
```

This *Table 8* demonstrate how Python can be used to address the complexity of spatial data visualization, including multidimensionality,scale, geometric complexity, spatial relationships, and spatial statistics.

Advanced Spatial Data Analysis

Traditionally spatial data has been stored and presented in the form of a map. Three basic types of spatial data models have evolved for storing geographic data digitally by Raster, Vector and image.

Raster Analysis

Raster consists of a matrix of cells organized into rows and columns where each cell contains a value representing information, such as temperature, Cold Weather Season Hot Weather Season, The Rainy Season of Retreating Monsoon.

Local Measures of Spatial Autocorrelation

Local measures of spatial autocorrelation are used to identify and visualize spatial patterns of similarity or dissimilarity within a spatial dataset. One commonly used measure of local spatial autocorrelation is the Local Moran's I statistic. The formula for Local Moran's I is as follows:

For a given feature i, let's assume it has a value zi and is located at a specific spatial location with coordinates (xi, yi). The Local Moran's I statistic for feature i is calculated using the following formula:

$$\left[I\_i = \backslash frac\{(z\_i - \backslash bar\{z\})\}\{S\wedge 2\}\backslash sum\_\{j\} w\_\{ij\}(z\_j - \backslash bar\{z\})\right] \tag{1}$$

Where:

(z_i) is the attribute value of feature i
(\bar{z}) is the mean of all attribute values in the dataset
(S^2) is the variance of all attribute values in the dataset
(w_ {ij}) is the spatial weight between features i and j

The summation is performed over all neighboring features j of feature i

After calculating the Local Moran's I statistic for each feature in the dataset, it can be used to identify spatial clusters of high values surrounded by high values (High-High clusters), low values surrounded by low values (Low-Low clusters), and high values surrounded by low values or vice versa (Outliers).

In Python, you can use libraries such as PySAL (Python Spatial Analysis Library) to calculate Local Moran's I and visualize the results. PySAL provides tools for spatial autocorrelation analysis and visualization, including the calculation of Local Moran's I.

```
import libpysalfrom libpysal import weightsfrom esda.
moran import Moran_Localimport geopandas as gpd# Read your
spatial dataset into a GeoDataFramegdf=gpd.read_file('C:/
Users/Sooree/Desktop/datascience/textbook/EthGIS/eth_basins.
shp') # Replace with the actual file path # Construct spatial
weightsw = weights.Queen.from_dataframe(gdf) # Calculate Local
Moran's Imoran_loc = Moran_Local(gdf['attribute_column'],
w) # Visualize the Local Moran's I resultsfrom splot.esda
import moran_scatterplot, plot_local_autocorrelationmoran_
scatterplot(moran_loc, p=0.05)plot_local_autocorrelation(moran_
loc, gdf, 'attribute_column')
```

In this example, you need to replace 'C:/Users/Sooree/Desktop/data science/textbook/Eth GIS/eth_basins.shp' with the actual file path to your spatial dataset and 'attribute column' with the name of the attribute column for which you want to calculate Local Moran's I.

The moran scatterplot and plot local autocorrelation functions from the 'splot. esd' module in PySAL can be used to visualize the results of the Local Moran's I analysis to calculate and visualize Local Moran's I using PySAL in Python.

Table 9 indicates the load spatial data from a file (''ethiopia_shapefile.shp'') that contains three columns x-coordinate, y-coordinate, and a variable of interest. We calculate the distance matrix using the pdist function from the scipy. spatial. distance module, and convert it to a square form using the square form function. We then create a spatial weights matrix using the weights. Finally, we calculate the

Table 9. Python code spatial data visualization malaria in Ethiopia

```
import geopandas as gpd
import matplotlib.pyplot as plt
# Load the shapefile for Ethiopia
ethiopia_map = gpd.read_file('ethiopia_shapefile.shp')
# Filter the data for the specific state, woreda, and zone
state = 'Oromia'
woreda = 'Bale'
zone = 'West'
filtered_data = malaria_data[(malaria_data['state'] == state) & (malaria_data['woreda'] == woreda) &
(malaria_data['zone'] == zone)]
# Merge the filtered data with the Ethiopia shapefile
merged_data = ethiopia_map.merge(filtered_data, on='ID', how='left')
fig, ax = plt.subplots(1, 1)
merged_data.plot(column='cases', cmap='OrRd', linewidth=0.8, ax=ax, edgecolor='0.8', legend=True)
ax.set_title(f'Malaria Incidence in {state}, {woreda}, {zone}')
plt.show()
```

Moran's I statistic using the essay. moran.Moran function from the pysal.explore the module, and print the results.

CONCLUSION

The chapter on spatial data visualization with Python introduces readers to a range of powerful visualization tools, including Matplotlib, Seaborn, Plotly, and Geopandas, which enable the creation of interactive and aesthetically appealing spatial visualizations. It emphasizes the significance of spatial data visualization in extracting insights from geographical data and covers fundamental concepts such as geographic coordinate systems, map projections, and spatial data formats to establish a strong foundation for readers. The introduction of Geopandas and Plotly Express equips readers with the ability to generate interactive maps, scatter plots, and choropleth maps with rich interactivity, facilitating the exploration and analysis of spatial data in a more engaging and intuitive manner. Ultimately, the chapter provides readers with the knowledge and practical skills to develop compelling spatial visualizations using Python libraries, empowering them to effectively communicate spatial data for diverse applications, from geographic analysis to urban planning and beyond.

Future recommendation, considering these future directions, the field of spatial data visualization in Python we can continue to evolve, addressing new challenges and opportunities in the realm of geospatial analysis and visualization. This notepad-like format provides a structured and concise outline of future recommendations for spatial data visualization with Python, capturing key areas for potential development and improvement.

REFERENCES

Anselin, L., Syabri, I., & Kho, Y. (2006). GeoDa: An introduction to spatial data analysis. *Geographical Analysis*, *38*(1), 5–22. doi:10.1111/j.0016-7363.2005.00671.x

Brown, J. E. (1971). Introduction to computers and their development. *Journal of the American Oil Chemists' Society*, *48*(5), 185–190. doi:10.1007/BF02883750

Coppock, J. T., Grelot, J. ., Yuju, H., Kanakubo, T., Morrison, J. L., & Rystedt, B. (1994). *Thematic Mapping from Satellite Imagery Visualization in Modern Cartography*.

Eberhard, K. (2023). The effects of visualization on judgment and decision-making: a systematic literature review. In Management Review Quarterly (Vol. 73, Issue 1). Springer International Publishing. doi:10.1007/s11301-021-00235-8

Fairbairn, D., Andrienko, G., Andrienko, N., Buziek, G., & Dykes, J. (2001). Representation and its relationship with cartographic visualization. *Cartography and Geographic Information Science*, *28*(1), 13–28. doi:10.1559/152304001782174005

Fornell, D. (2015). Understanding How Big Data Will Change Healthcare. In *Imaging Technology News* (*Vol. 55*, Issue 7, pp. 26–28). https://login.ezproxy.library.ualberta.ca/login?url=https://search.ebscohost.com/login.aspx?direct=true&db=rzh&AN=110150176&site=ehost-live&scope=site

Gete, Z. (2000). Implications of land use and land cover dynamics for mountain resource degradation in the Northwestern Ethiopian highlands. *Mountain Research and Development*, *21*(2), 184–191. doi:10.1659/0276-4741(2001)021[0184:IOLUAL]2.0.CO;2

Lansley, G., & Cheshire, J. (2016). An Introduction to Spatial Data Analysis and Visualisation in R. 1–121.

Lemenkova, P. (2020). Python libraries matplotlib, seaborn and pandas for visualization geospatial datasets generated by QGIS. *Analele Stiintifice Ale Universitatii "Alexandru Ioan Cuza" Din Iasi - Seria Geografie, 1*(1), 13–32.

Li, J. P., Haq, A. U., Din, S. U., Khan, J., Khan, A., & Saboor, A. (2020). Heart Disease Identification Method Using Machine Learning Classification in E-Healthcare. *IEEE Access : Practical Innovations, Open Solutions*, *8*, 107562–107582. doi:10.1109/ACCESS.2020.3001149

Rey, S. J., & Anselin, L. (2007). PySAL: A Python library of spatial analytical methods. *The Review of Regional Studies*, *37*(1), 5–27. doi:10.52324/001c.8285

Shadare, A. E., Musa, S. M., & Akujuobi, C. (2016). *Data visualization. December*.

ENDNOTES

1 https://datacatalog.worldbank.org/search/dataset/0039684
2 https://earthdata.nasa.gov/ and https://www.ncei.noaa.gov/
3 https://www.python.org
4 https://pandas.pydata.org/docs/

Chapter 5
Spatial Data Visualization With Python:
Techniques, Tools, and Real-World Applications

M. Shreenithi
Coimbatore Institute of Technology, India

D. Shanmugapriyaa
Coimbatore Institute of Technology, India

M. Sujithra
Coimbatore Institute of Technology, India

T. Rizanuma
Coimbatore Institute of Technology, India

B. Senthilkumar
https://orcid.org/0000-0001-8117-3314
Kumaraguru College of Technology, India

ABSTRACT

The field of spatial data visualization in Python is propelled by the significance of data science and geospatial analysis. Python is pivotal in providing a flexible toolkit for visualizing intricate spatial data. This guide explores the fundamentals, techniques, and practical applications of spatial data visualization, tracing its historical roots alongside the ascent of data science and geospatial analytics. Examining libraries like Matplotlib, Geopandas, Folium, and Plotly, the guide equips readers with essential tools for geospatial data visualization in diverse fields, including urban planning, environmental science, and healthcare. Emphasizing dynamic and interactive visualizations, it explores spatial data's pivotal role in space exploration, virtual and augmented reality, satellite navigation, industry

DOI: 10.4018/979-8-3693-1754-9.ch005

accuracy, and pandemic response. A focus lies on integrating IP geolocation for enhanced marketing accuracy, security, and compliance, showcasing how geospatial visualization improves public experiences in navigation, weather forecasting, public safety, healthcare.

EVOLUTION AND SIGNIFICANCE OF GEOSPATIAL DATA VISUALIZATION

The evolution of geospatial data visualization is a fascinating journey spanning millennia, reflecting humanity's ongoing efforts to understand and navigate our ever-changing world. From the rudimentary maps of ancient civilizations to complex geographic information systems (GIS) and web-based mapping tools, this evolution has been marked by technological advances. profound technology and a growing appreciation of the importance of geographic data. In this introduction, this book will explore the historical context and significance of this development, paving the way for a more detailed exploration of its main stages (Pundt, 2010).

Historical Overview

Geospatial data visualization or representation of geographic information management through maps, graphs, and other visual tools, has a long and rich history that has evolved significantly over time. Here is a brief historical overview of the development of geospatial data visualization:

- ***Ancient maps:*** The origins of geospatial data visualization can be traced back to ancient civilizations such as the Babylonians, Egyptians and Greeks, who created maps simple to navigate and understand their world. These early maps were often crude and lacked accurate scale.
- ***Renaissance Maps:*** During the Renaissance, cartography saw significant advances. Pioneers like Gerardus Mercator introduced map projections that preserved angles, making navigation more precise. This era also saw the birth of artistic and decorative maps (Koperski et al., 2019).
- ***Surveying and mapping:*** The 18th and 19th centuries marked a period of increasingly accurate geospatial data collection. Surveying, trigonometry, and the use of longitude and latitude coordinates contributed to more accurate mapping.
- ***Emergence of GIS:*** The advent of computers in the mid-20th century enabled the development of geographic information systems (GIS). GIS technology

has enabled the storage, analysis, and visualization of geospatial data in digital form (Pundt, 2010).

- ***Internet and Web Maps:*** With the development of the Internet, geospatial data has become widely available to the public. Companies like Google introduced interactive maps online, revolutionizing the way people interact with geographic information.

Importance and Applications

Geospatial data visualization is extremely important in today's world for several reasons:

- ***Decision Making:*** Geospatial data visualization facilitates informed decision making by presenting complex geographic information in an easy-to-understand and actionable format. It helps governments, businesses, and individuals make choices in resource allocation, urban planning, disaster management, and more.
- ***Environmental monitoring:*** Visualizing geospatial data is critical for monitoring and understanding environmental changes. It allows scientists and policymakers to track deforestation, climate change, land use and other important environmental factors.
- ***Urban Planning and Development:*** Urban planners use geospatial data to design efficient transportation systems, plan infrastructure, and manage urban growth. It helps optimize land use and ensure the sustainability of cities (Koperski et al., 2019).
- ***Emergency Response:*** During disasters, such as earthquakes, hurricanes, or wildfires, geospatial data visualization enables rapid response and evacuation planning. It helps first responders and relief agencies allocate resources effectively.
- ***Marketing and Retail:*** 4,444 Companies use geospatial data to analyse consumer behaviour and market demographics, allowing for targeted and selective marketing campaigns. Select locations for new stores.
- ***Navigation and Transportation:*** 4,444 Geospatial data powers GPS systems, ride-sharing apps, and navigation tools, improving travel and transportation efficiency.
- ***Natural Resource Management:*** Geospatial data visualization supports sustainable management of natural resources, including forestry, agriculture, and water resources.
- ***Healthcare:*** In epidemiology, geospatial data visualization can help track outbreaks and optimize the allocation of healthcare resources.

Role Of Python In Geospatial Visualization

Python has become a central and indispensable tool in the field of geospatial data visualization. Its flexibility, extensive library support, and integration capabilities make it the preferred choice for scientists, analysts, and researchers working with geospatial data.

Python Overview

Python plays a central role in geospatial visualization for several reasons:

- ***Data-driven transformation:*** The rapid growth of geospatial science data and analytics has changed the way we understand and interact with the world. Geospatial data, which provides location-specific information, is at the forefront of this transformation, enabling informed decision-making and insights into spatial relationships.
- ***Flexibility:*** Python's flexibility is a significant advantage. It can be seamlessly integrated into data science workflows, allowing management of various data formats including shape files, GeoJSON, and raster datasets. This makes it a powerful platform for merging and analysing disparate data sets.
- ***Extensive library ecosystem:*** Python has a comprehensive library ecosystem that is well suited for geospatial visualization. Libraries like NumPy and pandas make data manipulation easy, while specialized geospatial tools like GeoPandas, Folium, and Plotly are designed to manage spatial data and create interactive Visualization. work.
- ***Data processing and analysis:*** Python's geospatial library supports a variety of tasks, from basic operations like geocoding and spatial join to Advanced geospatial analysis. This functionality opens up new avenues for spatial research and analysis.

Benefits of Python in Geospatial Visualization

Python's role in geospatial visualization is based on several key advantages:

- ***Easy learn and use:*** Python is famous for its readability and simplicity. Its intuitive syntax makes it accessible to both novices and experienced programmers, reducing barriers to entry for analysing and visualizing geospatial data.

- ***Open Source:*** Python is open source and benefits from a large and active user community. This means it is constantly evolving and has a rich collection of user-contributed libraries, driving innovation and problem solving.
- ***Cross-platform compatibility:*** Python is compatible with many different operating systems, making it a flexible choice for users of different platforms.
- ***Interoperability:*** Python easily integrates with other programming languages and tools, enhancing its usefulness in geospatial and scientific workflows different data.

FUNDAMENTALS OF PYTHON FOR GEOSPATIAL DATA VISUALIZATION

The foundation of spatial data visualization is based on three basic pillars: Spatial data key concepts, Python programming basics, and an introduction to data visualization libraries.

Fundamental Spatial Data Principles

Spatial data includes information that is intrinsically linked to specific geographic locations on the Earth's surface. Key concepts include:

- ***Geospatial coordinates:*** Spatial data relies on coordinates such as latitude and longitude for precise location determination. These coordinates allow the representation of points, lines and polygons on the map.
- ***Geospatial Data Types:*** There are different types of geospatial data, including vector data (points, lines, polygons) and raster data (grid data, such as satellite images or elevation models). Understanding these types of data is essential for effective visualization.
- ***Projections:*** Projection is a method used to represent the Earth's three-dimensional surface on a two-dimensional plane, such as a map. Different projections all have their strengths and weaknesses, and choosing the right one is crucial for accurate visualization.

Data Visualization Libraries and Modules

Python's strength lies in its libraries, such as NumPy for numerical calculations, pandas for manipulating data, and specialized geospatial libraries such as GeoPandas for working with spatial data. Data visualization libraries in Python are essential for turning spatial data into meaningful images.

In the vast landscape of spatial data analysis, Python's powerful libraries serve as an important toolkit to bring your data to life through insightful Visualization. Among them, Matplotlib stands out as a versatile star, offering unparalleled functionality for creating static charts and maps of superior quality. Its unique ability to customize visual elements at a fine level makes it an essential tool for converting complex spatial patterns into visually appealing representations.

Seaborn is a Matplotlib-based extension that further refines spatial data visualization techniques. Since its inception, Seaborn has specialized in creating visually appealing statistical graphics with minimal code. This library excels at representing relationships between variables in spatial datasets, providing an elegant and straightforward approach to exploratory data analysis in a spatial context.

A dynamic force in the Python ecosystem, Plotly introduces a new dimension to spatial data visualization: interactivity. Known for its web-based features, Plotly allows you to create interactive maps and dashboards. This not only facilitates dynamic exploration of spatial data, but also adds a narrative layer to Visualization, increasing the storytelling potential of spatial insights.

GeoPandas seamlessly integrates geographic data functionality into the well-known Pandas framework and is tailored for advanced processing of geographic data. It bridges the gap between traditional data processing and the complex world of geospatial datasets. GeoPandas excels at overlay operations and choropleth map creation by simplifying complex spatial analysis and allowing users to discover hidden patterns in their data.

Folium is designed specifically for web-based applications and introduces a Python interface for creating interactive maps using Leaflet.js. Its strength is that it simplifies the process of creating dynamic web-based Visualization, making it an ideal choice for applications where user interaction is a priority. Folium can handle large data sets across marker clusters, keeping even large spatial data visually clear and informative.

Collectively, these Python libraries together form a robust ecosystem for spatial data visualization. Whether you're creating static graphics for publications or developing dynamic, interactive maps for web applications, these libraries provide a diverse and powerful set of tools. From the basic functionality of Matplotlib, to the interactive features of Plotly, to the advanced geospatial features of GeoPandas, Python's geospatial visualization libraries enable users to discover meaningful stories from geospatial data sets, uncover underlying patterns, and gain a better understanding of trends.

LITERATURE REVIEW

Geospatial data visualization is indispensable in interpreting and analysing spatial data. In the context of environmental monitoring and management, these Visualization are crucial for decision-making. As the amount and complexity of geospatial data continue to increase, effective visualization methods are essential for extracting meaningful insights. This literature review aims to examine the current landscape of geospatial data visualization, focusing on methods, tools, and challenges specific to environmental applications (Koperski et al., 2019).

Methods of Geospatial Data Visualization

The selection criteria emphasized recent publications within the last five years, relevance to geospatial data visualization, and a focus on environmental applications. This approach ensured that the review captured the most current and pertinent research in the field. Geospatial data visualization encompasses a variety of techniques tailored to different data types and analysis needs. Thematic mapping, for example, is widely used for visualizing spatial patterns and distributions. Choropleth maps are effective for displaying statistical data aggregated by area, providing a quick overview of trends. Heat maps, on the other hand, offer a visual representation of data density, highlighting areas of high and low concentration. Additionally, 3D visualization techniques provide a spatial perspective that can enhance understanding, particularly in terrain analysis and urban planning (Thomas et al., 2014).

Tools for Geospatial Data Visualization

Several software tools are available for geospatial data visualization, each offering unique features and capabilities. ArcGIS, a proprietary software, is renowned for its comprehensive mapping and spatial analysis functionalities, making it a popular choice among professionals. QGIS, an open-source alternative, provides similar capabilities and is favoured for its flexibility and cost-effectiveness. Tableau, a data visualization tool, stands out for its interactive and user-friendly interface, making it accessible to a wide range of users, including those with limited technical expertise (Pundt, 2010).

Challenges in Geospatial Data Visualization

Despite the advancements in geospatial data visualization, several challenges persist. Managing the complexity of spatial data, which includes issues related to volume, variety, and velocity, remains a significant challenge. Representing uncertainty and accuracy in spatial data is also a complex issue, as it requires careful consideration to avoid misleading interpretations (Koperski et al., 2019).

Figure 1. Python library and its description

Python Library	Description
GeoPandas	• Geopandas is like Pandas meet GIS. • It extends the datatypes used by pandas to allow spatial operations on geometric types. • Geometric operations are performed by shapely. It further depends on fiona for file access and matplotlib for plotting. • Here is the Quick Start Guide.
"MNSP"	• Classical Matplotlib, NumPy, SciPy and Pandas are usually handy for spatial data structures and algorithms, data visualization, multidimensional image processing and regular mathematical transformation.
Shapely	• With Shapely, you can create shapely geometry objects (e.g. Point, Polygon, Multipolygon) and manipulate them, e.g. buffer, calculate the area or an intersection etc.
Fiona	• Fiona can read and write real-world data using multi-layered GIS formats and zipped virtual file systems and integrates readily with other Python GIS packages such as pyproj, Rtree, and Shapely.
ipyleaflet	• If you want to create interactive maps, ipyleaflet is a fusion of Jupyter notebook and Leaflet. • You can control an assortment of customizations like loading basemaps, geojson, and widgets. • It also gives a wide range of map types to pick from including choropleth, velocity data, and side-by-side views.
Folium	• Just like ipyleaflet, Folium allows you to leverage leaflet to build interactive web maps. • It gives you the power to manipulate your data in Python, then you can visualize it with the leading open-source JavaScript library
Imageio	• Imageio is a Python library that provides an easy interface to read and write a wide range of image data, including animated images, volumetric data, and scientific formats.
PyQGIS	• QGIS, the free and open source community-driven GIS software. • PyQGIS is the Python environment inside QGIS with a set of QGIS libraries plus the Python tools with the potential of running other powerful libraries such as Pandas, Numpy or Scikit-learn.
Earth Engine	Earth Engine concepts illustrate important concepts about data types such as: • Image, The fundamental raster data type in Earth Engine. • ImageCollection, a stack or time-series of images. • Geometry, the fundamental vector data type in Earth Engine. • Feature, or a Geometry with attributes. • FeatureCollection, or a set of features. • Reducer, an object used to compute statistics or perform aggregations.

Figure 2. Python library and its description

GDAL/OGR	• The GDAL library is used for translating between GIS formats and extensions. • QGIS, ArcGIS, ERDAS, ENVI, and GRASS GIS and almost all GIS software use it for translation in some way.
RSGISLib	• The RSGISLib library is a set of remote sensing tools for raster processing and analysis. • To name a few, it classifies, filters, and performs statistics on imagery. • My personal favorite is the module for object-based segmentation and classification (GEOBIA).
ReportLab	• ReportLab library is one of the most satisfying libraries on this list. • I say this because GIS often lacks sufficient reporting capabilities. Especially, if you want to create a report template, this is a fabulous option.
MapClassify	• MapClassify implements a family of classification schemes for choropleth maps. • It is focused on the determination of the number of classes, and the assignment of observations to those classes.
RTree	RTree provides a number of advanced spatial indexing features such as - • Nearest neighbor search • Intersection search • Multi-dimensional indexes • Clustered indexes (store Python pickles directly with index entries)
WhiteBoxTools	• It is used for geospatial applications and workflow automation. • By combining the WhiteboxTools library with a high-level scripting language, such as Python, users are capable of creating powerful stand-alone geospatial applications and workflow automation scripts.
pydeck	• PyDeck is high-scale spatial rendering in Python. • Here is the QuickStart Guide.
GeoAlchemy 2	• GeoAlchemy 2 provides extensions to SQLAlchemy for working with spatial databases. • GeoAlchemy 2 focuses on PostGIS.
Spectral	• Spectral Python (SPy) is a pure Python module for processing hyperspectral image data. • It has functions for reading, displaying, manipulating, and classifying hyperspectral imagery. • It can be used interactively from the Python command prompt or via Python scripts.
Mapnik	• Mapnik is an open-source mapping toolkit for desktop and server based map rendering.

Figure 3. Python library and its description

Geemap	• Geemap is intended more for science and data analysis using Google Earth Engine (GEE)
GeoPlot	• A high-level geospatial data visualization library. • It's an extension to cartopy and matplotlib which makes mapping easy: like seaborn for geospatial. • Here is the QuickStart Guide.
ArcPy	• ArcPy library is for Esri ArcGIS. • ArcPy is meant for geoprocessing operations. • But it's not only for spatial analysis, it's also for data conversion, management, and map production with Esri ArcGIS.
SciKit-Image	• Scikit-Image is for image manipulation, e.g. histogram adjustments, filter, segmentation/edge detection operations, texture feature extraction etc.
SciKit-Learn	• SKLearn is a machine learning library for Regression, classification, dimensionality reductions etc. • Heavily used for Feature Extraction and data-preprocessing.
Descartes	• Descartes nables plotting of shapely geometries as matplotlib paths or patches. • Also it is a dependency for the geometry plotting functions of geopandas.
RasterStats	• RasterStats is used for zonal statistics. • Extracts statistics from rasters files or numpy arrays based on geometries.
Rasterio	• Rasterio is the go-to library for raster data handling. • It lets you read/write raster files to/from numpy arrays, offers many convenient ways to manipulate these array (e.g. masking, vectorizing etc.) and can handle transformations of coordinate reference systems. • Just like any other numpy array, the data can also be easily plotted, e.g. using the matplotlib library.
PySAL	• The Python Spatial Analysis Library contains a multitude of functions for spatial analysis, statistical modeling and plotting. It is intended to support the development of high-level applications.
xarray	• xarray labels the dimensions of the multidimensional numpy array and combines this with many functions and the syntax of the pandas library (e.g. groupby, rolling window, plotting). • Great for handling extensive image time series stacks.
PyProj	• PyProj library works with spatial referencing systems. • It can project and transform coordinates with a range of geographic reference systems. • PyProj can also perform geodetic calculations and distances for any given datum.

Trends In Geospatial Data Visualization

Emerging technologies such as virtual reality (VR) and augmented reality (AR) are increasingly being integrated into geospatial data visualization. These technologies offer immersive experiences that can enhance the understanding of spatial data, particularly in complex environments. Additionally, the integration of geospatial data with other data sources, such as social media and IoT devices, is enabling more comprehensive and insightful Visualization, providing a more holistic view of environmental patterns and trends.

Geospatial data visualization is a dynamic field with vast potential for advancing our understanding of complex spatial relationships. By leveraging advanced techniques and tools, researchers and practitioners can gain valuable insights into environmental patterns and trends, ultimately supporting informed decision-making. However, addressing the challenges associated with geospatial data visualization will require ongoing innovation and collaboration across disciplines, ensuring that Visualization remain effective and impactful in the face of evolving data complexities.

Geospatial data visualization is a crucial aspect of spatial analysis, providing insights into complex spatial relationships and patterns. While Python is a popular choice for geospatial data visualization, several other tools and languages are commonly used in this field, each offering unique capabilities and advantages.

WORKING WITH GEOSPATIAL DATA

Working with geospatial data involves analysing and visualizing information associated with a specific geographic location. This field combines traditional data with spatial context to provide insights into real-world phenomena. Geospatial applications range from mapping and navigation to environmental monitoring and urban planning. Integrating Geographic Information Systems (GIS) and specialized tools is essential to capture valuable data.

Understanding Spatial Coordinates and Projections

- ***Spatial Coordinates:*** Spatial coordinates represent locations on the Earth's surface using latitude and longitude values. Latitude measures the north south position, with 0° at the equator and ±90° at the poles. Longitude measures the east west position, with 0° at the Prime Meridian (Greenwich) and ±180° halfway around the globe.
- ***Coordinate Reference Systems (CRS):*** A Coordinate Reference System (CRS) is a framework for defining and interpreting spatial coordinates. Different CRSs use various units (e.g., degrees, meters) and datums (e.g., WGS 84, NAD 83) for precise location representation. Choosing the right CRS is crucial for accurate geospatial analysis and visualization.
- ***Common CRSs:*** WGS 84 (EPSG:4326): A widely used global CRS based on the World Geodetic System, using latitude and longitude in degrees.1UTM (Universal Transverse Mercator) Projections: A set of CRSs that divide the Earth into zones, ideal for regional and local mapping with meter-based coordinates.

Introduction to Geopandas for Vector Data

GeoPandas

GeoPandas is an opensource Python library that simplifies working with geospatial vector data. It extends the capabilities of pandas Data Frames by adding geospatial geometry support, making it easier to handle shapefiles, GeoJSON, and more.

```
import geopandas as gpd
```

GeoDataFrame A Geodata Frame is a core data structure in GeoPandas, combining tabular data (like a pandas Data Frame) with geospatial geometry information. It allows you to store and manipulate both attribute data and geometric shapes (points, lines, polygons) in a single object.

Loading Vector Data:

```
import geopandas as gpd
from shapely.geometry import Point

data = {
'Attribute1': [value1, value2, ...],
'Attribute2': [value1, value2, ...],
# Add more attributes as needed
}
gdf = gpd.GeoDataFrame(data, geometry=gpd.points_from_
xy(data['Longitude'], data['Latitude']))
gdf.crs = 'EPSG:4326'
```

GeoPandas provides functions for loading geospatial vector data from various formats, including Shapefile, GeoJSON, and more. This can be used gpd.read_file() to read and create GeoDataFrames from these formats.

Raster Data Manipulation with rasterio

```
gdf = gpd.read_file('path_to_shapefile.shp')
```

Raster Data Introduction

Raster data consists of a grid of pixels, where each pixel stores a value (e.g., elevation, temperature).It is commonly used in remote sensing, image analysis, and geographic information systems (GIS).

1. Rasterio Library:

Rasterio is a Python library for reading and manipulating geospatial raster data. It provides tools for opening and processing raster datasets, including GeoTIFF files.

```
import rasterio
```

2. Loading Raster Data:

You can use rasterio.open() to open raster datasets and load them into NumPy arrays. This step is crucial for accessing pixel values and performing raster operations.

```
with rasterio.open('data.tif') as src:
    data = src.read(1)
```

3. Raster Manipulation:

Rasterio offers functions for basic raster manipulations, such as cropping, resampling, and reprojecting, to prepare data for analysis or visualization.

```
window = ((min_row, max_row), (min_col, max_col))
cropped_data = src.read(1, window=window)
new_height = ...
new_width = ...
resampled_data = src.read(1, out_shape=(new_height, new_width),
resampling=Resampling.bilinear)
new_crs = 'EPSG:4326'  # Replace with your desired CRS
reprojected_data = src.read(1, out_crs=new_crs)
```

Data Preparation and Cleaning

Data preparation and cleaning are important steps to ensure the quality and reliability of datasets. This process includes identifying and handling missing values, correcting inconsistencies, and standardizing formats. Data cleaning ensures the accuracy of analysis and model training. Techniques such as outlier removal and normalization are used to refine the data to provide meaningful insights. Effective data preparation lays the foundation for sound analysis that reduces errors and improves overall data reliability for informed decision making.

1. Data Exploration

Data exploration involves examining geospatial data attributes, geometry types, and spatial distributions. Exploring data helps identify issues, trends, and patterns that inform subsequent analysis.

```
attribute_stats = gdf.describe()
print(attribute_stats)
gdf.plot(column='column_to_plot', cmap='colormap', legend=True)
```

2. Data Cleaning:

Data cleaning includes tasks like handling missing values in attributes or geometry coordinates. Correcting geometry errors, such as overlapping polygons or invalid shapes, is essential for accurate analysis.

```
gdf['Population'].fillna(0, inplace=True)
```

3. Data Transformation:

Data transformation involves aggregating, summarizing, or deriving new variables from existing geospatial data. For example, you can calculate area, density, or distance measures.

4. Data Conversion:

```
gdf['Area_sqkm'] = gdf.geometry.area / 1e6      # Convert to
square kilometers
gdf.to_crs("EPSG:4326", inplace=True)       # Convert to WGS
84
```

Data conversion deals with transforming geospatial data between different formats or coordinate systems (CRSs) to ensure compatibility with analysis tools and visualization platforms.

BASIC GEOSPATIAL VISUALIZATION AND ADVANCED VISUALIZATION TECHNIQUES

Basic Geospatial Visualization introduce foundational Python techniques for representing geographic data. Advanced geospatial visualization techniques, including choropleth maps, heatmaps, time series, and interactive maps, enable in-depth exploration of spatial data, revealing complex patterns, temporal dynamics, and interactive insights for diverse applications in geography, science, and decision-making (Wang et al., 2017).

Creating Simple Point Maps

Point maps are one of the simplest and most effective ways to visualize geospatial data. They represent individual data points as markers on a map.

Use Cases: Point maps are suitable for showing locations of specific events, businesses, or objects. Common use cases include plotting cities, landmarks, or sensor locations.

```
import geopandas as gpd
import matplotlib.pyplot as plt
import pandas as pdgdf.plot(marker='o', color='blue',
```

```
markersize=5)
plt.show()
```

Python Tools: You can create point maps in Python using libraries like GeoPandas and Matplotlib. GeoPandas provides a convenient way to manage and plot geospatial data, while Matplotlib handles the plotting.

Visualizing Spatial Relationships With Lines And Polygons

Visualizing spatial relationships involves displaying the connections, boundaries, or interactions between geographical entities. This can be done using lines (for connections) and polygons (for boundaries).

Use Cases: Line maps are often used for showing road networks, flight paths, or river systems.

Polygon maps are employed to represent regions, administrative boundaries, or thematic areas like land use.

```
gdf.plot(linewidth=1, edgecolor='red', facecolor='none')
plt.show()
```

Figure 4. Map with lines and polygons

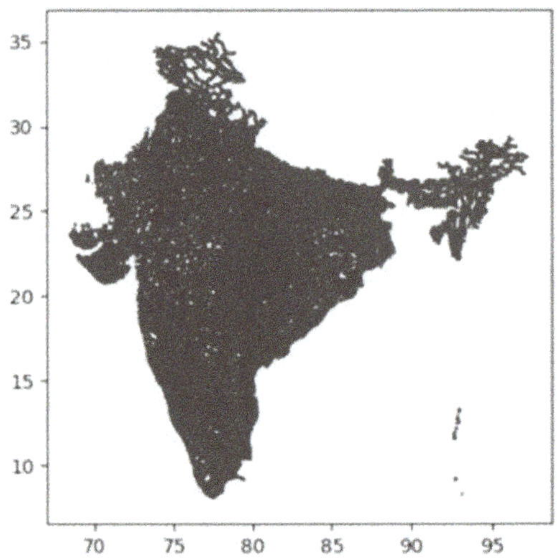

Python Tools: GeoPandas and Matplotlib are powerful tools for creating line and polygon maps. You can specify line styles, colours, and fill patterns to enhance Visualization.

Overlaying Multiple Layers

Introduction: Overlaying multiple layers means combining different geographical datasets on a single map to reveal relationships and patterns that may not be apparent in individual maps.

Use Cases: This technique is valuable when you want to compare and analyse multiple geographical features simultaneously. For instance, overlaying land use, population, and environmental data can provide insights into urban planning.

```
gdf1 = gpd.read_file('layer1.shp')
gdf2 = gpd.read_file('layer2.shp')
fig, ax = plt.subplots()gdf1.plot(ax=ax, color='blue')
gdf2.plot(ax=ax, color='red')
plt.show()
```

Figure 5. Map with multiple layers

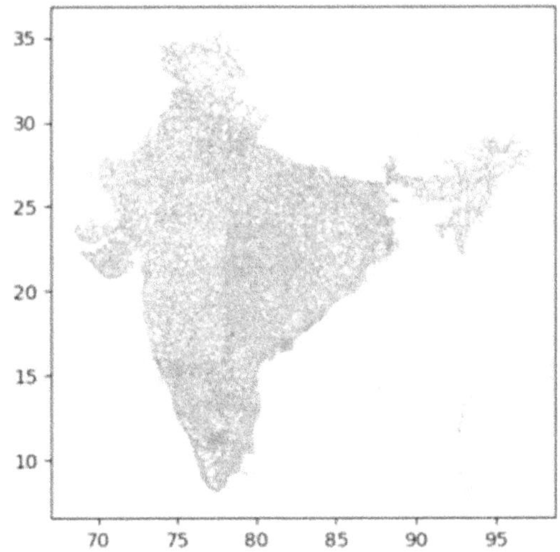

Python Tools: In Python, you can overlay layers by loading each dataset into a Geodata Frame and then using Matplotlib or GeoPandas to plot them on the same map. Controlling the order of plotting determines which layer appears on top.

Working With Categorical Data

Categorical data, representing discrete classes, plays a vital role in geospatial visualization. This introduction explores its significance and application in various domains.

Use Cases: Categorical data finds application in land use planning, epidemiology, ecology, and market analysis, facilitating informed decisions and insights.

Python Tools: Python libraries like GeoPandas, Matplotlib, Seaborn, Folium, and ColorBrewer empower geospatial practitioners to effectively visualize and analyse categorical data.

```
gdf.plot(column='Category', legend=True)
plt.show()
```

Figure 6. Map using categorical data

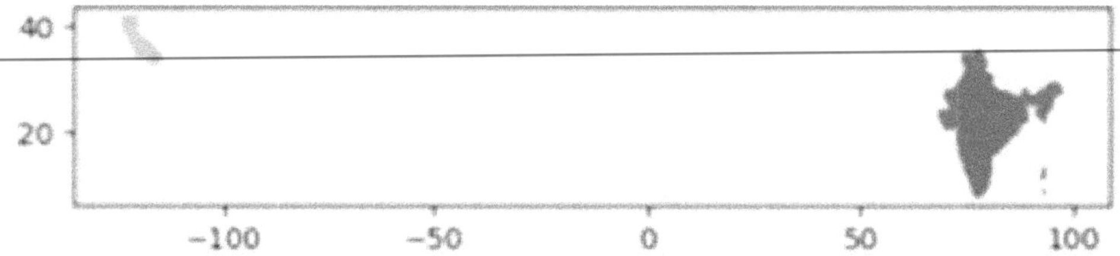

Creating Choropleth Maps

Choropleth maps are powerful for visualizing spatial patterns of continuous data, such as population density or economic indicators, by shading geographic areas based on data values.

Use Cases: Choropleth maps are widely used in demography, economics, and epidemiology to display regional variations in data. They are valuable for highlighting disparities and trends.

Figure 7. Choropleth maps

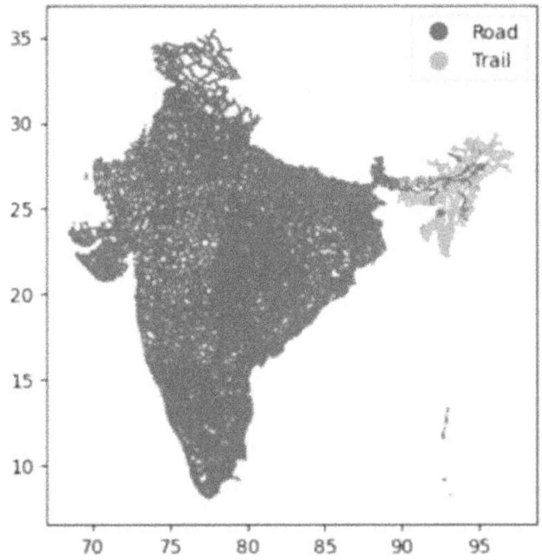

Python Tools: Libraries like GeoPandas and Matplotlib enable the creation of choropleth maps. You can choose colour schemes and classification methods to convey data effectively.

Heatmaps And Density Plots

Heatmaps and density plots provide a visual representation of data intensity or concentration across a geographic area. They are helpful for identifying hotspots and clustering.

Use Cases: Heatmaps are used in real estate to show property prices, in epidemiology to map disease incidence, and in ecology to study species distributions.

Python Tools: Seaborn, Matplotlib, and spatial libraries like rasterio are used to create heatmaps and density plots. They allow customization of colour gradients and contour lines.

```
sns.kdeplot(gdf['x_coordinate'], gdf['y_coordinate'],
cmap='YlOrRd', fill=True)
plt.show()
```

Time Series and Animations

Time series Visualization in geospatial data capture temporal changes. Animations add a dynamic dimension to data exploration, showing how spatial patterns evolve over time (Wang et al., 2017).

Use Cases: Time series maps are essential in meteorology, traffic analysis, and environmental monitoring. Animations help visualize processes like urban growth.

```
def animate(i):
    plt.clf()
    subset = gdf[gdf['timestamp_column'] <= i]
    subset.plot(column='data_column', cmap='coolwarm',
legend=True)
    plt.title(f'Time: {i}')
ani = animation.FuncAnimation(plt.figure(), animate,
frames=range(0, max_timestamp, step), repeat=False)
plt.show()
```

Python Tools: Libraries like Matplotlib, Plotly, and Geopandas support time series and animation creation. You can create frame by frame animations to convey complex spatiotemporal information.

Interactive Maps With Folium

Folium is a Python library for creating interactive maps. It allows you to add interactivity, popups, and tooltips to your maps for a richer user experience.

Use Cases: Folium is commonly used for web mapping, visualizing point of interest data, and sharing geospatial insights with an interactive audience.

Python Tools: Folium provides a straightforward way to create interactive maps with minimal code. You can add markers, polygons, and custom JavaScript to enhance interactivity.

```
import folium
m = folium.Map(location=[latitude, longitude], zoom_start=10)
folium.Marker([lat1, lon1], tooltip='Marker 1').add_to(m)
folium.Polygon(locations=[(lat1, lon1), (lat2, lon2), (lat3,
lon3)], color='blue', fill=True, fill_color='blue', fill_
opacity=0.4, tooltip='Polygon').add_to(m)
m.save('interactive_map.html')
```

Figure 8. Interactive maps

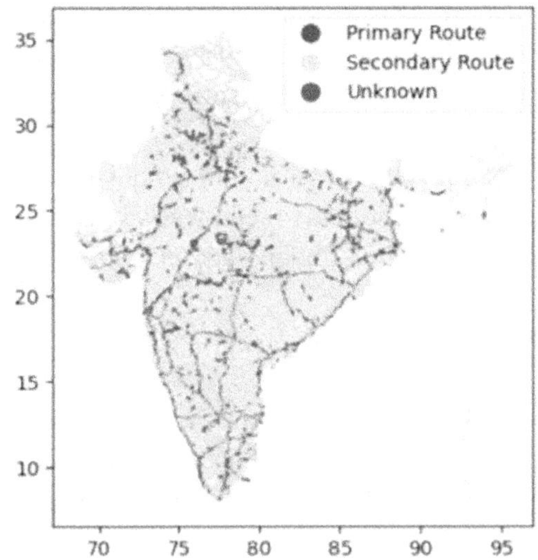

Static maps, while informative, possess limitations when it comes to conveying dynamic or time dependent information. This section of the chapter underscores the need for spatial Visualization that evolve and adapt, catering to real-time data or dynamic phenomena. It emphasizes the transition from fixed images to dynamic representations that reflect changes over time or in response to user interactions.

- ***Bokeh:*** Bokeh, a Python interactive visualization library, shines in generating dynamic and interactive plots and dashboards. It empowers users to create interactive maps with features like zooming, panning, and tooltips.
- ***Folium:*** Folium specializes in creating leaflet.jsbased interactive maps with Python. It facilitates the addition of markers, polygons, and popups, making it a go to choose for web-based mapping applications.
- ***Plotly:*** Plotly offers interactive plotting capabilities, enabling the creation of dynamic spatial Visualization that respond to user input. It is especially well-suited for creating web-based dashboards and applications.

HARNESSING SPATIAL DATA VISUALIZATION FOR ENHANCED DECISION MAKING AND GEOGRAPHIC INFORMATION SYSTEMS (GIS) INTEGRATION

The convergence of spatial data visualization across various domains, including Virtual Reality (VR), Augmented Reality (AR), satellite navigation technologies

(Satnav), and precision industries, is revolutionizing how we interact with and interpret location-based information. This transformative technology empowers users with immersive experiences, real-time insights, and data driven decision-making, fostering innovation and efficiency across diverse sectors. Geographic Information Systems (GIS) Integration introduces the vital synergy between Python and GIS software. It elucidates the interface mechanisms, data harmonization, and spatial analytical capabilities, empowering practitioners to seamlessly integrate and extract geospatial insights for informed decision-making and advanced research (Pundt, 2010).

Interfacing With Gis Software

Efficiently interfacing with GIS software is a paramount consideration for organizations and individuals engaged in spatial data management and analysis. Python offers multiple avenues for establishing a seamless connection with GIS systems:

Using Libraries: Python boasts a rich ecosystem of libraries, such as pyproj, geopandas, and fiona, which facilitate direct interaction with GIS platforms. These libraries empower users to effortlessly read, write, and manipulate geographic data in formats that align with GIS software requirements.

***Command Line Tools*:** Python serves as a versatile orchestrator for executing command line GIS utilities like ogr2ogr and gdal_translate. These utilities are integral for tasks involving data conversion and ensuring compatibility across different geospatial formats.

APIs: Some GIS software packages extend their functionality by offering Application Programming Interfaces (APIs). Python can harness these APIs, enabling programmatic communication with GIS systems. For instance, the ArcGIS API for Python enables seamless interaction with the renowned ArcGIS platform (Pundt, 2010).

Data Conversion And Compatibility

Data conversion stands as a foundational pillar of GIS integration, enabling harmonious coexistence between diverse data formats. Python assumes a pivotal role in this process:

- ***Format Conversion:*** Python wields the capability to perform conversions between a multitude of geospatial formats. Through utilities like ogr2ogr, Python orchestrates the transformation of data between formats like shapefiles, GeoJSON, KML, and others.

- ***Data Transformation:*** Python leverages libraries like shapely and pyproj to adeptly transform and reproject geographic data. This transformation ensures that data aligns seamlessly with the requisite coordinate systems and formats dictated by GIS software.
- ***Database Integration:*** Python facilitates the integration of geospatial data with databases frequently employed in the GIS domain. Widely used databases like PostgreSQL/PostGIS and SQLite readily interface with Python, empowering efficient data storage and retrieval (Pundt, 2010).

Spatial Analysis With Python

Spatial analysis involves extracting insights from geographic data through operations like buffering, overlay analysis, clustering, and more. Python provides several libraries and tools for conducting spatial analysis:

GeoPandas: Geo Pandas is an influential extension of the Pandas library designed to facilitate the processing and analysis of geospatial data in Python. By implementing the Geo Data Frame data structure, Geo Pandas seamlessly integrates geometric features into the familiar Pandas framework. It allows users to perform spatial joins, define queries and create visually appealing data Visualization. Geo Pandas focuses on simplicity and compatibility with Panda. It is the star of the Python constellation that reveals the complexity of spatial data structures.

Rasterio:Raster is the dominant tool in Python's arsenal for managing geospatial raster data. Its capabilities extend to reading, writing and coordinating various raster data, from satellite images to digital elevation models. A raster simplifies the interpretation of pixel values in a geospatial context and provides tools for tasks such as mosaicking and projection. By putting raster data into a Python command, Rasterio allows users to access and manipulate spatial data embedded in raster datasets.

PySAL (Python Spatial Analysis Library): PySAL is a Python ecosystem shell for spatial econometrics, spatial statistics, and spatial data visualization. This comprehensive library provides a wide range of tools for exploring and modelling spatial relationships in data. PySAL facilitates complex spatial analysis tasks, from spatial econometrics to analysing financial data with spatial dependencies to statistical tools for spatial pattern recognition. Its contribution extends to providing visualization capabilities for spatial patterns and relationships, which improves the interpretability of complex spatial data sets.

Network Analysis: Python and comprehensive toolkit include specialized libraries for network analysis, such as network, which gives Python the ability to navigate, optimize networks, and model transport networks. These libraries enable, for example, route mapping, identification of optimal routes between locations, and optimization of network performance. Python and its's network analysis capabilities provide a versatile and powerful tool, whether used for transportation, communication, or other domains requiring network optimization.

Geostatistics: Python is making a confident entry into geostatistics with packages such as SciPy, pykrige and GSLIB. SciPy, a core library for scientific computing, integrates geostatistical functions, while pykrige specializes in spatial interpolation. GSLIB completes the landscape with variogram analysers that capture spatial variability. Together, these packages allow Python users to perform tasks such as spatial interpolation, estimate values at unsampled locations, and analyse variograms to understand the spatial distribution of data. Python and its study of geostatistics strengthens his ability to handle various aspects of spatial data analysis.

Spatial Data Visualization In Virtual Reality (VR) And Augmented Reality (AR)

1. Enhancing Spatial Data Visualization

In the field of spatial data visualization, virtual reality (VR) offers a revolutionary approach, offering users an immersive experience in a rich three-dimensional environment. This technological leap goes beyond traditional two-dimensional representations and offers a new dimension to understanding geographic data. The applications of comprehensive visualization of spatial information in VR are extensive and include urban planning, archaeology and education, among others. Urban planning, a complex discipline that involves designing sustainable and liveable cities, can greatly benefit from VR-based visualization of spatial data.

Figure 9. Map using AR and VR

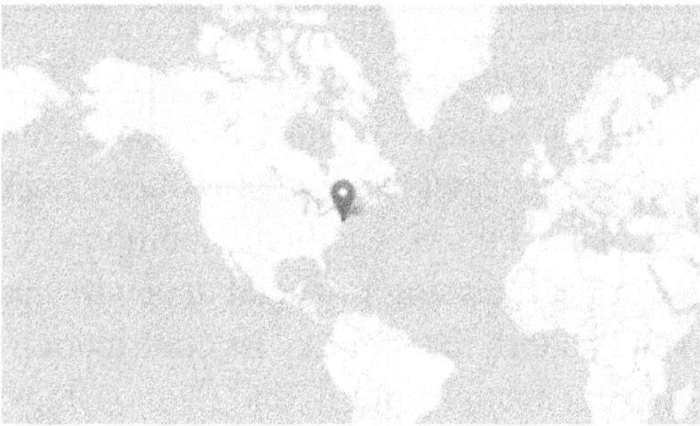

Designers can immerse themselves in intricately detailed virtual cityscapes and gain a holistic understanding of how different elements interact in a three-dimensional space. This immersion allows more intuitive exploration and analysis of spatial information, which paves the way for making informed decisions in the field of urban planning and development. The ability to virtually walk through a city plan, analyse traffic patterns and assess the impact of proposed changes is revolutionizing the planning process.

In archaeology, VR-based spatial data visualization offers an innovative way to study historical sites and artifacts. Scientists and archaeologists can practically reconstruct ancient landscapes, enabling a deeper understanding of past civilizations. The immersive experience allows users to navigate archaeological sites, examine artifacts up close and visualize historical context in a way that traditional methods cannot reproduce. This technology facilitates a more accurate interpretation of archaeological data and promotes the preservation and sharing of cultural heritage (Pundt, 2010).

Training settings also use immersive visualization of spatial data in VR. Complex spatial concepts that are often difficult to convey through traditional methods become accessible and engaging in a virtual environment. Students can study geographic phenomena, such as the topography of the Earth or the movement of tectonic plates, in a way that feels tangible and real. This hands-on, immersive learning improves understanding and retention of spatial concepts, making training more interactive and effective.

2. Transforming Spatial Awareness

Augmented reality (AR), on the other hand, overlay's location data over the real world, giving users a real-time view of their surroundings. This technology has transformative applications, particularly in areas such as navigation and situational awareness. While navigating, AR displays directional information and places of interest to the user, as well as a real-world view. These real-time instructions improve situational awareness and make navigation more intuitive and efficient. For example, pedestrians can follow virtual arrows on the sidewalk, and drivers can receive driving directions on the road ahead.

Integrating location data into the user's field of view revolutionizes navigation and reduces dependence on maps and screens. Situational awareness extends beyond navigation to a variety of industries, including emergency services, maintenance and field work. By incorporating relevant location data into the real world, AR empowers professionals to make informed decisions on the spot. For rescue workers, this means critical information about the environment during rescue operations. Service technicians can visualize hardware details on the machine, improving troubleshooting and efficiency (Wang et al., 2017).

3. Changing Decision Making, Education and Engagement

Integrating spatial data visualization with VR and AR not only provides an immersive experience, but also changes decision-making processes in various industries. Stakeholders from city planners to emergency responders can now use interactive data simulations that provide insight into possible scenarios and outcomes. This paradigm shift significantly improves the quality and efficiency of decision-making. In training scenarios, VR and AR become invaluable tools that allow users to practice tasks in a realistic virtual environment. This is particularly useful in occupations where spatial awareness and precise operations are critical, such as pilot training, medical procedures and complex machine operation (Wang et al., 2017).

The immersive nature of these techniques allows for hands-on training without the risks associated with real-world scenarios. This in turn improves agility and overall performance. In addition, visualization of location data in VR and AR increases engagement in educational environments. By transforming abstract concepts into concrete, immersive experiences, students are more likely to retain information and develop a deeper understanding of spatial relationships. Whether it's exploring historical events in VR or cutting through the layers of the Earth in AR, these technologies spark education by fuelling curiosity and discovery.

Spatial Data Visualization in Satellite Navigation (Satnav) Technologies

Location data visualization is an integral part of satellite navigation technology (Satnav) that improves navigation accuracy and real-time traffic management. Using real-time data such as satellite images, Satnav optimizes routes, reduces travel time and minimizes fuel consumption. It transforms decision making by dynamically adjusting routes based on real-time traffic updates. Satnav and its visualization of location data also improves safety by providing real-time information on road conditions and accidents, ensuring users reach their destination efficiently and safely.

1. Optimizing Navigational Accuracy

Satellite navigation systems (Satnav) have become an integral part of modern life, providing users with accurate and efficient navigation assistance. Central to the effectiveness of these systems is the use of location data visualization, a technology that optimizes navigation accuracy, improves real-time traffic management, and improves both safety and user satisfaction. The main purpose of Satnav systems is to guide users to their destination using the most efficient routes possible. Spatial data visualization plays a key role in achieving this goal, using real-time data, including satellite imagery and traffic data. These systems continuously analyse and interpret location data to provide users with up-to-date information on road conditions, traffic congestion and alternative routes. This ensures that users receive the most accurate and up-to-date navigation instructions and optimize their travel routes.

Real-time data, especially from satellites, allows Satnav systems to pinpoint the user and location. By viewing this location data, users can receive turn-by-turn directions that seamlessly guide them through their journey. This precision not only reduces driving time, but also reduces fuel consumption, contributing to environmental sustainability. As spatial data visualization evolves to include advanced technologies such as 3D mapping and augmented reality, users can expect an even more immersive and accurate navigation experience.

2. Real Time Traffic Management

Visualization of location data in Satnav technologies goes beyond individual navigation and plays an important role in real-time traffic management. These systems collect and analyse data from various sources such as GPS devices, traffic cameras and other connected devices. This real-time monitoring allows you to identify traffic congestion, accidents and road closures. By viewing this location data, Satnav systems can dynamically adjust navigation routes, steer drivers away from congested areas and suggest alternative routes.

Figure 10. Sat-Nav

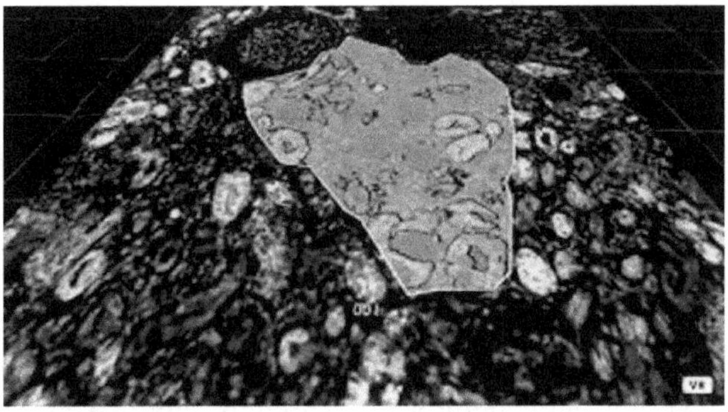

This real-time traffic management not only improves the individual travel experience, but also contributes to the overall traffic flow. The ability to effectively reroute drivers minimizes delays and reduces the environmental impact of traffic congestion, making urban traffic more sustainable. In addition, real-time traffic management facilitates communication between Satnav systems and traffic control centres. This exchange of location data allows authorities to quickly respond to changing traffic conditions and implement measures to effectively manage traffic flow, such as adjusting direction signs or closing roads. As a result, cities can optimize their transportation infrastructure based on real-time insights, improving overall mobility.

3. Improve security and user satisfaction

Satnav systems play a vital role in improving road safety as they provide real-time information on various factors that can affect the driver and the journey. Visualization of location data allows users to receive instant updates on road conditions, accidents and weather forecasts. This information is essential for making informed decisions while driving, which contributes to public safety.

For example, if an accident or adverse weather conditions occur on a planned route, Satnav systems can dynamically recalculate the route, directing users away from potential hazards. This proactive approach to safety helps reduce accidents and gives users a sense of security while traveling. User satisfaction is another important aspect affected by the effective use of location data in Satnav systems. By providing accurate and real-time navigation information, these systems ensure a hassle-free and efficient driving experience.

Users can confidently rely on their satnav for directions, knowing they will receive timely updates and alternative routes if needed. The result is not only a more pleasant driving experience, but also a positive perception of the Satnav system itself.

Spatial Data Visualization in Precision Industries

This section focuses on the important role of spatial data visualization in automating processes, optimizing heavy equipment operations, minimizing human error, and ensuring consistency between operations in the dynamic field of precision industries.

1. Automation in Precision Industries

The integration of spatial data into precision agriculture heralds a new era of automation. Imagine a fleet of autonomous harvesters moving across fields with unprecedented precision based on real-time spatial insights. This subsection describes transformative spaces, where spatial data plays a key role in automating tasks such as harvesting, planting, and monitoring. Through data-driven decision-making, farmers optimize yields, reduce labour costs, and maximize efficiency. The case studies in this section highlight examples where the synergy of spatial intelligence and automation has led to significant advances in agricultural practices.

2. Precision in Heavy Equipment Operation

The construction industry, which relies heavily on complex machinery, is undergoing a paradigm shift with the integration of spatial data. Real-time visualization of construction sites allows project managers to optimize heavy equipment operations. Excavators, cranes and bulldozers become interconnected hubs controlled by spatial insights, ensuring precise and efficient work. This section delves into how construction projects benefit from reduced downtime, accelerated timelines, and substantial cost savings through the strategic use of spatial data in heavy machinery management.

3. Minimizing Human Error

Human error, though inevitable, becomes a manageable factor with the integration of spatial data. This subsection explores how spatial insights act as a safeguard against inaccuracies caused by human judgment. In agriculture, for instance, the precise application of fertilizers and irrigation is guided by spatial intelligence, reducing the risk of overuse or underuse.In construction, decisions related to site planning and hazard mitigation are fortified by accurate spatial data, minimizing the potential for errors that could compromise safety and project timelines.

4. Consistency of Operations

Consistency is the basis of efficiency, and geospatial data ensures consistency of all operations in precision industries. This part of the study details how spatial insights can help maintain consistency across different processes. In agriculture, geospatial data facilitates consistent application of agricultural practices, resulting in consistent crop quality and yield. In construction, it ensures that each phase of a project is executed accurately, reducing variability and improving overall project performance.

From autonomous tractors navigating vast fields to construction sites coordinated with digital precision, the story is about a future where productivity and precision are not just goals, but characteristics inherent in spatially informed decision-making.

SHARING YOUR SPATIAL VISUALIZATION

Effectively sharing your spatial Visualization is essential to communicate insights and drive informed decision-making. Here are methods to share your Visualization:

Exporting Plots and Maps

After creating your spatial Visualization, you can export them in various formats for sharing:

- ***Image Formats:*** Export plots and maps as image files (e.g., PNG, JPEG, SVG) for easy sharing through emails, reports, or presentations. Tools like Matplotlib and GIS software often offer options for exporting images.
- ***Vector Formats:*** For scalability and high-quality printing, export maps in vector formats like PDF or EPS. This is useful when you need to include maps in printed materials.
- ***GIS File Formats:*** If you're working with GIS software, you can export your Visualization in common GIS formats like Shapefiles or GeoJSON for compatibility with other GIS systems (Pundt, 2010).

Building Interactive Web Based Dashboards

Interactive web-based dashboards are a powerful way to share spatial Visualization, allowing users to explore data on their own:

- ***JavaScript Libraries:*** Utilize JavaScript libraries like Leaflet, Mapbox, or D3.js to build custom interactive web maps and dashboards. These libraries offer extensive customization and integration options.
- ***Dashboard Tools:*** Platforms like Tableau, Power BI, or QlikView provide user-friendly interfaces for creating interactive dashboards that can include spatial Visualization. They are suitable for nontechnical users.
- ***Web Development:*** For maximum customization, consider developing a custom web application using web frameworks like Flask, Django, or JavaScript (e.g., React, Vue.js) to embed your spatial Visualization and interactivity (Shaito & Elmasri, 2021).

It can be argued that the theoretical foundations for creating interactive web-based spatial data visualization dashboards involve a multidisciplinary approach. Based on data visualization, human-computer interaction, spatial analysis and distributed systems, the development process tries to follow well-established principles that prioritize clarity, user interaction and effective communication of spatial data. As technology advances, this theoretical framework provides a solid foundation for creating web-based dashboards that not only present spatial data, but also enable users to explore, analyse, and gain meaningful insights from complex geographic data sets (Balla & Mester, 2017).

Figure 11. Representation of spatial data

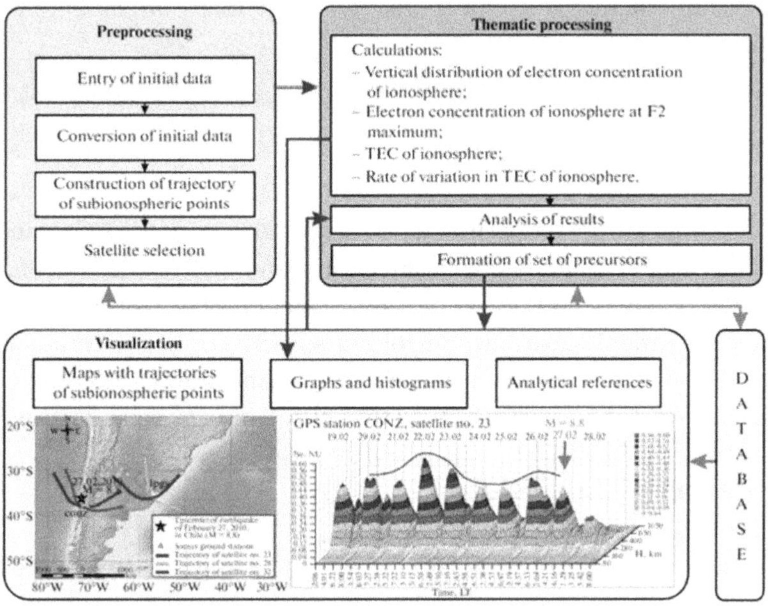

Preparing Visualization for Presentations and Reports

When sharing spatial Visualization in presentations or reports, consider the following:

- ***Storytelling:*** Frame your Visualization within a narrative to guide the audience through the insights you want to convey. Explain the context and significance of the data.
- ***Annotations:*** Use text labels, arrows, or callouts to highlight specific features or trends in your Visualization. Annotations provide context and clarity.
- ***Data Sources:*** Always provide information about the data sources, including dates, references, and any transformations or preprocessing performed. This ensures transparency and credibility.
- ***Captions and Legends:*** Include captions and legends to explain symbols, colours, and scales used in your Visualization. This helps the audience interpret the information correctly.
- ***High Resolution:*** Ensure that the Visualization are of high resolution and well formatted to maintain clarity when printed or projected in presentations.
- ***Accessibility:*** Make sure your Visualization are accessible to all users, including those with disabilities. Use alt text for images, provide alternative ways to access content, and consider colour choices for readability.

- ***Interactive Elements:*** If your presentation or report is digital, consider embedding interactive elements like clickable maps or charts to engage the audience.

REAL WORLD APPLICATIONS

This text delves into the practical utilization of spatial data visualization and analysis in Python across diverse domains. From urban planning to environmental science, epidemiology to logistics optimization, this chapter presents real-world case studies that showcase Python's instrumental role in solving complex spatial challenges. Professionals and researchers gain valuable insights into how Python empowers data driven decision-making in areas critical to societal and environmental wellbeing.

Urban Planning Case Study

Urban planning is a multifaceted discipline that requires sophisticated tools for data-driven decision making. Python has emerged as a versatile and indispensable tool in this field, giving urban planners the ability to analyse various data sets that are crucial to designing sustainable and livable cities. One of its key strengths is its ability to seamlessly handle demographics, transport modes and land data. Python's role in urban planning is illustrated by libraries such as GeoPandas and Matplotlib. GeoPandas extends the functionality of Panda to include geospatial data structures, making it easier for professionals to manage and analyse geospatial data.

By evaluating the impact of proposed infrastructure projects, Python facilitates evidence-based decision-making. Designing more efficient cities is possible with Python's spatial analysis tool, helping to create urban environments that are not only sustainable, but resilient to future challenges.

Figure 12. Urban Planning flow

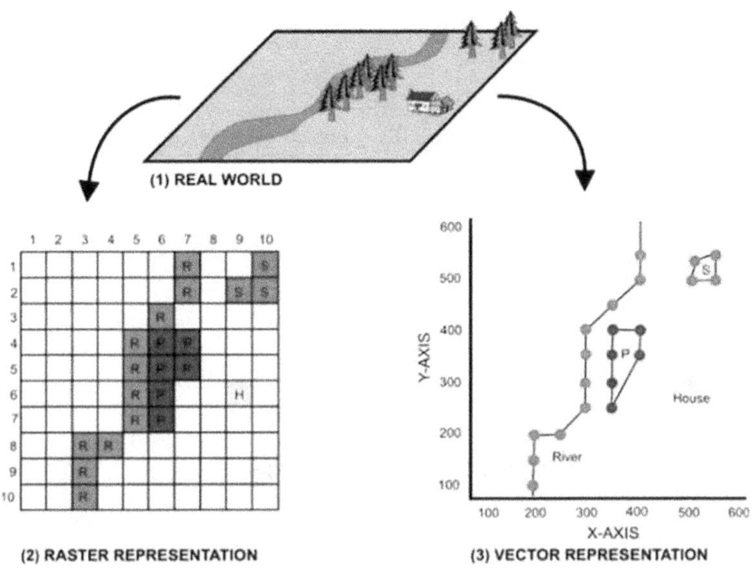

Environmental Analysis and Modeling

Python's reach extends to the field of environmental science, where scientists grapple with critical issues such as climate change, ecological conservation and agriculture. Python and its extensive libraries, including NumPy and SciPy, provide a strong foundation for sophisticated data analysis and modelling. These libraries allow environmental scientists to efficiently process large data sets, perform statistical analysis, and develop complex models.

Matplotlib and Folium further enhance Python's capabilities in environmental analysis, making it easier to visualize spatial data. Intuitive Visualization enable researchers to effectively communicate complex findings, bridging the gap between knowledge and decision makers. Python's adaptability enables the creation of complex environmental models, from climate simulations to habitat suitability analyses. It not only supports research, but also informs policy-making for proactive environmental management.

Figure 13. Layers of environment

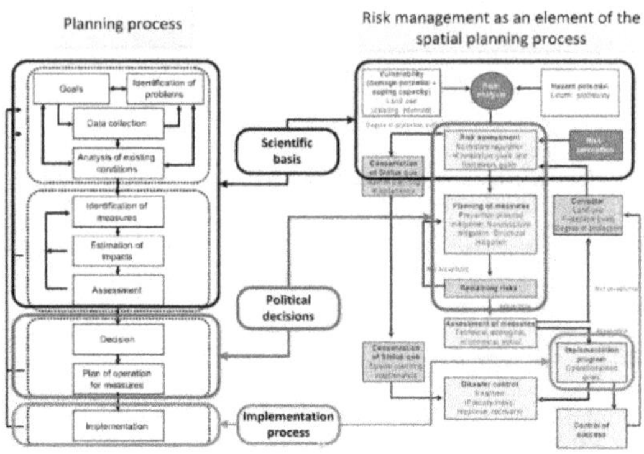

Optimization of Transport and Logistics

Efficient transport and logistics are indispensable for both companies and authorities. Python is a powerful ally in route optimization, fleet management and supply chain streamlining. Libraries such as Network and ORTools provide tools for route optimization and analysis of transport networks. Python-based geographic information systems (GIS) contribute greatly to the planning of public transport systems and the optimization of delivery routes. This application improves work efficiency, reduces costs and minimizes the environmental impact of tran sport operations.

Python's versatility enables transportation and logistics professionals to adapt and optimize their systems in response to dynamic and complex challenges. Taken together, these case studies highlight the versatility and effectiveness of Python and it's in solving complex spatial challenges in various industries (Koperski et al., 2019).

Figure 14. Transport optimization

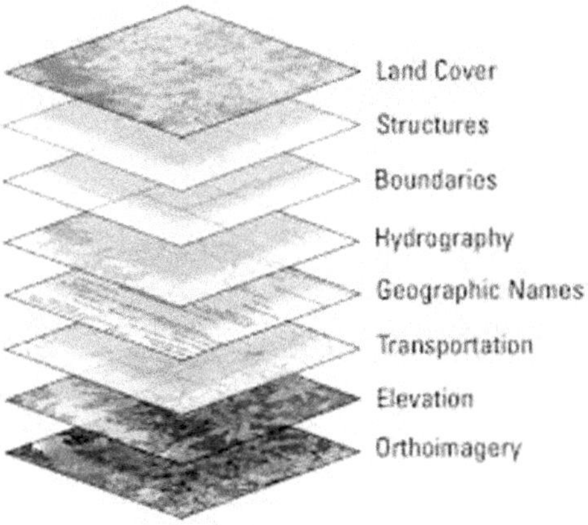

Spatial Data Visualization in Space Exploration

Space exploration relies heavily on spatial data visualization to ensure the success and safety of missions. This section discusses the use of spatial data visualization in two prominent space missions: Chandrayaan3 Lunar Exploration and Aditya L1 Solar Observation Missions. These missions highlight the importance of spatial data visualization in safe landings, rover operations, and solar dynamics understanding.

Chandrayaan3 Lunar Exploration

Chandrayaan3 is the third lunar exploration mission by the Indian Space Research Organisation (ISRO). Spatial data visualization plays a crucial role in Chandrayaan3's success by:

Terrain Mapping: Detailed lunar topographic maps are created using spatial data visualization techniques. These maps help in identifying safe landing sites for lunar landers.

- ***Rover Navigation:*** Visualizing spatial data from rover sensors allows mission operators to plan safe and efficient routes for lunar rovers.
- ***Scientific Analysis:*** Spatial data visualization tools assist in analysing data collected by scientific instruments onboard. This aids in lunar geology and resource exploration.

Figure 15. Chandrayaan-3

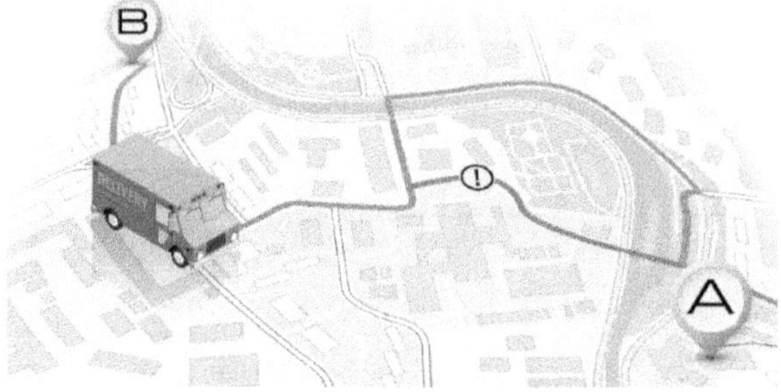

ADITYA L1 Solar Observation Missions

Aditya L1 is an Indian solar observation mission aimed at studying the Sun's outermost layer, the corona. Spatial data visualization is pivotal in Aditya L1's objectives:

- ***Coronal Imaging:*** Visualization tools enable scientists to observe and analyse the Sun's corona, helping understand solar dynamics and predict solar storms.
- ***Data Fusion:*** Spatial data from multiple solar observation instruments is fused to create comprehensive images of the Sun's surface and atmosphere.
- ***Predictive Models:*** Visualization aids in creating predictive models of solar activities, which are crucial for space weather forecasting and safeguarding space assets.

In both missions, spatial data visualization not only aids in mission planning and execution but also enhances our understanding of celestial bodies.

Spatial Data Visualization In Pandemic Response

The utilization of spatial data visualization has been instrumental in the global response to the COVID19 pandemic. This chapter explores the critical role of spatial data visualization in pandemic response, focusing on tracking, managing, and mitigating COVID19, allocating resources, enhancing public health communication, and supporting contact tracing and spread modelling (Shaito & Elmasri, 2021).

Tracking, Managing, and Mitigating COVID19

Spatial data visualization has been pivotal in tracking the spread of COVID19 and managing its impact:

Dashboard Creation: Public health agencies and organizations have developed interactive dashboards that display real-time COVID19 data, including infection rates, hospitalizations, and fatalities, at local, regional, and global levels. These dashboards provide policymakers and the public with a clear, visual understanding of the pandemic's progression.

Hotspot Identification: Spatial data analysis helps identify COVID19 hotspots, allowing authorities to deploy resources, such as testing facilities and healthcare personnel, to areas with the highest infection rates. This targeted response helps mitigate the virus's spread.

Figure 16. Covid-19 analysis

Travel and Mobility Data: Spatial data visualization incorporates travel and mobility data to track the movement of people, identifying potential sources of virus transmission and guiding travel restrictions and quarantine measures (Shaito & Elmasri, 2021).

Supporting Contact Tracing and Spread Modelling

Contact tracing and modelling the spread of the virus are essential for pandemic control:

Contact Tracing: Spatial data visualization supports contact tracing efforts by mapping the movements and interactions of infected individuals. This helps identify potential exposure points and facilitates timely notification and testing of contacts.

Spread Modelling: Epidemiologists use spatial data to create models that predict how the virus may spread in different scenarios. These models guide policy decisions, such as lockdowns and social distancing measures, to curb transmission (Shaito & Elmasri, 2021).

Integration of IP Geolocation and Spatial Data Visualization

In the ever-evolving digital landscape, the integration of IP geolocation and spatial data visualization offers a myriad of benefits, including tailored content delivery, marketing precision, enhanced security, compliance with geographic regulations, and data driven decision-making.

1. Personal user experiences

In the vastness of the Internet, delivering personalized and relevant content to users is the cornerstone of an engaging digital experience. IP geolocation is the basis for achieving this goal. By recognizing users' geographic location, websites and apps can adapt content to regional preferences, language and cultural nuances. This content personalization increases user engagement and satisfaction by creating an environment that resonates with the user and local context (Bella & Mester, 2017).

2. Target Marketing

Accuracy is key in digital marketing. Understanding the geographic location of users through IP location data allows marketers to refine their targeting strategies with unprecedented precision. Instead of using broad features, marketers can create campaigns that specifically target users in different regions. This targeted marketing approach not only increases the effectiveness of campaigns, but also reduces wasted advertising spend by directing resources where they are most likely to produce results.

3. High Security

As digital communication increases, so do the threats associated with online activity. Organizations face countless security challenges. IP tracking is an important tool for strengthening security measures by tracking the geographic origin of website traffic. One of the most important security benefits of an IP location service is its role in fraud detection and prevention. By analysing the geographic locations from which users access a website or application, organizations can identify unusual or suspicious usage patterns.

4. Findings and ethical implications

Organizations must prioritize transparency and provide clear information about how IP location data is collected, stored and used. Strong data protection measures such as encryption and anonymization help protect user privacy. In addition, organizations must comply with relevant data protection regulations, such as GDPR (General Data Protection Regulation) in Europe or CCPA (California Consumer Privacy Act) in the United States. These policies define guidelines for the lawful collection and processing of user data, including geographic location. In addition, ethical considerations include ensuring that geographic location IP is used responsibly and does not cause discrimination or prejudice (Shaito & Elmasri, 2021).

FUTURE TRENDS IN SPATIAL DATA VISUALIZATION

The field of spatial data visualization is evolving rapidly, driven by technological advancements, and changing societal needs. Here are some future trends to watch out for:

Emerging Technologies and Libraries

As technology continues to advance, new tools and libraries will emerge, enabling more sophisticated spatial data visualization techniques. These technologies may include:

- ***Web Based Visualization:*** Enhanced web-based frameworks and libraries for creating interactive and responsive spatial Visualization, making it easier to deliver data to a broader audience (Shaito & Elmasri, 2021).
- ***3D Visualization:*** Wider adoption of 3D spatial data visualization techniques for improved depth perception and understanding of geographical data.
- ***Realtime Data Streaming:*** The ability to visualize and analyse real-time spatial data from IoT devices, sensors, and satellites in a dynamic and interactive manner.

The Role of Machine Learning and AI

Machine learning and artificial intelligence are poised to play a significant role in spatial data visualization:

- ***Predictive Analytics:*** Machine learning algorithms will be used to predict spatial trends and patterns, helping users make data driven decisions.

- ***Automated Feature Detection:*** AI powered algorithms can automatically detect and classify features in spatial data, such as identifying objects in satellite imagery or anomalies in sensor data.
- ***Personalized Visualization:*** AI can tailor spatial Visualization to individual user preferences and needs, creating more personalized and effective displays of data.

Figure 17. Machine Learning in spatial data analysis

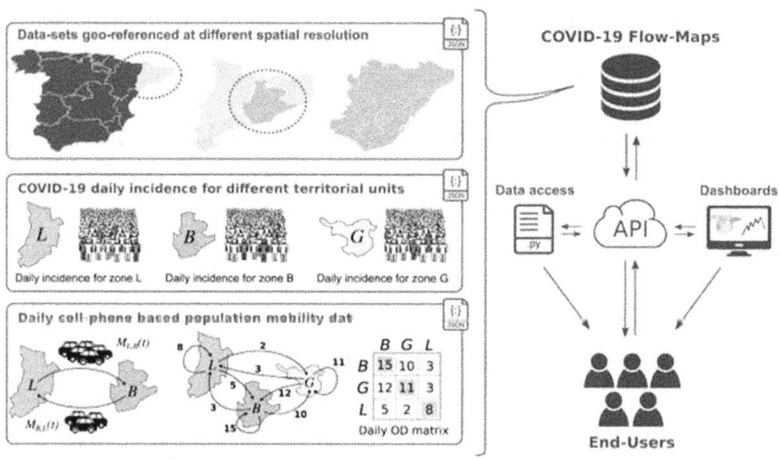

Ethical Considerations in Spatial Data Visualization

As spatial data becomes more accessible and powerful, ethical considerations will become increasingly important:

- ***Privacy Concerns:*** Protecting individuals' privacy in spatial data visualization, particularly in contexts like location-based services, will be a growing concern. Ensuring that data is anonymized and used responsibly will be critical.
- ***Bias and Fairness***: Addressing biases in spatial data and visualization algorithms to ensure equitable representation and decision making, especially in urban planning and resource allocation.
- ***Transparency and Accountability:*** Ethical standards and guidelines for spatial data visualization, including clear documentation of data sources and visualization methods, will become more standardized.

- ***Environmental Impact:*** Considerations of the environmental impact of data collection and visualization like satellite imaging and mapping, will gain prominence.

CONCLUSION

This chapters in this book provide a thorough and insightful journey into the field of spatial data visualization using Python, reflecting the impressive growth and importance of this field. The central role of Python as a flexible toolkit for visualizing complex spatial data is highlighted, providing a foundation for exploring fundamental principles, techniques, and real-world applications. Readers will have a comprehensive understanding of popular data visualization libraries to seamlessly analyse and interpret geospatial data. This chapter highlights the diverse applications of spatial data visualization, including urban planning, environmental science, business intelligence, and healthcare. This guide goes beyond static maps and considers dynamic, interactive Visualization to provide a holistic perspective on this topic. It focuses on the critical role of spatial data visualization in space exploration, virtual and augmented reality, satellite navigation, and pandemic response. We also discuss the integration of IP geolocation and spatial data visualization and their impact on marketing accuracy, security, and regulatory compliance.

REFERENCES

Balla, D. & Mester, T. (2017). *Possibilities of spatial data visualization with web technologies for cognitive interpretation*. Research Gate.

Fernando, J. (2018). *The Spatial-Perceptual Design Space: A New Comprehension for Data Visualization.*

Fiorinia, M. (2016). *AIS Data Visualization for Maritime Spatial Planning*. MSP.

Huang, B., Jiang, B., & Li, H. (2010). *An integration of GIS, virtual reality and the Internet for visualization, analysis and exploration of spatial data.*

Koperski, K., Adhikary, J., & Han, J. (2019). *Spatial Data Mining: Progress and Challenges Survey.*

Kwon, O. (2021). *Fostering Spatial Visualization Ability through Web-Based Virtual-Reality Program and Paper-Based Program.*

Pundt, H. (2010). *Visualization of spatial data for field-based GIS*. Research Gate.

Rogowitz B. (2021). *Data visualization: the end of the rainbow*. Research Gate.

Shaito, M. (2021). *Map Visualization using Spatial and Spatio-Temporal Data: Application to COVID-19 Data.* Research Gate.

Thomas, B. (2014). *Spatial augmented reality — A tool for 3D data visualization*. Research Gate.

Wang, D. (2017). *Spatial Temporal Data Visualization In Emergency Management: A view from data-driven decision.* Research Gate.

Chapter 6
Exploring Vector and Raster Data Formats for Geospatial Visualization With Python

Marsel Sonu M.
Christ University, India

Deepthi Das
Christ University, India

Arul Kumar Natarajan
https://orcid.org/0000-0002-9728-477X
Samarkand International University of Technology, Uzbekistan

Manimaran A.
https://orcid.org/0000-0002-5671-9466
VIT-AP University, India

ABSTRACT

The chapter uses Python to explore vector and raster data formats within geospatial visualization. It highlights their pivotal role across diverse environmental science, urban planning, and natural resource management domains. A nuanced comprehension of these formats is deemed essential for proficient geospatial visualization in Python, as they facilitate the storage and manipulation of spatial data. Vector data formats accurately represent points, lines, and polygons within a coordinate system. In contrast, raster data formats are tailored to depict continuous surfaces or grids of data. An array of libraries and tools are outlined for exploring and visualizing these geospatial data formats in Python, each serving distinct functionalities ranging from data manipulation to visualization. The chapter systematically introduces the concept of geospatial visualization, elucidates the disparities and application scenarios of vector and raster data formats, and subsequently elucidates various Python libraries and tools conducive to geospatial data manipulation and visualization.

DOI: 10.4018/979-8-3693-1754-9.ch006

OVERVIEW

Geospatial visualization is a potent instrument for analyzing and interpreting spatial data (McKitrick, 2023). Within the scope of this chapter, an exploration into the fundamental tenets of geospatial visualization is undertaken, leveraging Python as the primary tool. A particular emphasis is placed on comprehending two widely employed data formats: Vectors and Raster. In its essence, Vector data portrays geographic features as discrete, independent entities. Whether delineating cities, rivers, or administrative boundaries, each feature assumes the form of a distinct object, providing a nuanced representation of the Earth's surface. Conversely, Raster data paints a divergent portrait, representing geographic information as a grid of uniformly spaced cells or pixels reminiscent of a digital image. Embedded within this grid, each cell houses a value indicative of a specific attribute or measurement, constituting the fundamental elements of spatial analysis.

A foundational understanding of geospatial visualization is established by elucidating the dichotomy between Vector and Raster data formats. Through Python's versatile ecosystem, stakeholders are poised to embark on a journey to harness the transformative potential of spatial data analysis. This endeavor equips practitioners with the tools to extract insights, formulate informed decisions, and navigate real-world challenges with precision and efficacy (Feng & Koch, 2024).

VECTOR DATA FORMAT

Vector data is a powerful medium for representing spatial information by abstracting points, lines, and polygons. This versatile format allows for the precise delineation of geographic features, enabling the portrayal of intricate landscapes and artificial structures with remarkable fidelity. Figure 1 vividly demonstrates the symbiotic relationship between vector data and real-world imagery. The majestic mountain ranges are elegantly encapsulated within a polygon's confines, underscoring the vector data's capacity to encapsulate expansive geographical phenomena. Similarly, the river's meandering course finds expression as a line, gracefully tracing its path across the landscape.

In contrast, the quaint abode nestled amidst the verdant terrain is succinctly represented as a point, pinpointing its exact location on the map. With its inherent flexibility and granularity, vector data emerges as the quintessential tool for representing discrete features such as valleys, mountain ranges, and administrative boundaries, as exemplified by the diverse elements in the provided illustration. Moreover, its applicability extends beyond mere visualization, facilitating spatial analysis and informed decision-making across various domains.

Figure 1. Depicting a real-world image in Raster and Vector representation

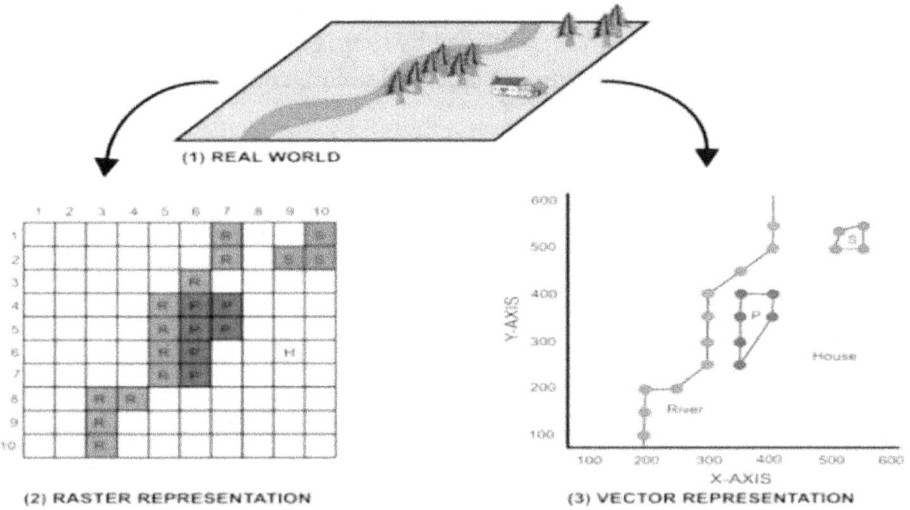

Figure 1 provides a visual depiction illustrating the distinction between raster and vector representations of real-world geographic features. This comparative visualization elucidates the fundamental disparities between the two data formats, facilitating a clearer understanding of their applications in geospatial analysis. In Python, Geopandas is the preeminent library for working with vector data. Renowned for its intuitive interface and robust functionality, Geopandas empowers users to easily manipulate, visualize, and analyze spatial datasets. Leveraging Geopandas, practitioners can unlock the full potential of vector data, ushering in a new era of spatial exploration and discovery (Anselin & Koschinsky, 2022).

Points

In the subsequent section, we delve deeper into vector data representation by examining specific geometric objects. The following code snippet demonstrates the utilization of Geopandas, a Python library renowned for its prowess in handling spatial data, to highlight the city of Bangalore as a point on a world map. Through step-by-step elucidation, the code illustrates importing necessary libraries, loading spatial datasets, creating visualizations, and ultimately showcasing Bangalore's geographic coordinates with vibrant prominence (Alam et al., 2022). This exemplar reinforces our understanding of point representation and elucidates the practical application of geospatial visualization techniques in Python. Points, the fundamental building blocks of vector data, embody simple geometric entities representing singular spatial locations. In geospatial visualization, points are powerful proxies for depicting specific features such as cities, landmarks, or sampling locations on the Earth's surface (Kirilenko, 2022). Defined by their x and y coordinates, each point assumes a unique position within the spatial domain, contributing to the intricate tapestry of geographic information.

In the provided example, Geopandas facilitates the seamless integration of spatial data manipulation and visualization, portraying Bangalore as a prominent point on the global stage. By harnessing Bangalore's latitude and longitude coordinates, the code orchestrates a compelling visual narrative, underscoring the city's geographical significance amidst the broader canvas of the world map. The resulting visualization serves as a testament to the efficacy of Python in unraveling spatial intricacies and illuminating spatial relationships with precision and clarity. Figure 2 shows the output of the Bangalore location highlighted on the world map.

```
Example: Using Geopandas to highlight Bangalore as a point on
the world map.
```

```
#Install GeoPandas
pip install geopandas

# Import the libraries
import geopandas as gpd
import matplotlib.pyplot as plt

# Load the built-in 'naturalearth_lowres' dataset from
GeoPandas.
world = gpd.read_file(gpd.datasets.get_path('naturalearth_
lowres'))

# Create a base map
world.plot(figsize=(15, 10), color='white', edgecolor='black')

# To plot Bangalore, we input the latitude and longitude of
Bangalore.
bangalore = gpd.GeoDataFrame(
{'city': ['Bangalore'], 'geometry': [gpd.points_from_
xy(x=[77.5946], y=[12.9716])[0]]}  # Coordinates for Bangalore)

# Highlight Bangalore on the map
bangalore.plot(ax=plt.gca(), color='red', markersize=100)
```

```
# Display the plot with a title
plt.title('Geospatial Visualization Example: Location of
Bangalore')
plt.show()
```

Figure 2. Bangalore Location Highlighted on World Map

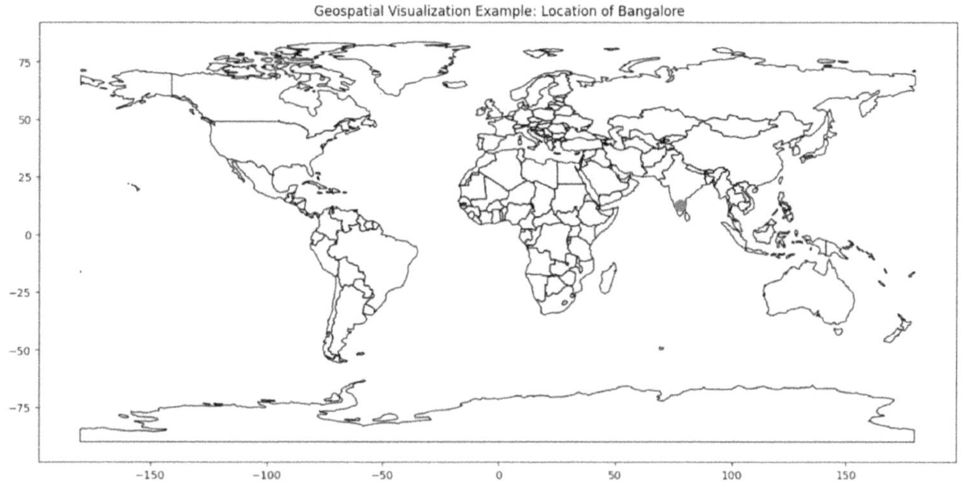

Lines

Within the domain of geospatial visualization, lines are indispensable for representing linear features across the Earth's surface, ranging from rivers and roads to administrative boundaries (Jan & Sachar, 2022). These geometric constructs, comprised of interconnected vertices joined by straight line segments, embody a wealth of spatial information encapsulating the intricate tapestry of our planet's topography (Vuckovic et al., 2021). This chapter embarks on a journey to unravel the intricacies of line representations in geospatial visualization, delving into their pivotal role in delineating spatial relationships and facilitating nuanced spatial analyses.

The provided code snippet showcases the utilization of Geopandas, a Python library renowned for its prowess in spatial data handling, to visualize a line connecting the cities of Bangalore and Sydney on a world map. Beginning with importing essential libraries, including Geopandas and Matplotlib, the code loads a low-resolution world map dataset from GeoPandas. Subsequently, the coordinates of Bangalore and Sydney are defined as Point objects, encapsulating their respective geographic locations. Leveraging these points, a GeoDataFrame is constructed, facilitating the integration of city markers onto the world map visualization. Additionally, a line, delineated by a LineString object connecting Bangalore and Sydney, is generated and overlaid onto the map, symbolizing the spatial relationship between the two cities. The resulting visualization offers a compelling depiction of the geographical connection between Bangalore and Sydney, exemplifying the diverse applications of lines in geospatial visualization.

```
Example: Using Geopandas to highlight a line from Bangalore to
Sydney on the world map.

# Importing necessary libraries
import geopandas as gpd
import matplotlib.pyplot as plt
from shapely.geometry import Point, LineString

# Load the built-in 'naturalearth_lowres' dataset from
GeoPandas, a low-resolution world map.
world = gpd.read_file(gpd.datasets.get_path('naturalearth_
lowres'))

# Coordinates of the cities
blr_coords = (77.5946, 12.9716)  # Bangalore, India
syd_coords = (151.2093, -33.8688)  # Sydney, Australia

# Create Point objects for the cities
blr = Point(blr_coords)
syd = Point(syd_coords)

# Create a GeoDataFrame with the points
cities = gpd.GeoDataFrame(geometry=[blr, syd],
index=['Bangalore', 'Sydney'])

# Create a line between Bangalore and Sydney
line = gpd.GeoDataFrame(geometry=[LineString([blr, syd])])

# Create the base map of the world
base = world.plot(figsize=(15, 10), color='white',
edgecolor='black')
```

```
# Plot the cities on the map
cities.plot(ax=base, marker='o', color='red', markersize=50)

# Plot the line on the map
line.plot(ax=base, color='blue')

# Show the map with a title
plt.title('Geospatial Visualization Example: Line from
Bangalore to Sydney')
plt.show()
```

The example output provides a visualization illustrating the connections between distinct geographical locations. The line representing the route from Bangalore to Sydney symbolizes diverse relationships, including travel routes, communication links, and cultural connections. Figure 3 illustrates a line connecting the cities of Bangalore and Sydney on a world map, showcasing the spatial relationship between these two geographic locations. This visualization depicts the geographical connection between the cities and exemplifies the practical application of lines in geospatial visualization, particularly in representing linear features across the Earth's surface.

Figure 3. Line from Bangalore to Sydney on the World Map

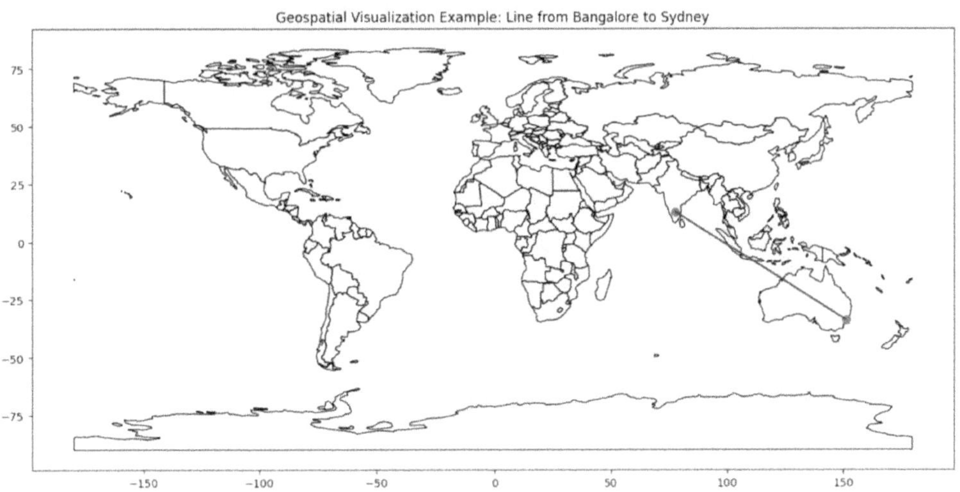

Polygons

Polygons serve as foundational elements in geospatial visualization, representing closed shapes that precisely delineate areas on the Earth's surface (Wei et al., 2024). These geometric entities find diverse applications, from depicting countries and states to demarcating parks and ecological reserves. Defined by a series of coordinates outlining their shape, polygons offer a robust spatial representation and analysis framework (Fan, 2021). This chapter delves into polygons in geospatial visualization, exploring their significance and practical applications. Through a detailed examination of examples and code demonstrations, readers will gain insight into the versatile nature of polygons and their role in spatial data visualization.

```
Example: Using Geopandas to highlight a Polygon (Bermuda
Triangle) on the world map.

# Importing necessary libraries
import geopandas as gpd
import matplotlib.pyplot as plt
from shapely.geometry import Polygon

# Load the built-in 'naturalearth_lowres' dataset from
GeoPandas, a low-resolution world map.
world = gpd.read_file(gpd.datasets.get_path('naturalearth_
lowres'))

# Coordinates of the Bermuda Triangle's vertices
miami_coords = (-80.1918, 25.7617)  # Miami, USA
bermuda_coords = (-64.7515, 32.3078)  # Bermuda
puerto_rico_coords = (-66.5901, 18.4655)  # Puerto Rico

# Create a Polygon geometry for the Bermuda Triangle
bermuda_triangle_geom = Polygon([miami_coords, bermuda_coords,
puerto_rico_coords])

# Create a GeoDataFrame with the polygon
bermuda_triangle = gpd.GeoDataFrame(index=[0],
geometry=[bermuda_triangle_geom])

# Create the base map of the world
base = world.plot(figsize=(15, 10), color='white',
```

```
edgecolor='black')

# Plot the Bermuda Triangle on the map
bermuda_triangle.plot(ax=base, color='green')

# Show the map with a title
plt.title('Geospatial Visualization Example: The Bermuda
Triangle')
plt.show()
```

The output of this visualization highlights a triangular area in the North Atlantic Ocean known as the Bermuda Triangle.

Figure 4. Visualization of the Bermuda Triangle in the North Atlantic Ocean

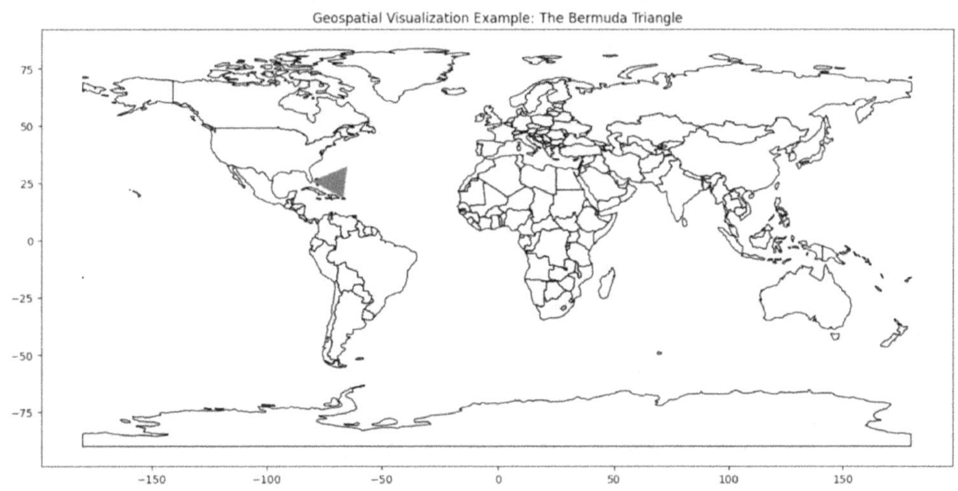

Figure 4 presents a visualization depicting the Bermuda Triangle in the North Atlantic Ocean. Using geospatial data and visualization techniques, this illustration highlights the triangular area encompassing Miami, Bermuda, and Puerto Rico. By delineating this enigmatic geographical phenomenon, the visualization contributes to a deeper understanding of spatial phenomena and their representation in geospatial visualization. The provided code snippet illustrates the utilization of Geopandas, a powerful Python library for geospatial data handling, to highlight the Bermuda Triangle polygon on a world map. Beginning with importing essential libraries such as Geopandas and Matplotlib, the code loads a low-resolution world map dataset from GeoPandas.

Subsequently, the coordinates of the Bermuda Triangle's vertices, encompassing Miami, Bermuda, and Puerto Rico, are defined. Leveraging these coordinates, a Polygon geometry representing the Bermuda Triangle is created, encapsulating the triangular area in the North Atlantic Ocean. The polygon is integrated into a GeoDataFrame, facilitating its visualization on the world map (Garcia et al., 2023). The resulting visualization offers a compelling depiction of the Bermuda Triangle, shedding light on this enigmatic geographical phenomenon.

RASTER DATA FORMAT

Raster data format is a cornerstone in geospatial visualization, portraying spatial information through a grid of cells or pixels. Each cell within the grid encapsulates a specific attribute, symbolizing diverse geographical features ranging from valleys and mountain ranges to individual structures like houses. Invaluable for representing continuous data with spatial variation, raster data finds its domain in Python-based geospatial analysis, with Rasterio as the predominant library. Within this framework, raster data inherits resolution, dictating the size of each cell within the grid. Higher resolution yields smaller cells, affording intricate detail but demanding higher memory and processing resources (Feng & Koch, 2024) (Tran et al., 2020).

Conversely, lower resolutions sacrifice detail for reduced file size and processing requirements. This chapter navigates the nuances of raster data format, shedding light on its pivotal role in spatial analysis and visualization. Concepts such as resolution and the composition of raster data bands are explored, unraveling their significance in real-world applications. Practical insights and code demonstrations equip readers to harness the power of raster data for diverse geospatial visualization tasks (Zhang et al., 2024)

Resolution

Raster data is a fundamental component in geospatial visualization, providing a structured grid format for representing spatial information. This example delves into the concept of resolution within raster data using Python's Rasterio library. Through the generation of synthetic raster data and subsequent resolution manipulation, the code illustrates how alterations in resolution affect the level of detail in spatial data representation. This exploration delves into the intricacies of resolution adjustment and its implications for geospatial analysis by presenting code demonstrations and visualizations (Garg, 2023).

The provided code initiates by crafting a synthetic raster image sized at 300x300 pixels, incorporating random data with Rasterio. Subsequently, the code showcases the process of resolution modification through data resampling utilizing Rasterio's Resampling module. In this instance, the resolution is decreased by a factor of 10, leading to a reduction in pixel count and, consequently, a decrease in image detail. Following the data resampling, the original and resampled images are juxtaposed using Matplotlib. While the original image depicts high resolution, the resampled counterpart reflects a lower resolution achieved via downscaling. This visual comparison vividly demonstrates how lower resolution yields a more pixelated image with diminished detail than the original high-resolution image. This example gives a more profound comprehension of resolution within raster data. Higher resolution facilitates more precise analysis by capturing finer details, whereas lower resolution proves advantageous in reducing data size and processing demands, particularly in scenarios where intricate detail is nonessential (Román et al., 2024).

```
Example: Using Rasterio to depict the change in resolution.

import numpy as np
import rasterio
from rasterio.enums import Resampling
import matplotlib.pyplot as plt

# Create a synthetic raster 300x300 raster with random data.
with rasterio.open(
'synthetic.tif',
'w',
driver='GTiff',
height=300,
width=300,
count=1,
dtype=np.float32
) as synthetic_raster:
data = np.random.rand(300, 300).astype(np.float32)
synthetic_raster.write(data, 1)

#Changing the resolution by reading the data back and
resampling it.
with rasterio.open('synthetic.tif') as dataset:

# decrease the resolution by a factor of 10.
```

```
data = dataset.read(
out_shape=(dataset.height // 10, dataset.width // 10),
resampling=Resampling.bilinear  # Bilinear is good for
downscaling as it averages nearby pixels.
)

#Plot the original data and the resampled data side-by-side.
fig, ax = plt.subplots(1, 2, figsize=(10, 5))

# Original image
ax[0].imshow(dataset.read(1), cmap='viridis')
ax[0].set_title('Original Image (High Resolution)')

# Resampled (lower resolution) image
ax[1].imshow(data[0], cmap='viridis')
ax[1].set_title('Resampled Image (Lower Resolution)')

plt.tight_layout()
plt.show()
```

Figure 5. Change in Resolution using Rasterio

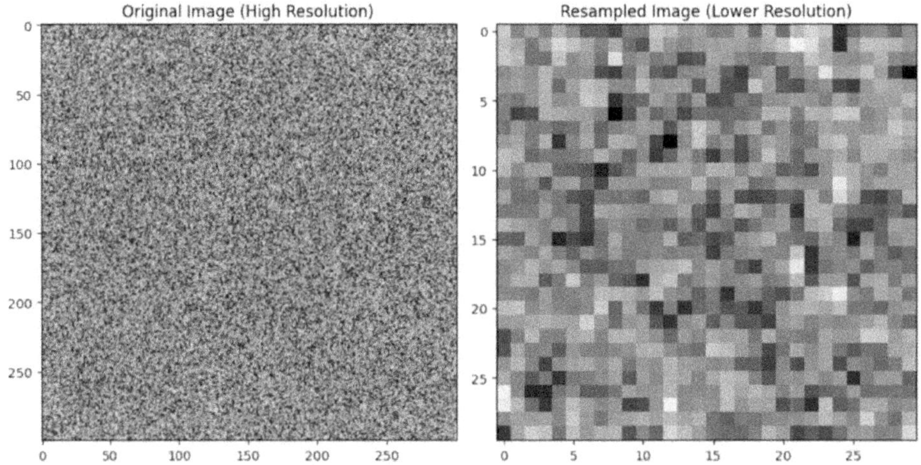

Output of Example: The resampled image appears significantly more pixelated than the original, clearly showing how a lower resolution results in a loss of detail. This visual difference illustrates the concept of resolution in raster data: higher resolution captures more detail, allowing for more precise analysis. In comparison, lower resolution can help reduce data size, which is best applicable for use cases that do not require fine-grained detail. Figure 5 showcases the change in resolution using Rasterio, a Python library for handling raster data. This figure illustrates the impact of resolution modification on spatial data representation through a series of visualizations. By comparing the original and resampled images, the figure elucidates how alterations in resolution influence the level of detail captured in raster data, providing valuable insights into the nuances of spatial analysis and visualization.

Bands

In geospatial analysis, raster data represents various attributes of the Earth's surface. One critical aspect of raster data is its ability to consist of multiple bands, each representing a distinct attribute or variable. For instance, in satellite imagery, different bands may capture data related to different wavelengths of light or specific environmental parameters such as temperature or vegetation. This capability enables the creation of composite images by combining multiple bands, providing valuable insights into different aspects of the Earth's surface. In this section, we explore how Rasterio, a popular Python library for working with raster data, facilitates the visualization of individual bands within raster datasets (Howari & Ghrefat, 2021).

The example demonstrates using Rasterio to visualize individual bands within a Raster dataset. Firstly, the file path to the GeoTIFF file containing the raster data is specified. Upon opening the file with Rasterio, a specific band is selected for visualization, represented by the variable band_number. In this example, the focus is on visualizing the temperature changes captured by the first band. The data from this band is read, and a visualization is created using Matplotlib and Rasterio's show function. The show function displays the raster data, applying a specified colormap (in this case, 'viridis') to represent temperature variations. A color bar is also included to provide a reference for interpreting temperature values. Finally, labels are added to the plot, such as the title indicating the band number and axes labels specifying the longitude and latitude. The resulting visualization provides insights into the temperature distribution across the geographic area covered by the raster dataset, showcasing the utility of visualizing individual bands for geospatial analysis and interpretation.

```
Example: Using Rasterio to depict the Bands.
import rasterio
from rasterio.plot import show
import matplotlib.pyplot as plt
```

```
# Path to your GeoTIFF file
file_path = 'HYP_HR.tif'  # replace with actual file path

# Open the file
with rasterio.open(file_path) as src:

# Let's say we want to visualize the temperature changes from
the first band
band_number = 1

# Read the specific band
temperature_band = src.read(band_number)

# Now, let's visualize this band with a color map representing
temperature
plt.figure(figsize=(10, 6)) show(temperature_band,
transform=src.transform, cmap='viridis')  # using the 'viridis'
colormap
plt.colorbar(label='Temperature (°C)', shrink=0.75)
plt.title(f'Temperature Distribution for Band {band_number}')
plt.xlabel('Longitude')
plt.ylabel('Latitude')
    plt.show()
```

Figure 6. Depicting Bands Using Rasterio

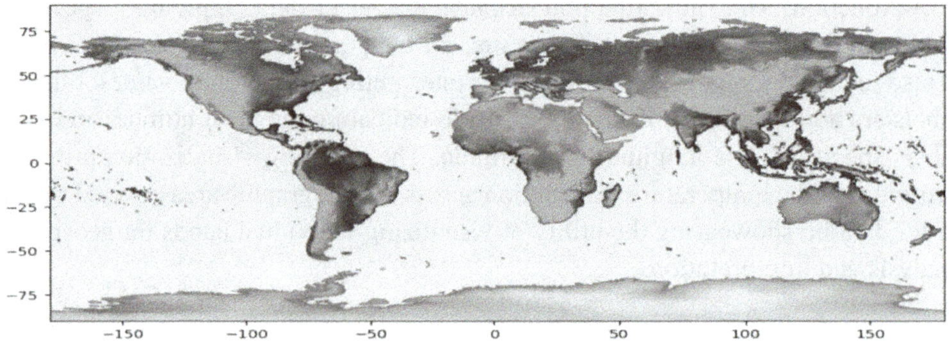

Output of Example: Each band would represent temperature data captured at different times, which would help visualize the changes. Figure 6 depicts the visualization of bands using Rasterio, a Python library designed for working with raster data. This illustration showcases the individual bands within a raster dataset, offering insights into various attributes or variables captured across the Earth's surface. By visualizing these bands, the figure provides a comprehensive understanding of the data content. It facilitates geospatial analysis and interpretation, highlighting the versatility and utility of raster data in spatial research.

VISUALIZATION TECHNIQUES

Python offers several visualization libraries, including Matplotlib and Seaborn, which are suitable for generating visually appealing maps and plots. These libraries enable the creation of choropleth maps, which color administrative boundaries based on population density values. For more complex geospatial data visualization tasks, the Cartopy library stands as a valuable resource. Cartopy specializes in mapping and visualizing geospatial data, providing a user-friendly interface for projecting, transforming, and plotting such data on various map projections. Additionally, Cartopy facilitates the integration of base maps and GIS data, creating detailed and accurate visualizations (Singh, 2023) (Kathar et al., 2023).

Geospatial visualization is a critical aspect of understanding and analyzing spatial data. Python offers diverse libraries for creating insightful and visually appealing maps and plots in this context. One such library is Cartopy, which specializes in advanced geospatial visualization tasks, providing tools to plot data on various map projections and integrate base maps and GIS data. In this example, we explore how to utilize Cartopy to visualize a theoretical path between two geographic locations, Bangalore and Sydney. The focus is understanding the code's functionality and examining the resulting visualization.

The output of this example is a visualization depicting the theoretical flight path from Bangalore to Sydney on a map. The plotted path, along with label starting and ending points, provides a clear representation of the geographic route between the two cities. This visualization aids in understanding the spatial relationship between the two locations and illustrates Cartopy's capability to create advanced geospatial visualizations.

```
Example: Using cartopy for advanced visualization

import matplotlib.pyplot as plt
import cartopy.crs as ccrs

def plot_theoretical_path(start, end, start_label, end_label):
```

```
"""
Plots a theoretical path on a map.

Parameters:
start (tuple): Coordinates for the starting point (longitude,
latitude).
end (tuple): Coordinates for the ending point (longitude,
latitude).
start_label (str): Label for the starting point.
end_label (str): Label for the ending point.
"""

# Create a map with a basic globe projection
ax = plt.axes(projection=ccrs.PlateCarree())
ax.stock_img()  # add a background image for context

# Plot the path with start and end coordinates
ax.plot([start[0], end[0]], [start[1], end[1]], color=red,
linestyle='--', lw=2, marker='o', transform=ccrs.PlateCarree())

# Mark the starting and ending points with labels
ax.text(start[0], start[1], start_label,
horizontalalignment='right', transform=ccrs.Geodetic())
ax.text(end[0], end[1], end_label, horizontalalignment='right',
transform=ccrs.Geodetic())

# Add coastlines for reference
ax.coastlines()

# Show the map
plt.show()

# Coordinates for Bangalore and Sydney
bangalore_coords = (77.5946, 12.9716)  # Bangalore
sydney_coords = (151.2093, -33.8688)  # Sydney

# Call the function to plot the path
plot_theoretical_path(bangalore_coords, sydney_coords,
'Bangalore', 'Sydney')
```

Output of Example: Tracing the flight path from Bangalore to Sydney (Figure 7). Figure 7 demonstrates the utilization of Cartopy, a Python library specialized in geospatial visualization, to depict the theoretical flight path from Bangalore to Sydney on a map. By leveraging Cartopy's capabilities, this visualization offers a detailed representation of the geographic route between the two cities, facilitating a nuanced understanding of spatial relationships and enhancing the visualization of complex spatial data.

Figure 7. Using Cartopy for advanced visualization of the theoretical flight path from Bangalore to Sydney on a map

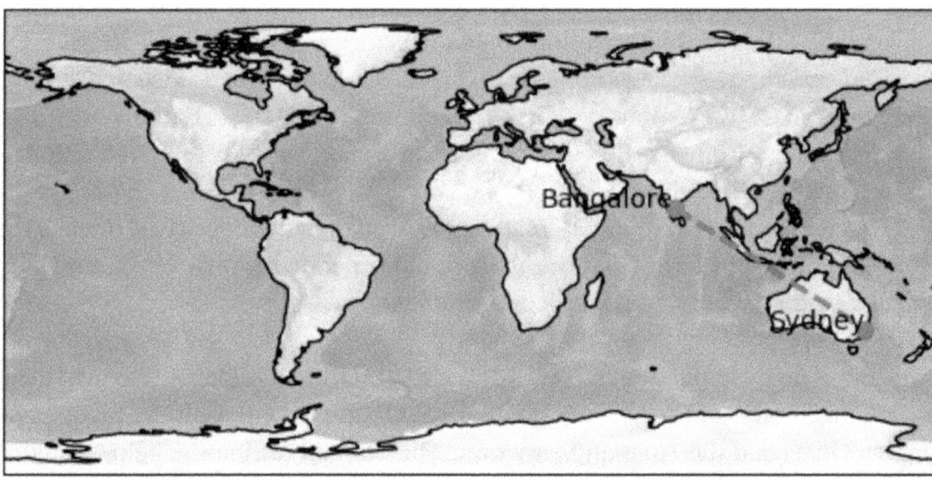

In conjunction with the libraries mentioned above, numerous additional Python packages serve specific purposes in geospatial data visualization. One such example is Folium, a library facilitating the generation of interactive maps through leaflet.js integration. Leaflet.js, a JavaScript library, is instrumental in producing interactive maps. Consequently, Folium amalgamates the versatility of Python with the robustness of JavaScript, yielding an efficient platform for interactive mapping. It furnishes an intuitive interface visualizing geospatial data, featuring embedded interactivity alongside functionalities like zooming, panning, and tooltips.

Folium library

Folium, a Python library, capitalizes on the capabilities of the Leaflet.js library, offering a straightforward method to generate interactive maps directly within the

Python environment. This facilitates the visualization of geospatial data on interactive maps, enhancing the comprehension of patterns, trends, and relationships within the data (Garcia, 2023).

Some of the key features include:

- *Interactive Maps:* Folium allows us to create interactive maps with various tilesets, such as OpenStreetMap, Mapbox, and Stamen Terrain. These maps can be zoomed in and out, panned, and customized with markers, polygons, and overlays.
- *Markers and Popups:* Markers are a common way to represent specific locations on a map. Folium provides an easy way to add markers to a map and customize their appearance. Additionally, we can attach popups to markers to provide additional information when clicked.
- *Choropleth Maps:* Choropleth maps are handy when visualizing data aggregated by regions, such as countries or states. Folium makes it straightforward to create choropleth maps by allowing us to assign colors to regions based on data values.
- *Heatmaps:* Heatmaps are a great way to visualize the density or intensity of data points on a map. Folium enables us to generate heatmaps by aggregating data and representing it with color gradients.

Below is an example of adding a marker (Figure 8) to an interactive map using the Folium library and marking a popular beach destination with the picturesque Western Ghats and the Arabian Sea views. The Stamen terrain tile helps visualize terrain maps, featuring hill shading and natural vegetation colors.

```
Code Snippet depicting the use of markers:
map = Folium.Map(location=['13.862637779043636',
'74.88275318203425'],
tiles='stamenterrain')
map.add_child(folium.ClickForMarker(popup='Must visit')) map
```

Figure 8. Depicting the use of markers

Figure 8 portrays the utilization of markers in geospatial visualization, exemplified through Python's Folium library. This visualization showcases the placement of a marker on an interactive map, representing a specific geographical location. By employing markers, the figure underscores the capability of Folium to enhance spatial understanding and facilitate the representation of critical points of interest on dynamic maps, thereby contributing to practical spatial analysis and visualization.

```
Code: Snippet representing heat map of a sample population
Density using Folium

import folium
from folium.plugins import HeatMap

# Create a Folium map centered around Bangalore
bangalore_map = Folium.Map(location=[12.9716, 77.5946], zoom_
start=10)
# Sample population density data for Bangalore and surrounding
areas population_density_data = [
    [12.9716, 77.5946, 1000],  # Bangalore (center) - higher
density
    [12.9507, 77.5848, 750],   # Surrounding area 1
```

```
    [12.9801, 77.6090, 800],    # Surrounding area 2
    # Add more data points for other areas

# Create a HeatMap layer using the data
HeatMap(population_density_data, radius=15, gradient={0.4:
'blue', 0.65: 'green', 1: 'red'}).add_to(bangalore_map)
# Display the map bangalore_map
```

Figure 9. The output of the Heatmap using Folium

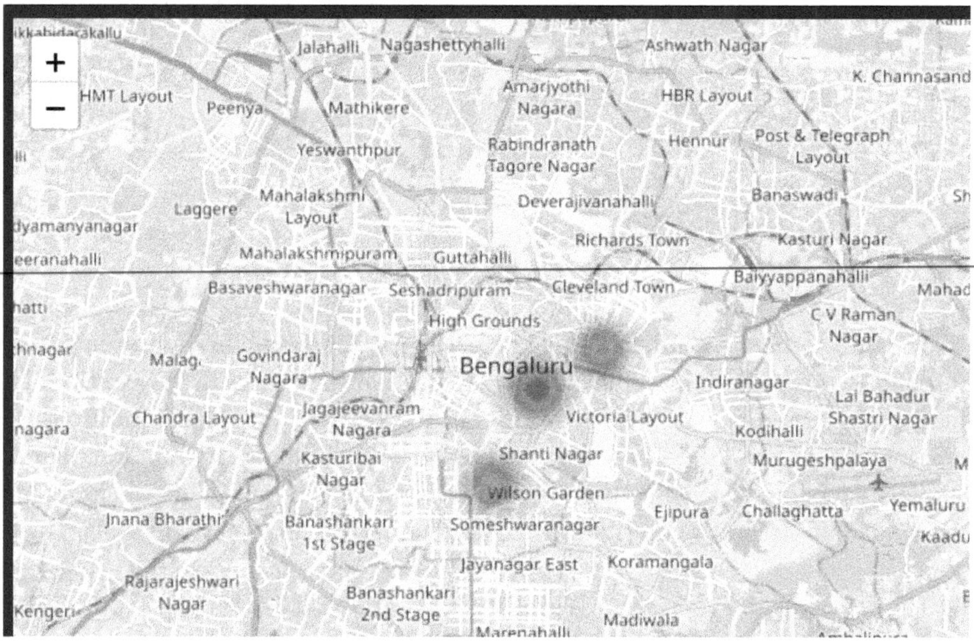

The output of the Heatmap using Folium is given in Figure 9. Figure 9 illustrates the output of a heatmap generated using the Folium library in Python for geospatial visualization. This visualization showcases the spatial distribution of data density or intensity across a map, providing insights into patterns and variations within the dataset. By depicting data aggregation through color gradients, the Heatmap aids in identifying areas of concentration or dispersion, thereby enhancing the understanding of spatial phenomena and supporting informed decision-making processes.

REAL-WORLD APPLICATIONS

Geospatial visualization using Python serves many real-world applications with significant implications across various sectors. In urban planning and development, Python's geospatial libraries like GeoPandas and Folium empower planners to analyze spatial data such as land use and transportation networks, facilitating informed decision-making regarding infrastructure allocation and development impact assessment. Moreover, in fields such as environmental monitoring, transportation and logistics optimization, public health surveillance, and natural resource management, Python's geospatial visualization capabilities enable stakeholders to address critical challenges by visualizing data patterns, tracking changes, and making data-driven decisions for sustainable resource management and efficient operations (Jamil, et al., 2024).

Urban Planning and Development

Geospatial visualization in urban planning and development plays a crucial role in understanding and managing the built environment. Python's geospatial libraries, such as GeoPandas and Folium, enable urban planners to visualize spatial data like land use, transportation networks, and population density. By overlaying various data layers, planners can identify areas with high demand for infrastructure, assess the impact of development projects, and optimize resource allocation.

Environmental Monitoring

Geospatial visualization using Python is widely used in environmental monitoring to analyze and visualize climate change, natural disasters, and habitat conservation data. With libraries like Matplotlib and Cartopy, researchers can create interactive maps, time series plots, and spatial analyses to monitor temperature, precipitation, air quality, and land cover changes. These visualizations help scientists and policymakers gain insights into the impact of environmental factors and make informed decisions for sustainable resource management.

Transportation and Logistics

Python's geospatial visualization capabilities are indispensable in the transportation and logistics industry. By integrating geospatial data with network analysis tools like NetworkX, transportation companies can optimize routes, track vehicles in real time, and identify bottlenecks in the supply chain. Visualization techniques like heatmaps and flow maps allow stakeholders to visualize patterns, identify inefficiencies, and improve operational efficiency.

Public Health

Geospatial visualization using Python has proven valuable in public health research and disease surveillance. By visualizing health-related data, such as disease outbreaks, hospital locations, and population demographics, public health professionals can identify hotspots, track the spread of diseases, and allocate resources effectively. Python libraries like Plotly and Bokeh enable the creation of interactive dashboards that provide real-time updates and facilitate data-driven decision-making in public health emergencies.

Natural Resource Management

Geospatial visualization is instrumental in managing and conserving natural resources. Python's geospatial libraries, combined with remote sensing data, enable the monitoring of forests, water bodies, and wildlife habitats. By visualizing changes in vegetation cover, water availability, and biodiversity, conservationists can identify areas at risk, prioritize conservation efforts, and monitor the effectiveness of interventions. These visualizations aid in preserving ecosystems and the sustainable use of natural resources.

CONCLUSION

In conclusion, this chapter delves into the pivotal role of Python in exploring vector and raster data formats for geospatial visualization. Through the lens of environmental science, urban planning, and natural resource management, it underscores the significance of understanding these formats for proficient spatial data manipulation and visualization. By systematically elucidating the disparities and application scenarios of vector and raster data formats, coupled with demonstrations of various Python libraries and tools, the chapter equips practitioners with the knowledge and skills to leverage spatial data effectively. Moreover, it highlights the real-world applications of geospatial visualization across diverse domains, ranging from urban planning and environmental monitoring to transportation logistics and public health, reaffirming the transformative potential of Python in addressing contemporary spatial challenges.

REFERENCES

Alam, M. M., Torgo, L., & Bifet, A. (2022). A survey on spatio-temporal data analytics systems. *ACM Computing Surveys*, *54*(10s), 1–38.

Anselin, L., Li, X., & Koschinsky, J. (2022). GeoDa, from the desktop to an ecosystem for exploring spatial data. *Geographical Analysis*, *54*(3), 439–466.

Fan, H., Zhao, Z., & Li, W. (2021). Towards measuring shape similarity of polygons based on multiscale features and grid context descriptors. *ISPRS International Journal of Geo-Information*, *10*(5), 279.

Feng, X., & Koch, J. (2024). Combining vector and raster data in regionalization: A unified framework for delineating spatial unit boundaries for socio-environmental systems analyses. *International Journal of Applied Earth Observation and Geoinformation*, *128*, 103745.

Garcia, C., Rabadi, G., Abujaber, D., & Seck, M. (2023). Supporting Humanitarian Crisis Decision Making with Reliable Intelligence Derived from Social Media Using AI. *Journal of Homeland Security and Emergency Management*, *20*(2), 97–131.

Garcia, E., Peyman, M., Serrat, C., & Xhafa, F. (2023). Join Operation for Semantic Data Enrichment of Asynchronous Time Series Data. *Axioms*, *12*(4), 349.

Garg, P. (2023). Understanding Geospatial Data. In Emerging Trends, Techniques, and Applications in Geospatial Data Science (pp. 1-14). IGI Global.

Howari, F. M., & Ghrefat, H. (2021). Geographic information system: spatial data structures, models, and case studies. In *Pollution Assessment for Sustainable Practices in Applied Sciences and Engineering* (pp. 165–198). Butterworth-Heinemann.

Jamil, A., Padubidri, C., Karatsiolis, S., Kalita, I., Guley, A., & Kamilaris, A. (2024). GAEA: A Country-Scale Geospatial Environmental Modelling Tool: Towards a Digital Twin for Real Estate. In *Environmental Informatics* (pp. 177–199). Springer.

Jan, A. F., & Sachar, A. (2022). Use of GIS in Road Infrastructure Development and Management. *International Journal of Innovative Research in Engineering & Management*, *9*(2), 338–342.

Kathar, S. P., Nagne, A. D., Awate, P. L., & Bhosle, S. (2023, December). Comparative Study of Supervised Classification for LULC Using Geospatial Technology. In International Conference on Soft Computing and its Engineering Applications (pp. 79-93). Cham: Springer Nature Switzerland.

Kirilenko, A. P. (2022). Geographic Information System (GIS) Making Sense of Geospatial Data. In *Applied Data Science in Tourism: Interdisciplinary Approaches, Methodologies, and Applications* (pp. 513–526). Springer International Publishing.

McKitrick, M. K., Schuurman, N., & Crooks, V. A. (2023). Collecting, analyzing, and visualizing location-based social media data: Review of methods in GIS-social media analysis. *GeoJournal*, *88*(1), 1035–1057.

Román, A., Heredia, S., Windle, A. E., Tovar-Sánchez, A., & Navarro, G. (2024). Enhancing Georeferencing and Mosaicking Techniques over Water Surfaces with High-Resolution Unmanned Aerial Vehicle (UAV) Imagery. *Remote Sensing*, *16*(2), 290.

Singh, R. (2023). Remote Sensing and GIS Applications in Soil Conservation. In *Soil and Water Conservation Structures Design* (pp. 367–413). Springer Nature Singapore.

Tran, B. H., Aussenac-Gilles, N., Comparot, C., & Trojahn, C. (2020). Semantic integration of raster data for earth observation: An RDF dataset of territorial unit versions with their land cover. *ISPRS International Journal of Geo-Information*, *9*(9), 503.

Vuckovic, M., Schmidt, J., Ortner, T., & Cornel, D. (2021). Combining 2D and 3D visualization with visual analytics in the environmental domain. *Information (Basel)*, *13*(1), 7.

Wei, S., Zhang, T., Yu, D., Ji, S., Zhang, Y., & Gong, J. (2024). From lines to Polygons: Polygonal building contour extraction from High-Resolution remote sensing imagery. *ISPRS Journal of Photogrammetry and Remote Sensing*, *209*, 213–232.

Zhang, H., Tang, G., Xiong, L., Yang, X., & Li, F. (2024). Geomorphology-oriented theoretical framework and construction method for value-added DEM. *Journal of Geographical Sciences*, *34*(1), 165–184.

Chapter 7
Geospatial Data Visualization With Folium

K. G. Suma
VIT-AP University, India

Gurram Sunitha
https://orcid.org/0000-0002-3305-8167
Mohan Babu University, India

J. Avanija
https://orcid.org/0000-0003-0605-7609
Mohan Babu University, India

Mohammad Gouse Galety
https://orcid.org/0000-0003-1666-2001
Samarkand International University of Technology, Uzbekistan

Chinthapatla Pranay Varna
Sree Vidyanikethan Engineering College, India

ABSTRACT

Geospatial data visualization is a powerful tool for understanding and analyzing spatial patterns and relationships. This chapter dives into the world of geospatial data visualization using Folium geospatial Python library. It discloses the significance and versatility of Folium for creating dynamic and accessible maps. This chapter provides hands-on experience of geospatial data visualization using Python geospatial library Folium. This chapter provides practical examples. Readers will gain insights into the power of Folium for creating interactive and visually compelling maps. Thereby, making geospatial data accessible and engaging

DOI: 10.4018/979-8-3693-1754-9.ch007

INTRODUCTION

Geospatial data visualization is a powerful tool for understanding and analyzing spatial patterns and relationships (Hu et al., 2022). This chapter provides hands-on experience of geospatial data visualization using Python geospatial library – *'Folium'* (Ajagbe et al., 2020). This library supports creation of compelling and informative geospatial visualizations (McClain, 2022). Mastering these libraries can enhance the ability of researchers, data scientists etc to communicate insights effectively (Rey et al., 2023). Geospatial data visualization is important in many application domains (Philippou, 2022). The hands-on experience provided in this chapter use *'Anaconda Navigator'* to setup Python working environment and *'Jupyter Notebook'* to execute Python code snippets.

Folium: A Powerful Tool for the Age of Accessible Geospatial Data

In today's data-driven world, geographic information plays a crucial role in various fields. From tracking environmental changes to analyzing disease outbreaks, visualizing geospatial data is more important than ever.

Folium' Relevance in the Current Context

- **Democratization of Geospatial Data**: Open-source data platforms and citizen science initiatives are generating vast amounts of geospatial data. Folium, with its user-friendly approach, empowers researchers, analysts, and even students to quickly visualize and explore this data, fostering new discoveries and insights.
- **Communication and Public Engagement**: Folium's interactive maps are a powerful communication tool. They can effectively convey complex geographical patterns and relationships to a broad audience. This is invaluable for raising awareness about environmental issues, public health concerns, and socio-economic trends.
- **Rapid Prototyping and Collaboration**: In fast-paced environments, Folium excels at rapid prototyping for geospatial visualizations. Its seamless integration with other Python libraries makes collaborative data exploration efficient, allowing teams to iterate on map designs quickly.

Folium's Contribution in Specific Areas

- **Disaster Management:** Folium can be used to create real-time evacuation maps during natural disasters, visualize flood zones, and track the spread of wildfires, aiding emergency response efforts.
- **Urban Planning:** Interactive maps created with Folium can help visualize traffic patterns, population density, and land-use changes, informing better urban planning decisions.
- **Public Health Surveillance:** By mapping disease outbreaks or tracking vaccination coverage, Folium can empower public health officials to identify trends and allocate resources effectively.

Folium, by makes interactive map creation accessible. It plays a significant role in the current era of readily available geospatial data. Its ease of use and focus on core visualization functionalities make it a valuable tool for researchers, analysts, and anyone who wants to effectively communicate the power of location-based information.

Setting Up the Environment

Readers are recommended to download and install *'Anaconda Navigator'* (Anaconda, 2023). Once the it is installed, *'Anaconda'* command prompt and *'Jupyter Notebook'* can be launched (Camara et al., 2021). Folium library has dependencies on various other Python libraries. Hence, it is recommended to create a new *'Anaconda'* environment to avoid any dependency conflicts between Python libraries. The following are the commands to create and activate new *'Anaconda'* environment. These commands are to be used in *'Anaconda'* command prompt.

```
conda create -n new_env_nameconda activate new_env_name
```

The next step is to install Folium in the new *Anaconda* environment (Getting Started with Folium, n.d.). The following command can be used to install *Folium* in *Jupyter Notebook* environment.

```
pip install folium
```

The library can also be installed through *Anaconda* command prompt using *'conda'* package manager or *'conda-forge'* package channel. The following are the commands to install *'Folium'*. These commands are to be used in *'Anaconda'* command prompt.

```
conda install foliumorconda install folium -c conda-forge
```

It is recommended to have *'Jupyter Notebook'* installed for an interactive and visual coding experience. All the code snippets discussed in this chapter are executed in *'Jupyter Notebook'* environment. Readers are expected to have a preliminary experience in working with *'Jupyter Notebook'* environment. All the code snippets discussed in this chapter are written in Python programming language. Hence, readers are also expected to have preliminary experience in basic Python programming, Pandas libraries, plotting libraries such as Plotly Express (Sunitha et al., 2023).

Creating a Basic Folium Map

This section discusses creation of a simple Folium map. Table 1 provides Python code snippet for creating a basic Folium map with specified geographical locations, zoom level and type of map tiles.

Table 1. Python code snippet for creating a basic Folium map

```
1  # importing Folium geospatial library
2  import folium as fl

3  # Creating a Folium figure with specific width and height
4  folium_fig = fl.Figure(width=750, height=500)

5  # Creating a Folium map centered at specific coordinates with an # initial zoom level, chosen map tile; Adding Folium map to Folium #figure; Creating a Folium Map Object
6  fl.Map(location = [13.628730, 79.419067],
7  tiles = 'OpenStreetMap',
8  zoom_start = 6).add_to(folium_fig)

9  # Visualizing Folium Figure
10 folium_fig
```

The code snippet in Table 1 uses *'Map'* class to create a Folium map (Folium API for Map, n.d.). Folium library is imported (Line 1). A Folium *'Figure'* is created with specific size (Line 4). A Folium Map is created with a specific location, tiles typle, and initial zoom level (Line 8). Further this map is added to Folium figure using *'add_to'* method (Line 8). Finally, the Folium figure is visualized (Line 10). Figure 1 shows the output for the code of Table 1.

Figure 1. Visualization of a basic Folium map

Figure Class

'Figure' is a class in Folium. It provides a way to organize and customize the layout of multiple maps or visualizations in a single container. It serves as a container for holding and organizing multiple maps within a single figure. It allows users to create a composite layout containing various subplots. Each subplot represents different maps and/or other visualizations.

Map Class

'Map' is a wrapper class in Folium library. It is the main container for creating interactive maps. It is used to initialize and configure a Folium map object. It provides parameters, methods, and attributes for customization, allowing users to configure a map's appearance, location, and zoom level.

Parameters

- *'location'* parameter takes a Python list or tuple consisting of latitude and longitude values. It represents the initial center of the map.
- *'zoom_start'* parameter is used to set the initial zoom level of Folium map. It determines how close or far the map is initially displayed. The value of *'zoom_start'* is an integer. The typical range for *'zoom_start'* is 0 to 18. Minimum value 0 provides a global view of the entire world. Maximum value 18 shows a highly detailed and close-up view of a specific location.

Appropriate range for a specific map may depend on the type of data being visualized and the level of detail to be displayed. *'zoom_start'* value needs to be adjusted accordingly to achieve desired initial zoom level in the Folium map.

- *'tiles'* parameter is used to specify the basemap or tile layer that is displayed on the map. Tiles are the individual images that make up the map background. Folium provides various visual styles in the form of tiles. Different tile layers can be selected for different needs. It can be configured with various options – 'OpenStreetMap, 'cartodb positron', 'Cartodb dark_matter' etc.

Attributes

- *'location'* returns the current center of the map.
- *'zoom'* returns the current zoom level of the map.
- *'tiles'* returns the current tiles or basemap type.

Methods

- *'add_child(child)'* adds a child element like markers, map, layers etc to the figure.
- *'add_to(parent)'* adds the map to a parent element, such as another Folium map or a Jupyter Notebook.

Creating Multiple Maps in a Folium Figure

This section discusses creation of multiple Folium maps as subplots in a single Folium figure (Folium API for Subplots in Figure, n.d.). Table 2 provides Python code snippet. The code snippet in Table 2 uses *'add_subplot'* method to add subplots (1x2) in a Folium figure (Lines 6 & 7); *'add_child'* method is used to add multiple maps as children (subplots) to a single Folium figure (Lines 9 & 10). Figure 2 shows the output for the code of Table 2.

Table 2. Python code snippet for creating multiple maps (subplots) in a Folium figure

```
# importing Folium geospatial library
import folium as fl

# Creating a Folium figure with specific width and height
folium_fig = fl.Figure(width=500, height=250)

# Adding subplots to Folium figure
map1 = folium_fig.add_subplot(1, 2, 1)
map2 = folium_fig.add_subplot(1, 2, 2)

# Creating maps and adding them as children(subplots) to figure
map1.add_child(fl.Map(location=[13.628730, 79.419067],
    tiles='OpenStreetMap', zoom_start=10))
map2.add_child(fl.Map(location=[13.628730, 79.419067],
    tiles='OpenTopoMap', zoom_start=10))

# Visualizing Folium Figure
folium_fig
```

Customizing Map Tiles

Map tile customization is essential in web mapping applications in order to tailor the appearance and functionality of map tiles based on specific needs. A few specific reasons are presented further. Map's visual style may have to be customized to align with the overall design of the application. This includes adjusting colors, fonts, and other visual elements to create a cohesive user experience. Map's appearance and functionality needs to be adapted for optimal display on different devices; especially on mobile devices with smaller screens. Customizing the map layout and controls ensures a seamless experience across various platforms.

Figure 2. Visualization of multiple maps (subplots) in a Folium figure

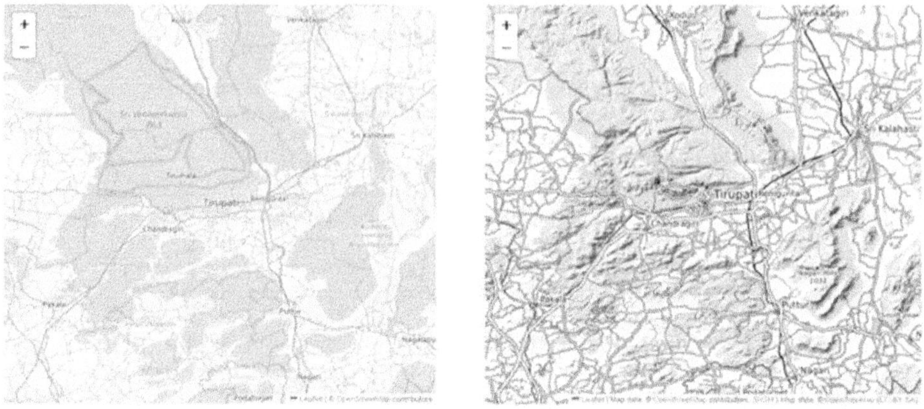

Folium is a simple tool connecting *Python* and *Leaflet* for maps. It is a streamlined interface to JavaScript Leaflet library. It has native backing for six tile types. It facilitates the inclusion of custom XYZ URLs and their attributions in map creation. This design underscores its dependence on external tile provider sources for map rendering.

Tile providers are services that offer map tiles, which are small square images representing portions of a map at different zoom levels. These tiles can be seamlessly stitched together to create a complete map. Some of the popular tile providers are OpenStreetMap, Mapbox, Stamen Design, Esri ArcGIS, Thunderforest etc (Tile Providers: OpenStreetMap, n.d.; Tile Providers: Stamen Maps, n.d.; Tile Providers: Thunderforest, n.d.). These providers offer different map styles, themes, and additional features. Each of these providers have specific URL templates. To use their tiles, specific URLs need to be built. Also, providers like Thunderforest give access to their tiles through an API key. API keys may be obtained by registering on their website.

XYZ tiles are a popular way to represent and serve map data on the web (Tile Providers: xyzservices, n.d.). The term *"XYZ"* refers to three components used to address a specific tile in a map:

X: The horizontal coordinate (column) of the tile in the map grid.

Y: The vertical coordinate (row) of the tile in the map grid.

Z: The zoom level indicating the level of detail. A higher Z value corresponds to a closer view of the map.

These tiles are organized in a pyramid structure, with each level of the pyramid representing a different zoom level. At each zoom level, the map is divided into a grid of tiles, and each tile is uniquely identified by its X, Y, and Z coordinates.

XYZ tiles are commonly used in web mapping applications. They allow efficient tile-based rendering and display. They are compatible with various mapping libraries and services. Thereby, making it easy to integrate maps into websites and applications. When working with XYZ tiles in Python, libraries like Folium, Leaflet, or Mapbox may be used to load and display these tiles. Additionally, various online map tile providers, such as OpenStreetMap, Mapbox, and others, offer XYZ tiles as a standard way to access map data.

Table 3 shows Python code snippet for customizing tiles in a Folium Map using *'xyzservices'*. The *'xyzservices.providers'* package is imported (Line 4). A Folium Map is created by setting *'tiles'* parameter to map tile type *'xyz.OpenTopoMap'* (Line 8). The *'build_url()'* method builds specific XYZ URLs for the map tile type. On using map tiles from external providers, attribution needs to be mentioned by setting *'attr'* parameter. Figure 3 shows the output for the code of Table 3.

Table 3. Python code snippet for customizing tiles in a Folium map

```
1  # importing Folium geospatial library
2  import folium as fl

3  # importing xyzservices
4  import xyzservices.providers as xyz

5  # Creating a Folium figure with specific width and height
6  f = fl.Figure(width=800, height=600)

7  # Creating a map with customized map tiles from xyzservices
8  fl.Map(location=[13.61, 78.3],
   tiles=xyz.OpenTopoMap.build_url(),
   attr='OpenTopoMap',
   zoom_start=6).add_to(f)

9  # Visualizing Folium Figure
10 folium_fig
```

Figure 3. Visualization of customized map tiles in a Folium map

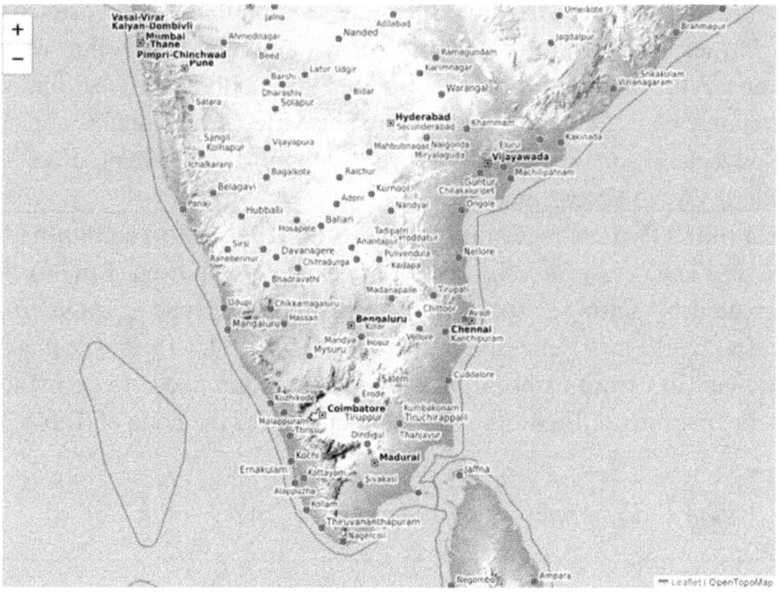

Adding Markers

Markers can be added to a Folium map to pinpoint significant locations. Table 4 shows Python code snippet for adding markers to a Folium Map using *'Marker'* and *'CircleMarker'* methods (Folium API for Marker, n.d.). It shows how to how to create markers. Also it demonstrates how to customize markers. Marker icons may be customized by using *'Icon'* method (Line 12). Marker sizes and colors may be adjusted by setting parameters *'radius', 'color', 'fill', 'fill_color' etc* (Line 14). Figure 4 shows the output for the code of Table 4.

Table 4. Python code snippet for adding customized markers to a Folium map

```
# importing Folium geospatial library
import folium as fl

# importing xyzservices
import xyzservices.providers as xyz

# Creating a Folium figure with specific width and height
folium_fig = fl.Figure(width=600, height=500)

# Creating a map with customized tile map from xyzservices
map_object = fl.Map(location=[13.61, 78.3],
tiles=xyz.OpenTopoMap.build_url(),
attr='OpenTopoMap',
zoom_start=6).add_to(folium_fig)

# Adding marker at specific location
fl.Marker(location=[16.505329, 80.661209]).add_to(map_object)

# Customizing Marker Icons
fl.Marker(location=[13.628730, 79.419067], icon=fl.Icon(color='pink',
icon='info-sign')).add_to(map_object)

# Adjusting Marker Sizes and Colors
fl.CircleMarker(location=[12.9716, 77.5946],radius=20, color='blue',
fill=True, fill_color='blue').add_to(map_object)

# Visualizing Folium Map Object
map_object
```

Adding Interactive Features

A Folium Map can be made interactive by adding features like popup, tooltip, click events etc. Adding interactive features to a Folium map can be incredibly useful. Interactive features like zoom, pan, and click functionality allow users to explore a map at their own pace. Interactive elements empower users to filter data based on their interests. This allows them to focus on specific aspects of the map and personalize their exploration experience. Interactive features can make complex geospatial information easier to grasp by allowing users to visualize data dynamically. This is particularly beneficial for conveying trends, changes over time, or relationships between different spatial variables. Animation and time-series data visualization can be incorporated into Folium maps to create engaging narratives and showcase the evolution of spatial patterns over time. Interactive maps can be used to compare scenarios, identify potential problem areas, and support data-driven decision making in various fields like urban planning, resource management, and disaster response.

Table 5 demonstrates Python code snippet for creating a Folium Map with interactive features. A marker can be made interactive by setting its *'popup'* or *'tooltip'* parameters (Lines 10 and 12). These features contribute to providing additional context and information to users interacting with the map. Popup feature is applied to markers to display additional information when the user clicks on them. Tooltip feature is used to show additional information when hovering over a marker. Click events enhance the interaction between the map and the user. The *'ClickForMarker'* method adds a click event to the map (Line 14). It creates new markers at the locations wherever user clicks on the map (Folium API for Click for Marker, n.d.). Also, popup content can be associated with the markers. Here, the popup content is set to display the latitude and longitude of the clicked location. Figure 5 illustrates the output for the code in Table 5.

Figure 4. Visualization of Folium map with customized markers

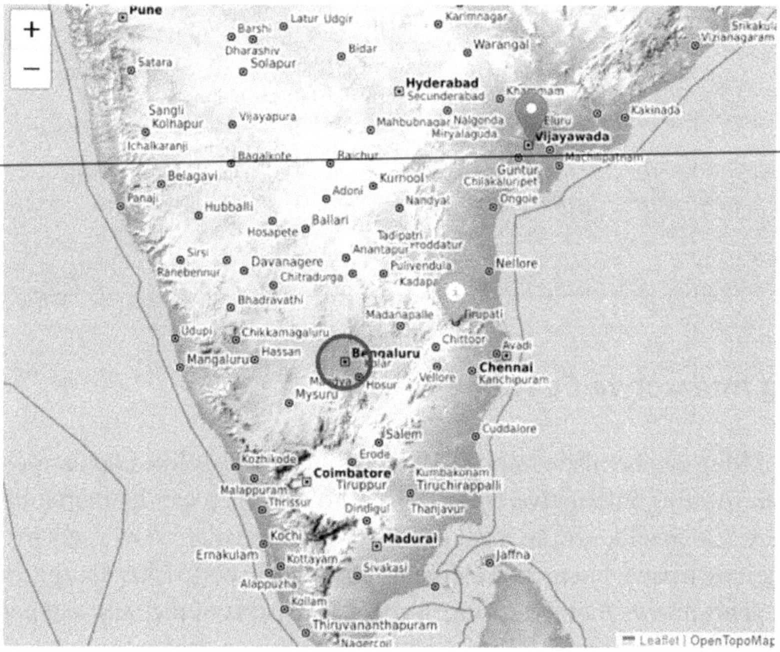

Table 5. Python code snippet for creating a Folium map with interactive features

```
# importing Folium geospatial library
import folium as fl
# importing xyzservices
import xyzservices.providers as xyz

# Creating a Folium figure with specific width and height
folium_fig = fl.Figure(width=600, height=500)

# Creating a map with customized tile map from xyzservices
map_object = fl.Map(location=[13.61, 78.3],
tiles=xyz.OpenTopoMap.build_url(), attr=’OpenTopoMap’,
zoom_start=6).add_to(folium_fig)
# Adding marker with popup text at specific location
fl.Marker(location=[13.628730, 79.419067], popup=’Holy City’, icon=
fl.Icon(color=’pink’, icon=’infosign’)).add_to(map_object)
# Adding marker with popup text at specific location
fl.Marker(location=[16.505329, 80.661209], tooltip=’HiTech City’,
icon=fl.Icon(color=’pink’, icon=’info-sign’)).add_to(map_object)
# Adding click event to create markers on user clicks
fl.ClickForMarker(
‘<b>Lat:</b>${lat}<br/><b>Lon:</b>${lng}’).add_to(map_object)
# Visualizing Folium Map Object
map_object
```

Figure 5. Visualization of Folium map with interactive features

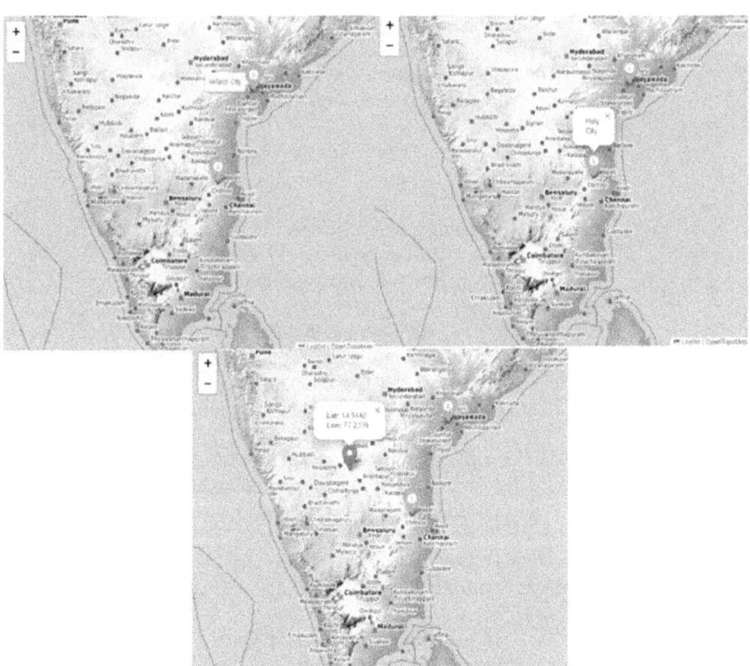

Map Layer Control

Creating multiple layers in a Folium map allows to overlay different types of information on the same map (Map Layer Control, n.d.). Thereby, providing a more comprehensive and interactive visualization. Each layer can represent distinct data or styles. Users can selectively show or hide these layers using the *'LayerControl'* widget (Folium API for Map Layer Control, n.d.). It allows to switch between various tile layers or feature groups. *'LayerControl'* is commonly used with tile layers. But it also supports feature groups and other types of layers in Folium. *'LayerControl'* is dynamic. It updates map's display dynamically as the user interacts with the control.

The default layer in the Python layer control is *'OpenStreetMap'*. It is chosen because of its global availability as a free and open-source map dataset. Its regular updates ensure the use of latest information.

Table 6 shows Python code snippet for creating a Folium Map with multiple layers. Multiple layers are created using *'TileLayer'* method (Folium API for Tile Layer, n.d.). Two tile layers (layer 1 and layer 2) are created with different map styles – layer 1 with Topographic Map using the *'OpenTopoMap'* tile provider; layer 2 with Ocean Base Map using the *'Esri. OceanBasemap'* tile provider (Lines 10 and 11).

Table 6. Python code snippet for creating a Folium map with multiple layers

```
1   # importing Folium geospatial library
2   import folium as fl

3   # importing xyzservices
4   import xyzservices.providers as xyz

5   # Creating a Folium figure with specific width and height
6   folium_fig = fl.Figure(width=600, height=400)

7   # Creating a map with specific location
8   map_object = fl.Map(location=[12.9716, 77.5946],
    zoom_start=6).add_to(folium_fig)

9   # Creating two tile layers with different map styles
10  layer1 = fl.TileLayer(name='Topographic Map',
    tiles=xyz.OpenTopoMap.build_url(),
    attr='OpenTopoMap').add_to(map_object)

11  layer2 = fl.TileLayer(name='Ocean Base Map',
    tiles=xyz.Esri.OceanBasemap.build_url(),
    attr='Ocean Base Map', show=False).add_to(map_object)

12  # Creating a LayerControl widget to switch between layers
13  fl.LayerControl(collapsed=False).add_to(map_object)

14  # Visualizing Folium Map Object
15  map_object
```

Figure 6. Visualization of Folium map with multiple layers

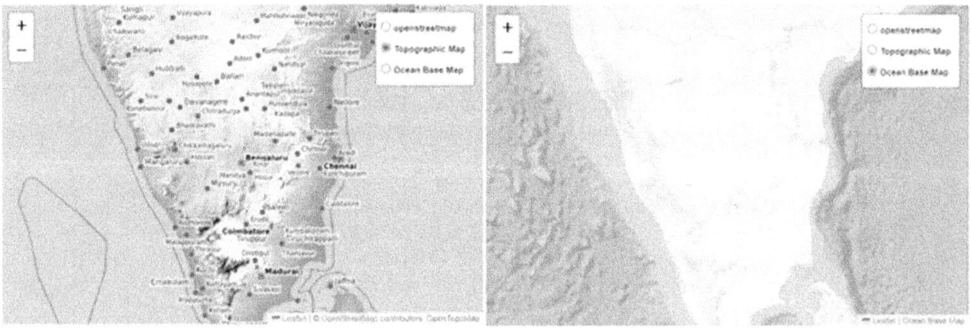

The *'show=False'* parameter set for layer 2 initially hides this layer. When the map with multiple layers is visualized, layer 1 will become the default layer. A LayerControl widget is added to the map using *'LayerControl()'* method. This widget allows to switch between layers of the map (Line 13). The *'collapsed=False'* parameter ensures that the widget control is expanded by default. Figure 6 illustrates the output for the code in Table 6.

Heatmaps

'HeatMap' is a powerful plugin in Folium (Heat Map in Folium, n.d.; Folium API for HeatMap, n.d.). It allows visualization of the intensity of data points on a map. It is particularly useful when there are large set of data points and the concentration or distribution of values across a geographical area. Table 7 shows Python code snippet for creating a Folium Map with heatmaps. A sample data is created (Line 7). Intensity of data of each geographical location is recorded along with its latitude and longitude values. The intensity values are represented using a gradient color scheme. Higher intensities are displayed with warmer colors. A heatmap layer is added to Folium map using *'HeatMap'* method. HeatMap is interactive, and users can zoom in and out to explore the distribution of data points. Appearance of heatmap can be customized by setting the radius of influence, blur, gradient color scheme etc. Figure 7 illustrates the output for the code in Table 7.

Table 7. Python code snippet for creating a Folium map with heatmap

```
# importing Folium geospatial library
import folium as fl

# importing HeatMap from Folium plugins
from folium.plugins import HeatMap

#Sample data-list of tuples of format (latitude, longitude, intensity)
# intensity values should be on scale [0, 1]
heat_data = [(12.9716, 74.5946, 1),
(13.9716, 72.5946, 0.4),
(11.9716, 75.5946, 0.9)]

# Creating a Folium figure with specific width and height
folium_fig = fl.Figure(width=600, height=400)

# Creating a Folium map
map_object = fl.Map(location=[12.9716, 74.5946],
zoom_start=6).add_to(folium_fig)

# Adding a HeatMap layer
# Customizing radius, blur, gradient of heatmaps
HeatMap(heat_data, radius=15, blur=10,
gradient={0.4: 'blue', 1: 'red'}).add_to(map_object)

# Visualizing Folium Map Object
map_object
```

Figure 7. Visualization of Folium map with heatmap

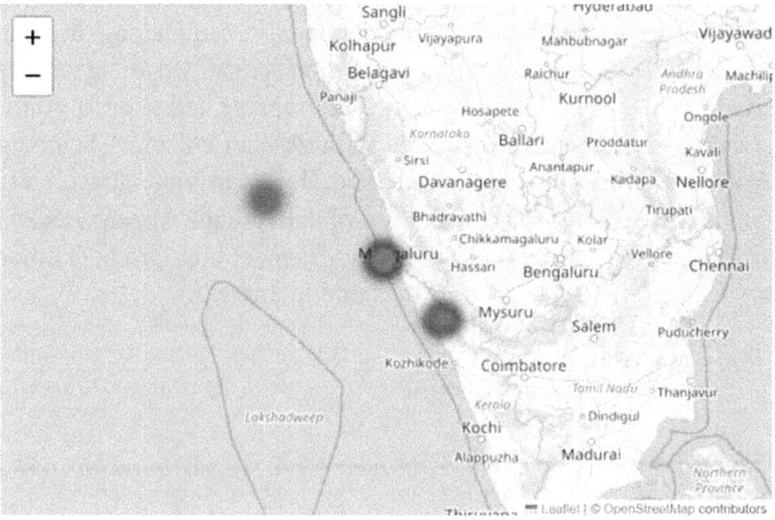

Choropleth Maps

Table 8 shows Python code snippet for creating a Folium Map with choropleth map. Sample population density data of a few Indian states is created (Lines 7 and 9). A choropleth layer is added to the Folium Map (Line 17). The geographical data used in this example code is of Indian states (Indian State GeoJSON Data, 2020). A choropleth layer is created using *'Choropleth'* method. Its parameters are as follows.

- *geo_data* – GeoJSON file path or URL that contains the geographical boundaries for the choropleth map.
- *data* – DataFrame or GeoDataFrame containing the data to be mapped. In this case, it represents the population density data for different states.
- *columns* – list of column names in the data DataFrame that correspond to the state (or geographical entity) and the value to be visualized (population density, in this case).
- *key_on* – path to the key in the GeoJSON file that matches the state names in the columns. In this example, it is set to 'feature.properties.NAME_1', suggesting that the GeoJSON file has a property named NAME_1 under properties that corresponds to the state names.
- *fill_color* – color palette or colormap used to fill the polygons. Here, 'Paired' represents a qualitative colormap from the Matplotlib colormap collection.
- *fill_opacity* – opacity of filled polygons, ranging from 0.0 (completely transparent) to 1.0 (completely opaque). In this case, it is set to 0.9.
- *line_opacity* – opacity of the polygon boundaries or borders, ranging from 0.0 (completely transparent) to 1.0 (completely opaque). Here, it is set to 1.0, making the boundaries fully opaque.
- *legend_name* – name displayed for the legend associated with the choropleth map. In this case, it is set to 'Population Density'.
- *nan_fill_color* – color used for polygons where data is missing (NaN values) in the data. Here, it is set to 'blue'.
- *nan_fill_opacity* – The opacity of the fill for polygons where data is missing, ranging from 0.0 to 1.0. In this example, it is set to 0.1.

Figure 8 illustrates the output for the code in Table 8.

Table 8. Python code snippet for creating a Folium map with choropleth map

```
# importing libraries
import folium as fl
import pandas as pd

# importing xyzservices
import xyzservices.providers as xyz

# Sample population density data (in Crores)
data_dict = {'State': ['Andhra Pradesh', 'Arunachal Pradesh', 'Assam',
'Bihar', 'Chhattisgarh'],
'Population_Density': [5.31, 0.15, 3.57, 12.67, 3.01]}

# Creating a Pandas DataFrame from the population density data
population_density = pd.DataFrame(data_dict)

# Load GeoJSON file path for Indian states geometries
geojson_path = 'D:/india_state_geo.json'

# Creating a Folium figure with specific width and height
folium_fig = fl.Figure(width=800, height=600)

# Create a Folium map
map_object = fl.Map(location=[20.5937, 80.9629], zoom_start=5,
tiles=xyz.Esri.OceanBasemap.build_url(),attr='Ocean Base Map')

# Adding Choropleth layer to Folium Map
fl.Choropleth(
geo_data=geojson_path,
data=population_density,
columns=['State', 'Population_Density'],
key_on='feature.properties.NAME_1',
fill_color='Paired',
fill_opacity=0.9,
line_opacity=1.0,
legend_name='Population Density',
nan_fill_color='blue',
nan_fill_opacity=0.1,
).add_to(map_object)

# Visualizing Folium Map Object
map_object
```

Figure 8. Visualization of Folium map with choropleth map

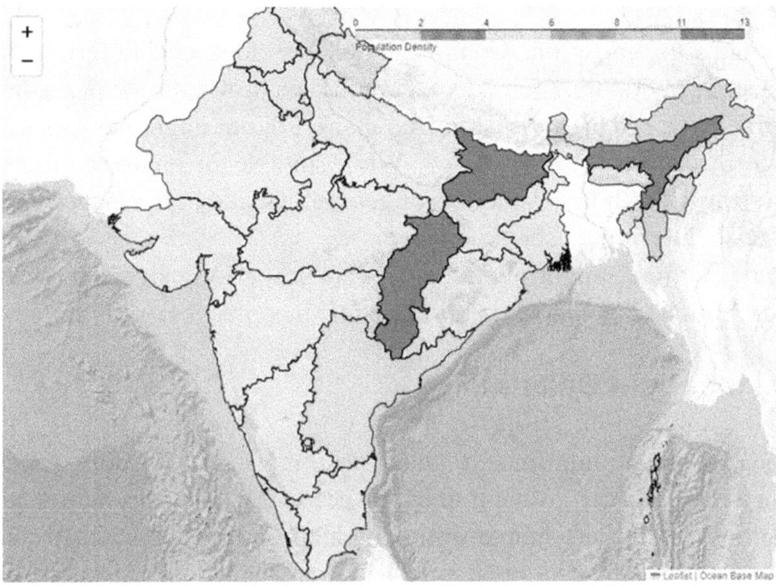

Folium: A Powerful Tool With Trade-offs

Folium is a popular Python library for creating interactive web maps. It offers a user-friendly interface for those familiar with Python, allowing for quick and visually appealing map creation. However, like any tool, Folium has its limitations and challenges that users should be aware of.

Strengths of Folium

- Ease of Use: Folium integrates seamlessly with other Python libraries like Pandas and GeoPandas, making it simple to visualize geospatial data. Its concise syntax allows rapid prototyping.
- Interactive Features: Folium maps are interactive, allowing users to zoom, pan, and click on features for detailed information. This makes data exploration intuitive and engaging.
- Customization Options: While primarily focused on basic map elements, Folium offers customization options for markers, popups, and tiles. Users can tailor maps to specific needs.

Limitations and Challenges of Folium

- Limited Customization: Compared to more advanced libraries like Plotly or Deck.gl, Folium offers less granular control over map aesthetics and interactivity. For highly customized maps, Folium might not be sufficient.
- Performance with Large Datasets: While suitable for smaller datasets, Folium can struggle with large amounts of geospatial data. Performance can become sluggish, impacting user experience.
- Limited Offline Support: Folium maps are primarily designed for web display. There's limited support for creating static map images or offline viewing.

Finding the Right Balance

While Folium has limitations in deep customization, its core functionality aligns perfectly with the growing need for accessible and informative geospatial visualizations. The library empowers individuals and organizations to unlock the potential of geospatial data for better decision-making, communication, and public engagement.

CONCLUSION

Folium excels at creating interactive and informative web maps for smaller datasets. It is a great choice for quick visualizations, educational purposes, and proof-of-concept projects. However, for highly customized maps, complex interactions, or visualizing large datasets, other libraries might be more suitable.

This chapter explored geospatial data visualization utilizing Folium geospatial Python library. It provides practical insights into configuring environments, creating diverse maps, and customizing visual elements. Discussion has been presented from foundational concepts to advanced features, to create clear and dynamic geospatial visualizations. Folium's simplicity and robustness, facilitates effective communication of complex geospatial data. The techniques explored, such as interactive markers, heatmaps, and choropleth maps, underscore the versatility of Python mapping with Folium. This chapter equips readers with valuable skills to navigate the realm of geospatial data visualization in a straightforward manner.

REFERENCES

Ajagbe, S. A., Oladipupo, M. A., & Balogun, E. O. (2020). Crime belt monitoring via data visualization: A case study of folium. International Journal of Information Security. *Privacy and Digital Forensic*, *4*(2), 35–44.

Camara, G. S., Camboim, S. P., & Bravo, J. V. M. (2021). Using Jupyter Notebooks for Viewing and Analysing Geospatial Data: Two Examples for Emotional Maps and Education Data. *The International Archives of the Photogrammetry, Remote Sensing and Spatial Information Sciences*, *46*, 17–24. doi:10.5194/isprs-archives-XLVI-4-W2-2021-17-2021

Choropleth Map in Folium (n.d.). Python. https://python-visualization.github.io/folium/latest/reference.html #module-folium.features

Folium API for Choropleth Map (n.d.). Python. https://python-visualization.github.io/folium/latest/getting_started.html

Folium API for Click for Marker (n.d.). Python. https://python-visualization.github.io/folium/latest/reference.html#folium.features.ClickForMarker

Folium API for Heat Map (n.d.). Python. https://python-visualization.github.io/folium/latest/reference.html#folium.plugins.HeatMap

Folium API for Map Layer Control (n.d.). Python. https://python-visualization.github.io/folium/latest/reference.html#folium.map.LayerControl

Folium API for Map (n.d.). Python. https://python-visualization.github.io/folium/latest/reference.html

Folium API for Marker (n.d.). Python. https://python-visualization.github.io/folium/latest/reference.html#folium.map.Marker

Folium API for Subplots in Figure (n.d.). Python. https://python-visualization.github.io/folium/latest/advanced_guide/subplots.html

Folium API for Tile Layer (n.d.). Python. https://python-visualization.github.io/folium/latest/reference.html#folium.raster_layers.TileLayer

Getting Started with Folium (n.d.). Python. https://python-visualization.github.io/folium/latest/getting_started.html

Heat Map in Folium (n.d.). Python. https://python-visualization.github.io/folium/latest/user_guide/plugins/heatmap.html

Hu, N., Meng, X., Liu, H., & Li, Y. (2022, June). Exploring Geospatial Data Visualization Based on Python. In *Proceedings of the 2022 2nd International Conference on Control and Intelligent Robotics* (pp. 858-862). 10.1145/3548608.3559322

Idrees, B. A., Naif, S. S., Galety, M. G., & Arulkumar, N. (2022). An optimized algorithm for selecting stable multipath routing in MANET using proficient multipath routing and glowworm detection techniques. In *Sustainable Advanced Computing: Select Proceedings of ICSAC 2021* (pp. 511-523). Singapore: Springer Singapore. 10.1007/978-981-16-9012-9_41

Indian State GeoJSON Data. (2020). Kaggle. https://www.kaggle.com/datasets/sauravmishra1710/indian-state-geojson-data

McClain, B. P. (2022). *Python for Geospatial Data Analysis*. O'Reilly Media, Inc.

Philippou, A. (2022*). Data visualisation of environmental risks in real estate properties* [Bachelor's thesis, University of Twente].

Pradhan, M. R. (2023). PIRAP: Medical Cancer Rehabilitation Healthcare Center Data Maintenance Based on IoT-Based Deep Federated Collaborative Learning. *International Journal of Cooperative Information Systems*.

Rey, S., Arribas-Bel, D., & Wolf, L. J. (2023). *Geographic data science with python*. CRC Press. doi:10.1201/9780429292507

Sunitha, G., Abouali, A., Galety, M. G., & Sriharsha, A. V. (2023). Dynamic Programming with Python. In *Advanced Applications of Python Data Structures and Algorithms* (pp. 102–122). IGI Global. doi:10.4018/978-1-6684-7100-5.ch005

Sunitha, G., Sriharsha, A. V., Yalgashev, O., & Mamatov, I. (2023). Interactive Visualization with Plotly Express. In *Advanced Applications of Python Data Structures and Algorithms* (pp. 182–206). IGI Global. doi:10.4018/978-1-6684-7100-5.ch009

Tile Providers: OpenStreetMap (n.d.). OpenStreet Map. https://www.openstreetmap.org/

Tile Providers: Stamen Maps (n.d.). Stamen Maps. https://maps.stamen.com/

Tile Providers: Thunderforest (n.d.). Thunder Forest. https://www.thunderforest.com/

Tile Providers: xyzservices (n.d.). ZYX Services. https://martinfleischmann.net/xyzservices-a-unified-source-of-xyz-tile-providers-in-python/

Chapter 8
Applying Geospatial Data to Choose the Optimal Route During the Road Design Stage

Ehyaudin Musema Ahmed
https://orcid.org/0009-0003-7368-8109
Debre Birhan University, Ethiopia

Mohammed Nuru Mohammed
Debre Birhan University, Ethiopia

Mohammad Gouse Galety
https://orcid.org/0000-0003-1666-2001
Samarkand International University of Technology, Uzbekistan

ABSTRACT

This chapter addresses the challenge of road route selection by employing a systematic geospatial data integration approach during the design phase. The methodology revolves around the acquisition and analysis of datasets, such as satellite imagery, DEMs, GIS, and other important data sets. Leveraging Python programming, the chapter endeavors to develop a tailored geospatial framework for optimizing road route selection, with a focus on evaluating socio-economic, engineering and environmental impacts and aligning with sustainability objectives. Resulting in a robust methodology determining optimal alignments based on terrain, environmental, and socioeconomic factors using MCA. The chapter extend to providing data-driven recommendations for the development of resilient infrastructure, ensuring that road design aligns with sustainability principles to foster economic prosperity, environmental conservation, and societal well-being a paradigm shift in road expansion strategies.

DOI: 10.4018/979-8-3693-1754-9.ch008

INTRODUCTION

Governments, carriers, and operators worldwide are looking for opportunities to invest in improving transportation infrastructure, including railway tracks, roads, and highways. It is becoming increasingly difficult to locate adequate routes for new transportation infrastructure for a variety of reasons, including continued urban development, greater public awareness of potential social and environmental consequences, and, most importantly, rising cost pressures (Tischler, 2017).

Route options are delineated as approximate alignments situated within the route corridor, serving as a basis for comparative analysis to ascertain the optimal route. It is imperative that the 'footprint' of each route option is meticulously defined to facilitate a comprehensive assessment of its feasibility, encompassing engineering, social, and environmental aspects, along with preliminary cost estimation. This evaluation should be conducted in consideration of topographical and geometric constraints. Additionally, beyond the incorporation of intermediate towns and nodal points, control points may arise due to geographical factors, such as a significant river crossing point or a col designated as a mountain pass.

Ethiopia, a nation of diverse landscapes and rapid urbanization, faces a critical challenge in developing and maintaining an efficient road infrastructure network. Road infrastructure is crucial for fostering economic growth, improving connectivity, and enhancing the overall quality of life for its citizens. However, Ethiopia's unique geographical diversity, characterized by the Ethiopian Highlands, lowland plains, and varying climatic conditions, presents complex challenges for road planning and design. This background section provides a comprehensive overview of the context, challenges, and opportunities in Ethiopian road infrastructure planning, drawing from a selection of pertinent literature.

Challenges in Ethiopian Road Infrastructure Planning

The geographical diversity of Ethiopia necessitates adaptive and innovative road infrastructure planning. The rugged Ethiopian Highlands are marked by steep slopes and challenging terrain, demanding careful route selection and engineering solutions to minimize construction costs and environmental impact. Conversely, the lowland plains and valleys pose different challenges, such as flooding and soil instability, that require specialized planning and design considerations (Teshome & Beyene, 2017). Rapid urbanization compounds these challenges, increasing the demand for efficient road networks to connect growing cities and support economic activities (Yitayew, 2020).

Objectives of Road Design and Route Selection

After choosing the best initial route (Corridor), minor tweaks are allowed for engineering/environmental reasons.

Decision-makers typically explore at least 3 options for connecting start and end points, meeting design standards. Upgrading existing routes might not need 3 options, but considering alternatives is still recommended.

Feasibility of each route option is checked using maps, desk studies, and potentially satellite imagery. In some cases, on-site surveys might be needed for final confirmation.

Importance of Geospatial Analysis and Python Programming

Leveraging geospatial data (satellite imagery, GIS, etc.) offers a scientific and efficient solution for road design decision-making in Ethiopia. This data provides valuable insights into the landscape.

Studies highlight the value of geospatial data (terrain, population, etc.) for optimizing road design through data-driven route selection. Geospatial analysis and Python programming become key tools, offering a powerful combination for informed decision-making.

Geospatial analysis, using tools like elevation models and GIS, helps engineers assess complex terrain and plan feasible routes. It also allows for integrating various factors (environmental, social, etc.) through Multi-Criteria Analysis (ERA, 2013).

Geospatial analysis aids decision-making by considering environmental, social, and engineering factors. It optimizes road alignment and minimizes impact through tools like Multi-Criteria-Analysis. Python streamlines data processing and integrates with GIS for efficient workflows.

Python's versatility extends beyond automation. Its open-source nature promotes collaboration and scales for projects of all sizes, making it valuable for both custom road design tools and complex analyses.

Combining geospatial analysis with Python programming provides a powerful solution for optimal route selection in road design. This approach enhances spatial understanding, automates workflows, and fosters collaboration, leading to more sustainable and efficient road networks.

ROUTE SELECTION IN ROAD DESIGN

Route selection is critical in road design, balancing safety, cost, and environmental impact. Engineers consider topography, regulations, and community needs to create efficient and sustainable road networks (Dr J Rolt et al., 2005). While challenges are inevitable, good engineering practices can mitigate them.

Key Prospects in Route Selection

This chapter defines a route corridor as a flexible area (not a fixed width) examined for potential road options. The size considers existing roads, road purpose, terrain, and social/environmental factors (ERA, 2013).

This chapter defines a route as a specific path within a corridor, and a corridor as a flexible zone that can hold many potential routes. This highlights the adaptability of transportation planning. Corridors provide a framework for various routes, showcasing both planning flexibility and the importance of corridors in supporting entire transportation networks.

Evolution of Route Selection Methods in Road Design

Route selection, a crucial step in road design, has undergone a significant transformation throughout history. Early methods focused primarily on engineering efficiency, while contemporary practices encompass a broader range of social, environmental, and technological considerations. This section explores a glimpse on this evolution, highlighting the key transitions in route selection methodologies.

Early Route Selection (Pre-1900s)

Traditionally, route selection relied on human expertise and on-site surveys as illustrated in Figure-1 and Figure-2 below Engineers identified the shortest or most level path between origin and destination, often prioritizing construction ease and minimizing costs. Social and environmental factors were largely disregarded during this period.

Figure 1. Engineer conducting reconnaissance survey on site 1900

Figure 2. Surveyors conducting rural side reconnaissance

The Rise of Engineering-Centric Methods (1900s)

The 20th century saw the introduction of more sophisticated techniques. The development of road design standards and manuals provided a framework for route selection. These methods incorporated factors like topography, soil conditions, and drainage. However, they still primarily focused on engineering feasibility and construction costs (Syamsunur et al., 2011).

The Shift Towards Multi-Criteria Analysis (Late 20th Century)

As environmental and social concerns gained prominence, route selection methodologies evolved to incorporate a broader range of factors. Multi-criteria analysis (MCA) emerged as a popular approach. MCA considers various criteria, such as environmental impact, social disruption, land acquisition costs, and safety, alongside traditional engineering considerations. This shift towards a holistic approach required more complex data collection and analysis techniques like Geographic Information Systems (GIS).

Furthermore, the availability of satellite images at high resolution in digital format, its repetitive coverage and multi-spectral capabilities help scientists to monitor land use change over time. Remote sensing technology makes it possible to generate land use/cover maps at regional or global scales (Jameel, 2017).

Integration of Public Participation (Early 21st Century)

The importance of public involvement in infrastructure projects grew in the late 20th and early 21st centuries. Public participation GIS (PPGIS) emerged as a valuable tool for incorporating community input into route selection. PPGIS platforms allow stakeholders to visualize potential routes, identify areas of concern, and provide feedback. This participatory approach fosters transparency and helps to minimize social disruption (Heldt, 2022).

The Rise of Artificial Intelligence (AI) and Big Data (Mid-21st Century)

The emergence of AI and big data analytics is further be able to transform route selection. Machine learning algorithms can analyze vast datasets, including historical traffic patterns, real-time sensor data, and environmental information. (Balkaya & Tutumluer, 2021).

This allows for the optimization of routes for factors like traffic flow, noise pollution, and construction efficiency. AI-powered systems can continuously learn and improve based on new data, potentially leading to more dynamic and adaptable route planning.

Future Trends

Route selection in road design has undergone a significant evolution. From a purely engineering-centric approach to a holistic method considering social, environmental, and technological factors, the process has become increasingly complex and data-driven. As technology continues to advance, we can expect further innovations in route selection methods, leading to the development of more sustainable, efficient, and resilient road networks.

The future of route selection is likely to see a continued integration of sophisticated technologies. Real-time data from connected vehicles and infrastructure will inform route optimization strategies. Advanced analytics will be used to assess climate change impacts and design resilient infrastructure. The emphasis on sustainability will drive the development of methods that minimize environmental impact and promote eco-friendly construction practices.

A Multifaceted Approach

Selecting the optimal road route alignment requires a comprehensive approach that considers various aspects beyond the aforementioned factors. Here's other key factors involved:

1. Social Impact Assessment (SIA):

A study by (H.J.Jani, 2018) emphasizes the importance of SIA in road projects. They propose a framework for integrating SIA into the early stages of planning, ensuring potential disruptions to communities, cultural heritage, and livelihoods are identified and mitigated.

Among the key considerations are;

- Displacement of communities and businesses
- Noise and air pollution impacts on residents
- Accessibility and equity of access to the road for different social groups
- Cultural heritage preservation and potential disruptions to sacred sites

The sustainable livelihoods approach is a way of thinking about the objectives, scope, and priorities for development activities (Serrat, 2017).

It is based on evolving thinking about the way the poor and vulnerable live their lives and the importance of policies and institutions. It helps formulate development activities that are;

- People-centered
- Responsive and participatory
- Multilevel
- Conducted in partnership with the public and private sectors
- Dynamic
- Sustainable

World Bank and African Development Bank policies prioritize avoiding displacement in projects due to its negative social and economic impacts. If resettlement is necessary, it must be fair, minimize disruption, and improve livelihoods. Clear guidelines and a participatory plan (RAP) are required. The RAP should address livelihoods, compensation, and well-being, ensuring displaced people have a voice (ERA, 2013).

Other Socio-Cultural Issues that needs to be considered in the Social Impact Assessment such as;

I. Cultural and archaeological sites
    - Archaeological sites that could be affected by each route option
    - Number and type of cemeteries that could be affected by each route option
    - Local Administration Office
    - Local Cultural and Tourism Office
    - Local Administration Officials and Elders

II. Religion, ethnic groups and languages used
    - Major ethnic groups in and round the project areas with their percentage share and settlement with respect to each route option
    - Look for possibility of conflicting interests among residents or ethnics groups in the vicinity of each route option.

2. The Environment and Climate Change Considerations:

A paper by (HenningTheuns, 2017) explores how to integrate climate change considerations into road infrastructure life cycle. He proposes a methodology for assessing vulnerability to extreme weather events and designing roads for resilience which sustains through the asset management.

Among the Other considerations included are;

- Increased risks of flooding, landslides, and wildfires due to climate change
- Greenhouse gas (GHG) emissions associated with construction and operation of the road.
- Carbon sequestration potential of existing landscapes along possible routes.

Transportation infrastructure like roads, railways, and waterways bring undeniable benefits to society, but they come at a cost to wildlife habitat and the corresponding bio-diverse ecosystem. This cost manifests in several ways.

Direct wildlife mortality increases dramatically as roads and railways fragment habitats, creating deadly barriers and raising collision rates. Beyond immediate deaths, habitat fragmentation is a major threat. Transportation infrastructure severs once-contiguous natural areas, isolating populations and disrupting movement pattern (IENE, 2022).

Roads fragment wildlife habitat, harming species with large territories. Habitat loss due to construction disrupts animal populations and isolates them in smaller areas. Additionally, roads alter local environments and increase human activity near wildlife corridors, further disrupting behavior (IENE, 2022).

Road-related habitat fragmentation significantly harms biodiversity. To create a sustainable transportation system, we need a holistic approach that considers social and ecological factors. Collaboration among ecologists, planners, and engineers is essential.

Ecologists and engineers must collaborate to minimize road impact on wildlife. Strategies include wildlife crossings (Figure 3), habitat restoration, and route planning to reduce footprint and fragmentation. A holistic approach is key to balancing human mobility with wildlife well-being for future generations (IENE, 2022).

Figure 3. Wildlife crossings in a highway

3. Public Participation:

A study by (Xiao & Hao, 2023) highlights the importance of public participation in road infrastructure projects. They propose strategies for effective public engagement to gather local knowledge and concerns. Which also considers;

- Transparency and open communication with stakeholders throughout the planning process
- Utilizing various methods for public participation (e.g., surveys, workshops, town hall meetings)
- Addressing public concerns and incorporating them into the decision-making process.

A Public Consultation and Disclosure Plan (PCDP) should be prepared by the client or its consultant at ESIA stage and is therefore not required for route selection.

However, stakeholder consultation is a critical element of the environmental and social assessment, and can contribute significantly to baseline data collection studies if issues become apparent during consultation that were not evident during desk study and field surveys (ERA, 2013).

Consultations should be undertaken with the following parties:

- EPA, including regional Environmental Procedures Authority representatives
- relevant district offices of ERA
- any other community representation groups
- industry and local transport providers, where appropriate
- any relevant NGOs (national and international) operating in the project area. It is important to ensure that all potentially-affected groups are consulted (ERA, 2013).

4. Technological Innovations:

Planning new transportation routes involves addressing environmental concerns, aligning with existing infrastructure, and considering social, financial, and political factors. The convergence of geospatial imaging, softcopy photogrammetry, regional significance analysis, and alignment optimization has streamlined this process. (Gipps et al., 2001)

Geospatial imaging provides cost-effective, data-rich information, while softcopy photogrammetry and regional significance analysis extract essential planning details. The automation of alignment generation creates economical, high-quality routes. Together, these technologies revolutionize alignment planning, transforming previously cumbersome tasks into efficient, cost-effective processes (Gipps et al., 2001).

5. Financial Feasibility:

Financial feasibility remains a cornerstone consideration in selecting an optimal road route alignment. However, the evaluation extends beyond simply identifying the most cost-effective construction option. A recent study by (Asfoor & Halit, 2017) emphasizes the importance of Life Cycle Cost Analysis (LCCA) in transportation infrastructure projects. This approach considers not just initial construction costs, but also maintenance expenses, potential rehabilitation needs, and the road's long-term economic benefits. Additionally, exploring public-private partnerships (PPPs) can be a viable strategy for financing road projects, as highlighted by (Raymond E. Levitt, 2019). PPPs leverage private sector expertise and funding, potentially alleviating the entire financial burden from the public sector. While cost remains a crucial factor, a comprehensive financial feasibility assessment should encompass both short-term construction expenses and long-term economic considerations.

- Considerations:
- Construction and maintenance costs of different route options
- Potential for public-private partnerships and funding opportunities
- Long-term economic benefits of the road project (e.g., job creation, increased access to markets)

6. Regulatory Compliance:

According to the World Bank Operational Policies (OP 4.01) issued in 1999 and revised in 2004, the Bank requires environmental assessment (EA) of projects proposed for Bank financing to help ensure that they are environmentally sound and sustainable, and thus to improve decision making. The borrower is responsible for carrying out the EA.

According to African Development Bank (AfDB) Environmental Policy; A strong and diversified economy shall be recognized as a just means to enhance the capacity for environmental protection; however, all development-related decision-making processes shall integrate economic, social and environmental considerations. Nonetheless, lack of financial resources shall not constitute an impediment to the promotion of community-based natural resource protection and management (ERA, 2013).

Wildlife Policy Protecting the wildlife resources and their habitats, maintaining the balance of nature for posterity in accordance with international wildlife conventions and agreements to which the country is a signatory (ERA, 2013).

Among the Relevant Multilateral Agreements Convention which enforces implementation compliances concerning;

- the Protection of World Cultural and Natural Heritage ratified 1972.
- International Plant Protection Convention.

- Convention on International Trade in Endangered Species (CITES), ratified in 1989.
- Vienna Convention on Ozone Layer Protection (1990);
- Montreal Protocol for Substances Depleting the Ozone Layer (1990);
- United Nations Convention on the Law of the Sea. Framework Convention on the Law of the Sea. Framework Convention on Climate, ratified in 1994.
- Convention on Biological Diversity, ratified in 1994.
- African Convention on the Conservation on Natural Resources.
- Convention on desertification ratified 1997.
- Convention on Biodiversity (Rio convention) (1997);
- Framework Convention of United Nations on Climate Change (1997);
- Convention on the Control of Trans-boundary Movement of Hazardous Substance

Precise procedures need to be followed during project implementation to confirm compliance with the requirements with the general consideration of;

- Environmental regulations regarding land use, emissions, and endangered species
- Social safeguards outlined in national and international policies
- Obtaining necessary permits and approvals from relevant authorities

All in all, MCA methodologies, as discussed previously, are valuable tools for weighing these diverse factors. By integrating social impact assessments, climate change considerations, public participation, technological innovations, financial feasibility, and regulatory compliance into a robust MCA framework, stakeholders can reach informed decisions on the optimal road route alignment.

Traditional Approaches vs. Geospatial Approach

Traditional Approaches of route selection relied on manual methods, involving on-site assessments, topographical maps, and desk studies. Decision-making was often based on limited data, leading to potential inefficiencies and increased costs.

Geospatial Approach runs around the advent of geospatial technologies, including Geographic Information System (GIS) software, satellite imagery, and digital elevation models (DEM), has revolutionized route selection. The geospatial approach integrates advanced tools to visualize and analyse the impact of various factors on potential road alignments. This data-driven approach enhances precision, efficiency, and the overall effectiveness of route selection.

Factors Influencing Route Selection

After defining the corridor, several social, environmental, and engineering factors are considered to identify and evaluate potential routes. A detailed assessment allows for objective comparisons and the selection of a preferred route. This chosen route then guides the design and implementation process.

In essence, the route selection process is intricately woven with the scrutiny of diverse factors, emphasizing the need for a thorough analysis of socio-economic dynamics, environmental implications, and engineering specifications. This multifaceted examination ensures that the chosen route aligns seamlessly with overarching project goals and sets the stage for a well-informed, strategic decision-making process (ERA, 2013).

Role of Advanced Tools in Route Selection

Advanced tools play a crucial role in modern route selection, providing a comprehensive and data-driven decision-making framework. To mention some, Geographic Information System (GIS) software facilitates the integration of spatial data, allowing engineers to visualize and analyse factors like terrain, population density, and environmental conditions. This aids in making informed decisions during the route selection process.

High-resolution satellite imagery provides a detailed view of the landscape, assisting in the identification of potential route options and the evaluation of environmental and topographical factors. Digital Elevation Models (DEM) offer a three-dimensional representation of the terrain, enabling a detailed analysis of slope, elevation changes, and overall topographic difficulty.

In essence, the use of advanced tools in route selection enhances the precision, efficiency, and sustainability of road design. These tools empower engineers to make informed decisions that align with economic goals, environmental conservation, and societal well-being.

Sustainability and Environment a Contemporary Approach

A recent study by (González Del Campo & Gazzola, 2020) explores innovative methods for environmental impact assessments (EIA) in road projects. They examine cutting-edge tools using data collection, modeling, and analysis to create more sustainable road designs and accurate EIAs.

(Althoey et al., 2023) work examines next-generation materials for low-carbon built environment using a life-cycle assessment approach. The study not only explores the environmental impacts of these materials but also assesses their sustainability throughout their life cycle. This research is crucial in the ongoing effort to reduce the carbon footprint of road infrastructure, providing insights into materials that align with broader environmental goals.

The basic framework of Environmental Procedures describes the need to carry out an Initial Environmental Examination (IEE) as an impact screening process that allows road projects to be classified according to whether they are Schedule I or Schedule II activities[1].

The principal difference between the two categories relates to the trigger level for the need to carry out an ESA during design. The ESA (or ESIA as it is more commonly known – Environmental and Social Impact Assessment) is therefore not the subject of this Route Selection Manual; the focus here is on the production of an IEE for route selection purposes.

The Recommended Sequence of Environmental and Social Assessment Activities for Route Selection is depicted in Figure-4.

Figure 4. Recommended sequence of environmental and social assessment activities for route selection

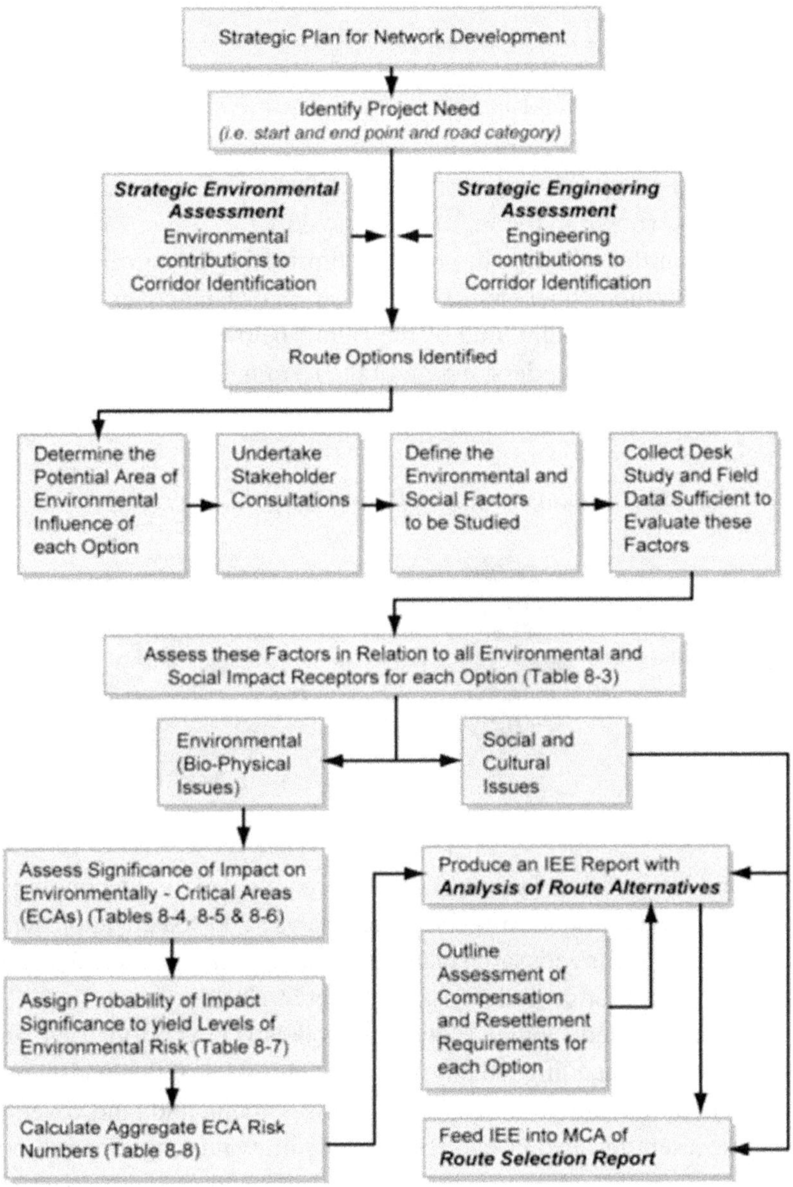

GEOSPATIAL DATA IN ROAD DESIGN

Types of Geospatial Data

Geospatial data is a cornerstone in the field of road design, providing valuable insights into the landscape and enabling informed decision-making. Geospatial data encompasses a diverse range, including vector and raster data. Vector data, such as points, lines, and polygons, represent discrete features like roads and boundaries (Moon, 2015). As shown in the Figure-5 below-left;

Raster data, on the other hand, presents information in a grid format, suitable for depicting continuous phenomena like elevation or land cover (David & Zeiler, 2004) as depicted in the right side of the figure below. The fusion of these data types allows for a holistic understanding of the terrain, essential for route selection and design optimization.

Figure 5. Vector data demonstration (left) and raster data (right)

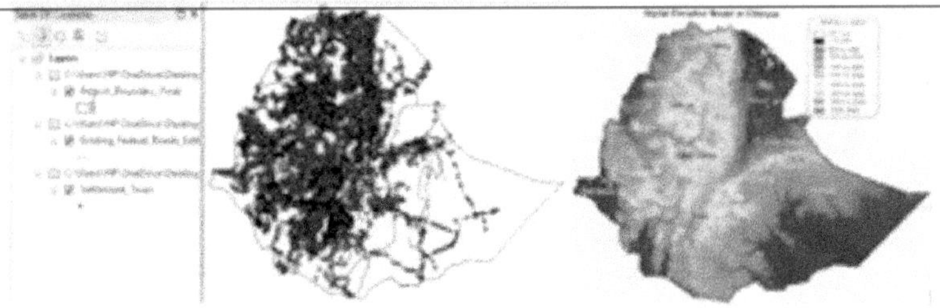

Geographic Information System (GIS) emerges as a pivotal technology in road design, offering a dynamic platform for spatial analysis and visualization. It facilitates the integration of various geospatial datasets, providing engineers and planners with a comprehensive tool for decision-making (Li et al., 2015).

Satellite imagery and Digital Elevation Models (DEM) offer rich sources of information for understanding the physical characteristics of the landscape. Aiding in route identification, environmental impact assessment, and land cover analysis.

DEMs, representing terrain elevation, assist in evaluating slope, topographic difficulty, and overall landscape morphology (Lillesand et al., 2015). Combined, these geospatial tools empower engineers to make informed decisions regarding optimal route selection and alignment design.

Geographic Information System (GIS) in Road Design

The integration of geospatial data in decision-making processes is fundamental for achieving sustainable and effective road design. Geospatial data contributes significantly to decision-making by providing a visual and analytical foundation (Li & Dragicevic, 2018).

It aids in the identification of optimal route alignments, assessment of environmental impact, and consideration of socio-economic factors.

Satellite Imagery and Digital Elevation Models (DEM)

DEMs, representing terrain elevation, assist in evaluating slope, topographic difficulty, and overall landscape morphology (Lillesand et al., 2015). Combined, these geospatial tools empower engineers to make informed decisions regarding optimal route selection and alignment design.

Importance of Geospatial Data in Decision-Making

The integration of geospatial data in decision-making processes is fundamental for achieving sustainable and effective road design. Geospatial data contributes significantly to decision-making by providing a visual and analytical foundation (Li & Dragicevic, 2018).

APPLYING PYTHON IN GEOSPATIAL ANALYSIS

Python has gained prominence as a go-to programming language in road design, offering a range of functionalities for data processing, analysis, and visualization.

Python Programming Language in Road Design

Its simplicity, readability, and extensive libraries make it an ideal choice for engineers and planners involved in the road design phase. Python facilitates the automation of repetitive tasks and the execution of complex algorithms, enhancing the overall efficiency of the road design process.

Research by (Grinberg, 2018) highlights Python's ascendancy in the field of geospatial analysis, underscoring its applicability in road design. The study emphasizes Python's flexibility in handling geospatial data and its role in streamlining workflows, ultimately contributing to more streamlined and effective road design processes.

Integration of Python With Geospatial Tools

Python seamlessly integrates with various geospatial tools, creating a synergy that amplifies the capabilities of both. This section explores how Python collaborates with geospatial tools, such as Geographic Information System (GIS) software, to enhance the analytical and visualization aspects of road design.

Studies by (Lawhead, 2013) elucidate the integration of Python with GIS tools, elucidating how Python scripts can automate GIS processes and analyses. This integration not only expedites tasks but also introduces a level of customization, allowing engineers to tailor analyses to specific project requirements, a crucial aspect in the intricate process of road design.

Advancements on Geospatial Technologies, Python Analysis, and Road Design Methodologies

In recent years, the scholarly landscape has witnessed a surge in research at the intersection of geospatial technologies, Python programming for geospatial analysis, and advancements in road design methodologies. This literature review synthesizes key findings from prominent works in these domains, providing a comprehensive overview of current research trajectories.

Recent research by (Ammapa et al., 2022) highlights the use of Geographic Information Systems (GIS) with remote sensing data for road network planning. This approach improves decision-making by analysing terrain, land use, and environmental factors, leading to better road design strategies.

Python programming is becoming increasingly important for geospatial analysis (McClain, 2022). Scripting in Python automates tasks, manipulates data, and allows for algorithm development. Its open-source nature fosters collaboration and tools like GDAL and Fiona streamline geospatial workflows.

Despite advancements in planning, materials, and management, a gap exists between user needs and road conditions. This highlights the ongoing challenge of effectively using technology and concepts to improve all aspects of road infrastructure.

Dealing with the development of novel physical experiments, computational models, and theoretical analyses, and the use of reliable models, characterization techniques, and evaluation tools to achieve traffic operation efficiency, safety, and energy savings. High-quality reviews and original research papers that present current research gaps, theoretical frameworks, methodologies, and practical approaches are welcome.

Potential topics include but are not limited to the following:

- Material characterization and innovative experimental techniques

- Multi-scale and multi-physics modelling of pavement materials
- Sustainable and innovative materials for road infrastructure
- Conception and implementation of novel road design approaches
- Road infrastructure construction techniques and strategies
- Rehabilitation, preservation, and maintenance strategies and approaches
- Pavement management systems
- Design and operation of new generation transport systems

Research is focusing on future advancements in road design using data and machine learning (Verbytskyi, 2023). This includes using real-time traffic data and dynamic factors to create adaptive road designs that can respond to changing needs, leading to more sustainable and resilient infrastructure.

Research highlights Python's advantages for geospatial analysis in road design (McKinney, 2017). Its extensive libraries like NumPy and Pandas allow for complex data manipulation. Python's open-source nature fosters collaboration and innovation in road design methods.

This review explored the intersection of three rapidly evolving fields: geospatial technologies, Python programming for geospatial analysis, and advancements in road design methodologies as summarized herein below;

1. Geospatial Technologies: A Data-Rich Landscape

Geospatial technologies encompass a range of tools and techniques for capturing, managing, analyzing, and visualizing data with a geographic component. Recent advancements include:

- **High-Resolution Satellite Imagery:** Increased satellite constellations provide frequent, high-resolution imagery for detailed infrastructure assessments and road network and/or route planning as seen in figure 6 below showing different values of high-resolution optical satellite imagery.

Figure 6. The value of high-resolution optical satellite imagery

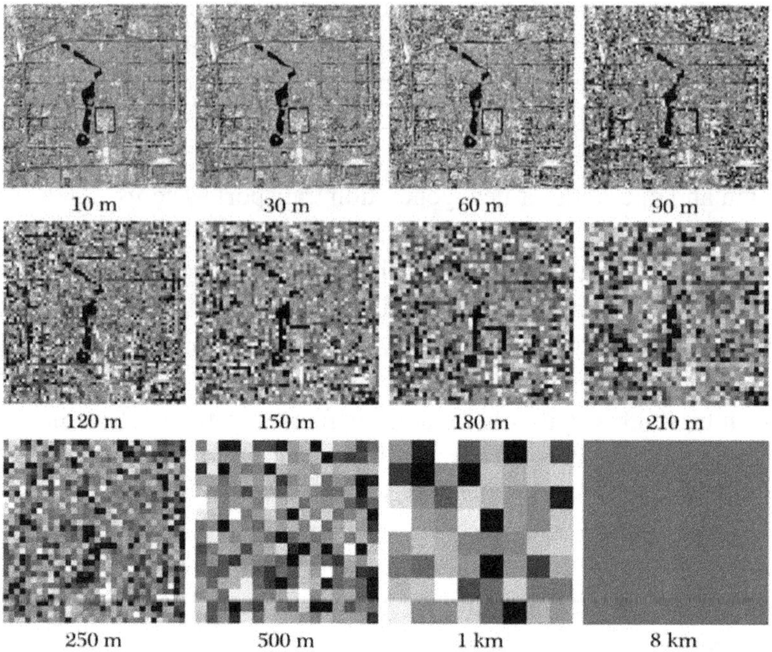

- **LiDAR (Light Detection and Ranging):** LiDAR offers precise 3D point cloud data, crucial for terrain analysis, road safety assessments, and vulnerability mapping in disaster-prone areas.

Which stands for Light Detection and Ranging, is a remote sensing method that uses light in the form of a pulsed laser to measure ranges (variable distances) to the Earth. These light pulses combined with other data recorded by the airborne system generate precise, three-dimensional information about the shape of the Earth and its surface characteristics as shown in the figure 7 below.

Its versatility and high resolution give it applications in archaeology, climate monitoring, city planning, meteorology, mining, and much more.

Figure 7. LIDAR derived topographic map black and white 3D

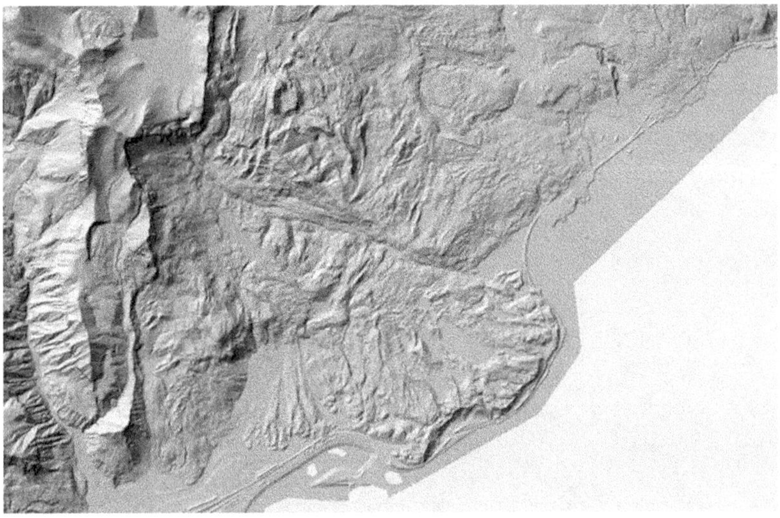

- **Real-Time Sensor Integration:** Integrating real-time traffic data from sensors with geospatial platforms enables dynamic traffic management and route optimization for road networks as shown in figure 8 below.

Figure 8. Real-Time sensor integrated map

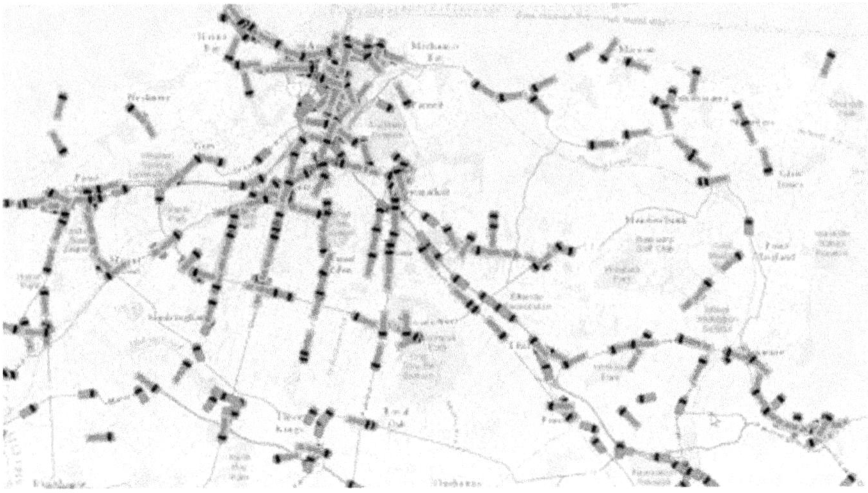

2. Python: A Powerful Tool for Geospatial Analysis

Python has become a dominant force in geospatial analysis due to its:

- **Open-Source Nature:** Free and readily available libraries like GeoPandas, Rasterio, and NetworkX facilitate efficient data manipulation, spatial analysis, and network modeling for road design as depicted in Figure-9, Figure-10 and Figure-11 Respectively.

Figure 9. Geo-Pandas Map integration

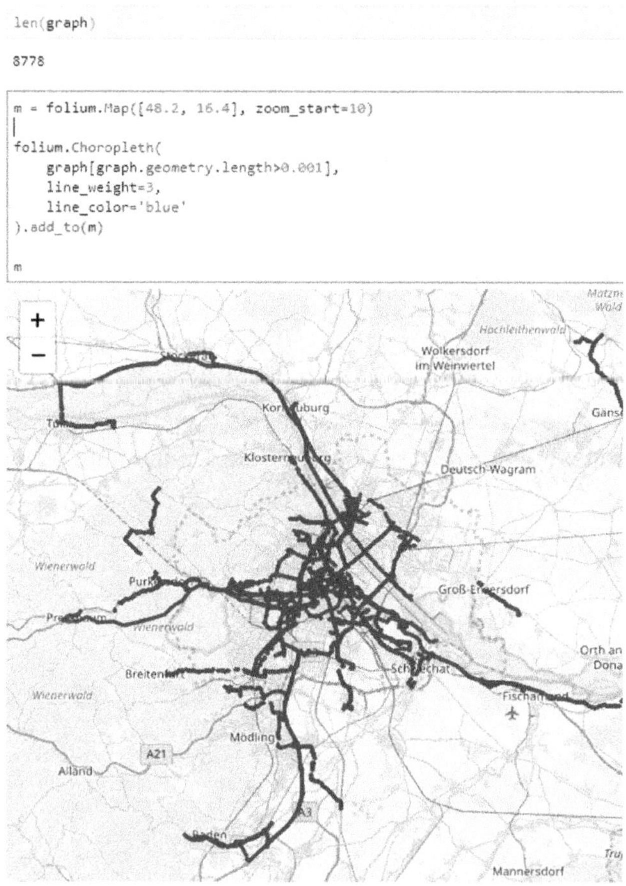

Figure 10. Rasterio map

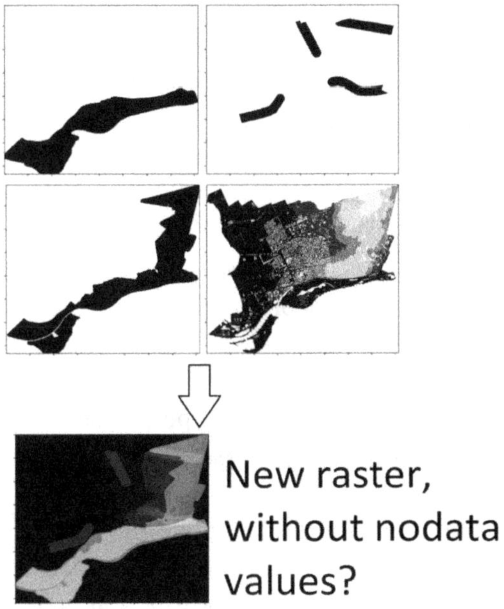

Figure 11. Network X map integration

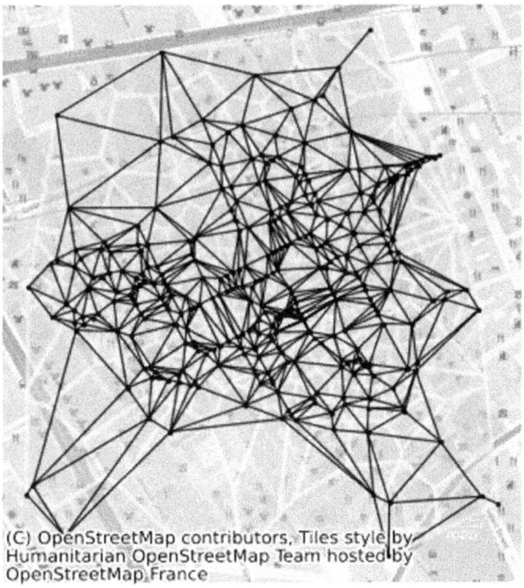

- **Extensive Community Support:** A vast developer community ensures continuous library updates, tutorials, and online resources for tackling complex geospatial problems in road planning.
- **Data Science Integration:** Python's seamless integration with data science libraries like NumPy and SciPy allows for advanced statistical analysis and machine learning applications on geospatial data for informed road design decisions.

3. Advancements in Road Design Methodologies

Road design methodologies are incorporating new technologies and user-centric approaches:

- **Sustainable Road Design:** Emphasis is placed on using recycled materials, green infrastructure, and permeable pavements to minimize environmental impact during road construction.
- **3D Modeling and Simulation:** Utilizing 3D modeling software and simulation tools allows for virtual prototyping of road designs, optimizing traffic flow and safety features before construction as depicted in Figure 12 of Auto-Desk Civil 3D extraction.

Figure 12. Civil 3D software extraction simulating a road project and surrounding geography

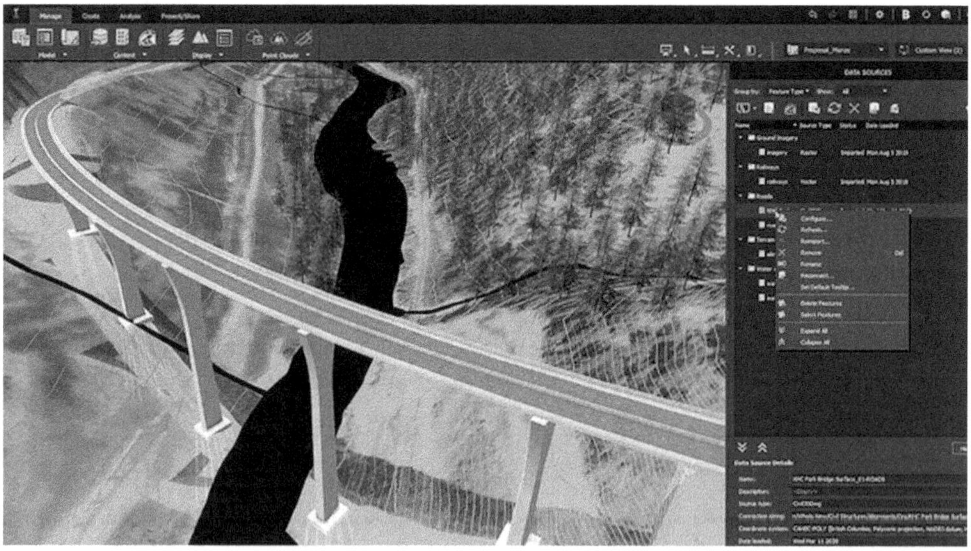

New geospatial tools, Python programming, and improved road design methods are revolutionizing infrastructure planning. This requires collaboration between geospatial scientists, programmers, and civil engineers to address the complexities of modern road design.

Ethical Considerations in Geospatial Analysis and Python Programming

While advancements in geospatial tools, Python programming, and road design methods are revolutionizing infrastructure planning (as previously mentioned), responsible use is crucial. Integrating these technologies in road design offers vast opportunities, but ethical considerations demand attention. This section explores these ethical dimensions, emphasizing responsible practices for geospatial analysis and Python programming.

Data Privacy and Security

One of the primary ethical considerations in geospatial analysis is the safeguarding of data privacy and security. As geospatial datasets often contain sensitive information, the responsible handling, storage, and sharing of such data become paramount. Ensuring compliance with data protection regulations and employing encryption measures when handling geospatial data in Python programming are essential steps in upholding ethical standards.

Transparency and Reproducibility

Ethical geospatial analysis and Python programming require a commitment to transparency and reproducibility. Researchers and practitioners should provide clear documentation of their methods, codes, and data sources to facilitate reproducibility and enable scrutiny by the scientific community. This transparency fosters trust and accountability, aligning with ethical principles in research and development.

Informed Consent and Community Engagement

When utilizing geospatial data which is confidential/not available in open source and that may impact local communities, obtaining informed consent and engaging with community stakeholders become ethical imperatives. Open communication, community involvement, and an understanding of the potential social implications of geospatial analysis contribute to ethical decision-making in road design. Python programming should align with these considerations by incorporating mechanisms for community input and feedback.

Bias and Fairness in Algorithmic Decision-Making

Geospatial analysis often involves the use of algorithms, and ethical concerns arise when these algorithms exhibit biases or perpetuate unfair outcomes. Python programming allows for the development and implementation of these algorithms, necessitating a critical examination of potential biases. Addressing algorithmic fairness and bias mitigation measures should be integral to the ethical deployment of geospatial analysis tools.

Environmental and Cultural Sensitivity

Respecting environmental and cultural sensitivities is a crucial ethical consideration in geospatial analysis. Python programming should be employed with a mindfulness of the potential impact of infrastructure development on the environment and cultural heritage. Collaborative efforts with environmental experts and cultural stakeholders can ensure that geospatial analysis aligns with ethical standards.

In recent years, the scholarly landscape has witnessed a surge in research at the intersection of geospatial technologies, Python programming for geospatial analysis, and advancements in road design methodologies. This literature review synthesizes key findings from prominent works in these domains, providing a comprehensive overview of current research trajectories.

Geospatial tech is booming, especially for urban planning and infrastructure. Li et al. (2021) showed how Geographic Information Systems (GIS) improve decision-making in road design. Combining remote sensing data with GIS helps analyze terrain, land use, and environmental factors for better road planning.

Python scripting (Smith & Johnson, 2020) automates tasks and analysis in geospatial studies. Its open-source nature fosters collaboration, and tools like GDAL and Fiona simplify data handling, making geospatial workflows more efficient.

Advancements of new geospatial tools and Python programming are transforming infrastructure planning, especially for roads. (Li et al., 2021) show how GIS improves decision-making, while (Smith & Johnson, 2020) highlight Python's role in automating tasks. Research by (Jones et al., 2022) explores using data and machine learning to optimize road design for changing traffic patterns, ensuring sustainable infrastructure.

Combining geospatial data, Python programming, and advanced design methods creates a powerful approach for road infrastructure planning. This requires collaboration between geospatial scientists, programmers, and engineers. As research progresses, embracing these synergies is key to overcoming challenges in modern road design.

ROUTE IDENTIFICATION AND SELECTION PROCESS

As far as decision-making is concerned for new roads, there are essentially four steps:

- Project identification
- Selection of the corridor
- Identification of route options within the corridor
- Selection of the preferred route option using an MCA (Multi-Criteria Analysis)

Figure 13 below summarises the processes and data sources that combine to make up the corridor selection and route selection stages of a new road construction project.

This chapter clarifies the role of corridor and route selection within the road design process, emphasizing its placement before detailed design. Traditionally, Ethiopia has over-designed routes during selection, leading to wasted resources. This highlights the importance of focusing on selection first and detailed design later.

However, route selection shouldn't prioritize options with economic, environmental, social, or engineering drawbacks (ERA, 2013). This suggests combining route selection with a preliminary design phase can eliminate the need for a separate feasibility study. Route selection can assess viability, while preliminary design provides more detailed cost estimates for a confident economic analysis.

Figure 13. Flow chart for corridor and route selection

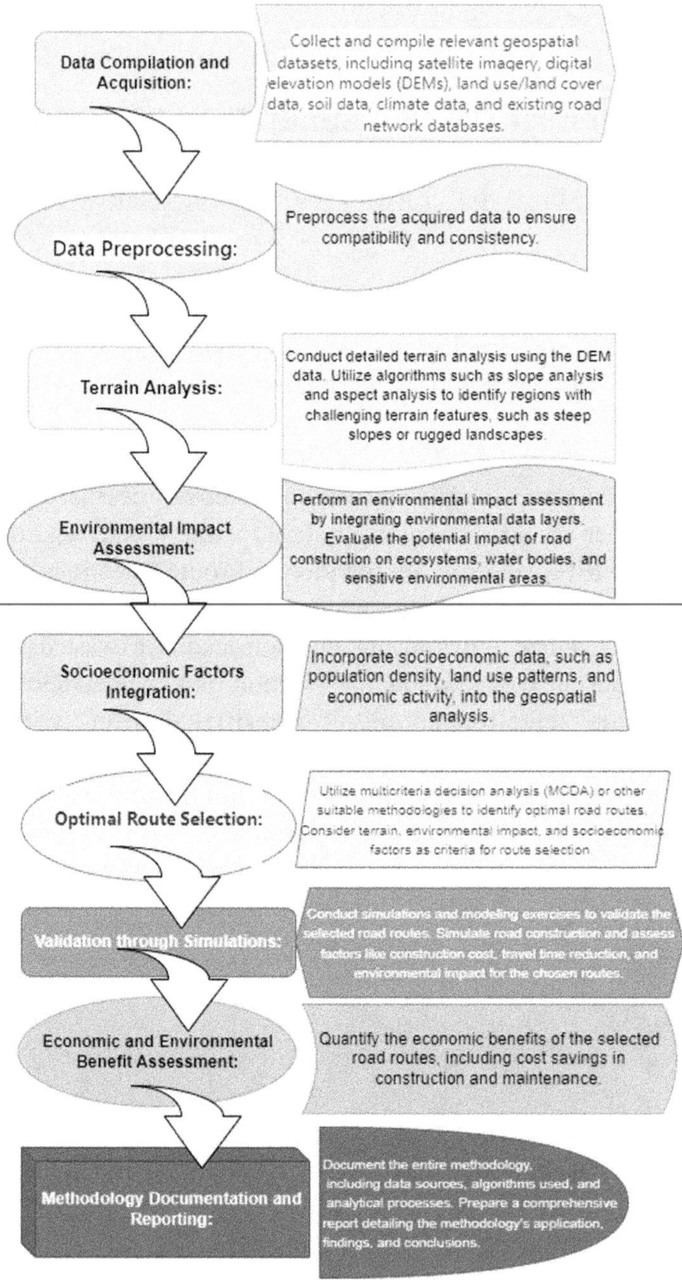

Model for Route Selection

Recommended model for the steps to be followed during route selection shall comprise of the following which the process model is also depicted in Figure 14;

Route Selection Steps

i. Know the termini points
ii. Identify and locate
iii. Conduct preliminary and reconnaissance surveys
iv. Select a corridor.
v. Possible centre-lines
vi. Examine each of the alternative alignment
vii. Final design

Figure 14. Process model for best route selection

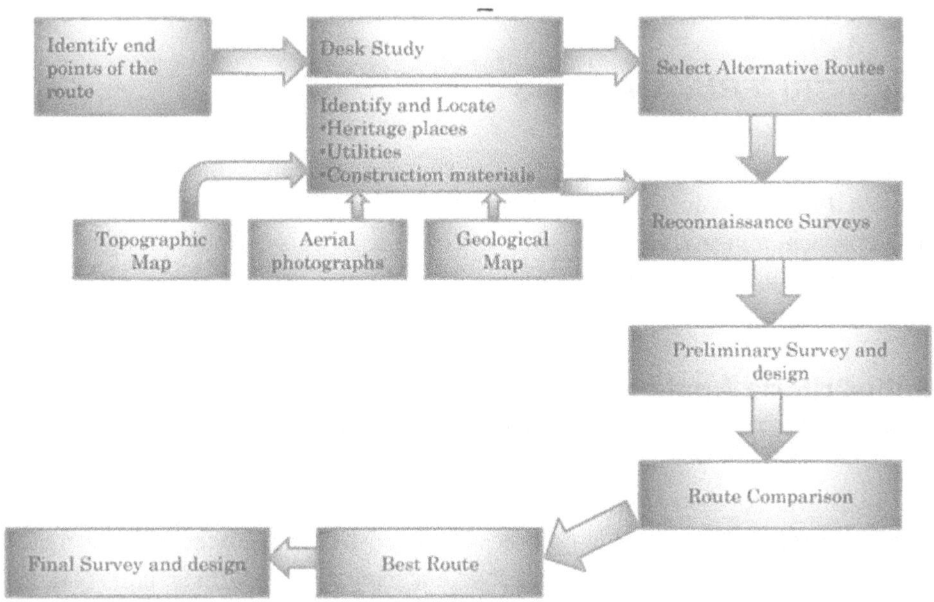

Identification of Route Options

In customary practice, the determination of route choices adheres to a systematic sequence. This involves;

- First establishing control points through which route options must pass, encompassing entities like towns, villages, and intersections within the road network.
- Subsequently, considerations extend to recognizing topographical and environmental constraints that influence the viability of route alternatives, such as identifying low points in mountainous terrains to optimize elevation changes or recognizing regions of environmental significance that impose impassable restrictions.
- Moreover, the identification of appropriate locations for significant river crossings or fan passages is prioritized.
- Following this preliminary phase, the delineation of three initial route options ensues, employing topographical maps or satellite imagery. This process is underpinned by a comprehensive evaluation encompassing environmental, social, engineering, geological, and hydrological facets as a primary criterion having the respective objective weights which has been discussed in the next sub-topic herein below.

MULTI-CRITERIA-ANALYSIS

Multi-Criteria Analysis (MCA) is a systematic decision-making methodology employed in the evaluation of various alternatives for major road investments. In this approach, a thorough assessment of road projects is conducted, considering a multitude of characteristics encompassing economic, social, environmental, and other pertinent factors.

Multi-Criteria Analysis (MCA) Methodologies in Road Design: A Look at Recent Literatures

Multi-Criteria Analysis (MCA) methodologies are increasingly being employed in road design to navigate the complex decision-making process. These methods consider various factors beyond just cost, ensuring a more holistic and sustainable approach. Here's a review of recent literature on the topic:

Application of MCA methods in making decisions about transport infrastructure: This 2018 study by (Brucker & Macharis, 2011)1 explores using a combination of MCA methods for route selection in a multimodal transportation network. They highlight the benefits of combining the ε-constraint method with D-CRITIC (Distance Correlation-based Criteria Importance Through Inter-criteria Correlation) and a modified TOPSIS (Technique for Order Preference by Similarity to Ideal Solution) approach. This combined approach offers a more robust solution for identifying Pareto-optimal solutions (those that cannot be improved in one criterion without worsening another) in complex transportation projects.

A framework for evaluating sustainable transportation infrastructure projects using multi-criteria decision analysis: This research by (Wang, 2014) proposes a comprehensive MCA framework for sustainable transportation infrastructure projects. Their framework incorporates economic, social, and environmental criteria, along with public participation, making it relevant for contemporary road design considerations.

A review of multi-criteria decision-making methods for sustainable infrastructure: This 2019 literature review by (Navarro et al., 2019) examines various MCA methodologies used in sustainable infrastructure construction. They analyze the strengths and weaknesses of different methods like the Analytical Hierarchy Process (AHP) and Multi-Attribute Utility Theory (MAUT), providing valuable insights for selecting the most suitable approach for a specific road design project.

These are just a few examples, and numerous other recent studies explore the application of MCA methodologies in road design. It's important to note that the specific MCA method chosen will depend on the unique challenges and priorities of each project.

Procedure for MCA in Road Design

The MCA evaluation process involves assigning ranks (i.e. 1st, 2nd, 3rd etc.) to each project based on different characteristics, such as economic viability, social impact, and environmental sustainability. This ranking process is iterated for each characteristic, ensuring a comprehensive and nuanced understanding of the projects' strengths and weaknesses.

To establish a coherent basis for decision-making, weights are allocated to individual characteristics, reflecting their relative importance. This weight allocation is a crucial step as it guides the computation of an overall score for each project. The overall score is determined by multiplying the rank of a project for each characteristic by its respective weight. This computation yields a holistic measure of the project's desirability, taking into account all relevant criteria (ERA, 2013).

It is important to emphasize that the purpose of the weighting procedure is not to evaluate the absolute value of any characteristic, such as the environment, but rather to facilitate objective comparisons between different projects. The focus is on relative assessments, acknowledging that different criteria carry varying levels of significance in the decision-making process.

Multi-Criteria Analysis plays a pivotal role in the decision-making landscape for road investments, offering several advantages. It enables decision-makers to make informed choices by considering a diverse set of factors, fostering a holistic evaluation that includes economic, social, and environmental dimensions. The systematic nature of MCA enhances transparency in decision-making, providing a structured framework for justifying and communicating decisions to stakeholders.

Moreover, MCA can be tailored to the specific needs and priorities of different regions, allowing for adaptability in decision-making processes. By considering multiple criteria, MCA aids in identifying and addressing potential risks associated with road investments, contributing to the development of more resilient and sustainable projects.

Establishing a Criteria for the MCA

Multi-Criteria Analysis (MCA) entails the systematic evaluation and ranking of route alternatives, leveraging a comprehensive set of criteria that encompass economic, social, environmental, and various other considerations. The efficacy of methodologies like MCA hinges upon the subjective values held by individuals engaged in the formulation of criteria. As a result there is a very wide variation in the formulation and characteristics of the ranking used. Here in below ARC-Map Extraction showing the proposed route alignments is depicted in Figure-15;

Figure 15. ARC-Map extraction showing the proposed route alignments

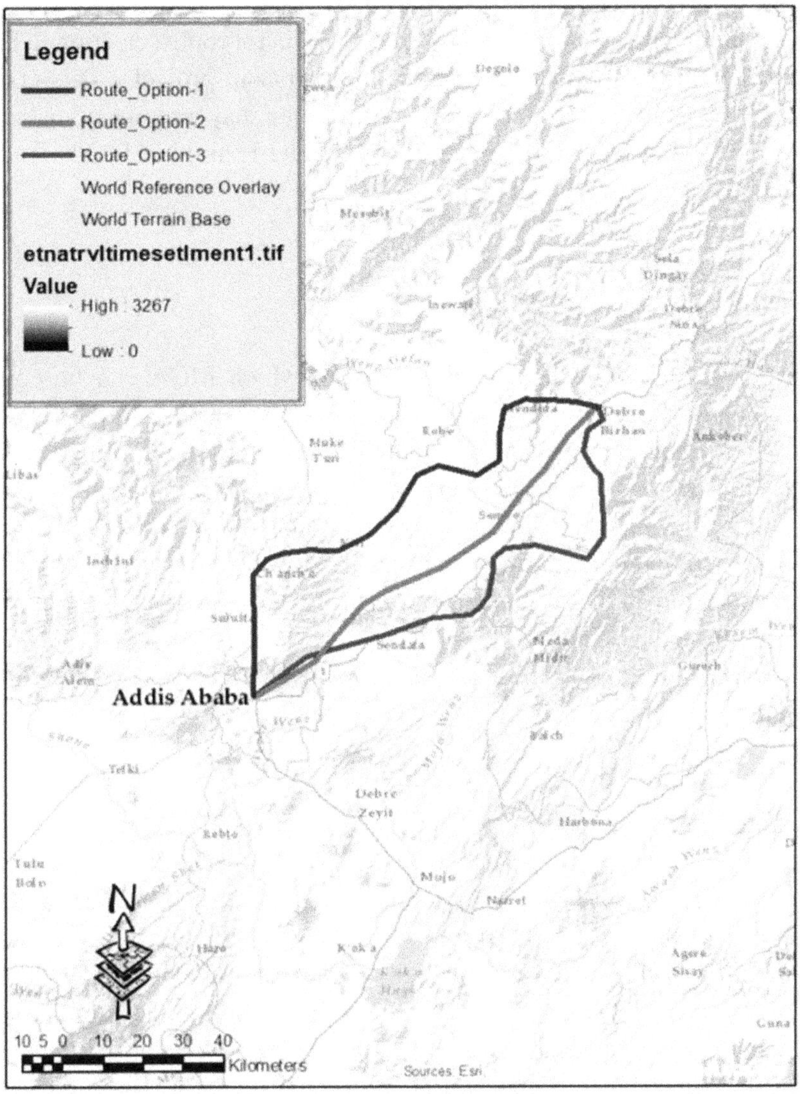

For each criterion the different alternatives has to be assessed and put into rank order. This process is then repeated for the other criteria. Weights are then assigned to each criterion and an overall score is obtained. (The score for each criterion is the product of the rank and the respective weights). The overall score gives a measure of the overall desirability of the alternative project route.

Thus, considering the comparison would be executed using the python programming language it is reasonably taken the under listed criterions for analysis that are possibly be obtained in terms of the respective relevant Geo-spatial Shape files.

This chapter recommends a set of general criteria for route selection in Ethiopia, considering international practices and Ethiopian road projects. This criteria set is suitable for Multi-Criteria Analysis (MCA) but may require adjustments for specific situations. The details are available in the Ethiopian Roads Authority's Route Selection Manual (2013).

CRITERIA RATIONALES

There is a considerable range of criteria from which MCA can be conducted, drawing from available reference sources and from recent road studies in Ethiopia. These criteria can be sub-divided into quantified and non-quantified. Thus, one can effectively make a long-list of MCA criteria while categorizing in different baskets.

For his chapter purpose a route option for Addis Ababa to Debre Birhan Road project has been taken arbitrarily only for simulation purpose while the MCA is formulated to compare the three alternatives using the underlisted principal criteria:

1. Socio-Economic (Administrative partly subsumed here)
2. Environmental
3. Engineering

Socio-Economic Criteria

Socio-economic factors are critical for Multi-Criteria Analysis (MCA) in road design. They ensure project benefits community well-being and contribute to sustainable development. The first facet within the socio-economic considerations is Community Access (Road Shape File), a criterion that demands an in-depth examination of how the proposed road alignment interfaces with existing settlements. This evaluative process seeks to quantify the impact on local communities by determining the number of settlements traversed by the road shape file, thereby shedding light on potential enhancements or disruptions to community accessibility.

Socio-economic factors include considering how the route intersects administrative areas (region, zone) to avoid potential implementation challenges. It also considering the social impact of resettlement estimating the number of displaced households for each route, considering the social impact of resettlement (ERA, 2013).

Economic factors include a thorough assessment of the economic activities surrounding each route option. This involves identifying major sectors like industry, mineral resources, agriculture, animal husbandry, and trade. By understanding the existing economic landscape, engineers can predict the potential impact of the road on the surrounding region.

Road construction can disrupt existing land use patterns. Socio-economic factors consider how the route would sever these patterns, affecting areas like cultivated land, grazing fields, forests, wetlands, and settlements. This analysis helps understand the environmental and social consequences of each route option.

Dealing with the Socio-economics benefit aspect the following needs to be taken into account;

- **Location**: The location of the corridor may have a large influence on the surrounding area, and it is the location of this entire project area that needs to be described here. A map will be required showing population centres and other geographical features.
- **Population**: What is the total population within the project area? How does this vary within the area? What are the main ethnic groups?
- **Economy**: What are the main drivers of the local economy within the project area?
- **Natural resources**: What natural resources are located within the project area? These might include rivers for irrigation or power development, forested areas, agricultural areas, minerals.
- Future economic growth goes beyond immediate resources. Factors include anticipated population growth, potential for tourism development due to the new road, and the road's own impact on stimulating economic activity. Climate change and existing development plans can also significantly influence the project's economic viability.

Social Implications of Road Projects: A Discussion

Road projects are a double-edged sword. While they can improve connectivity, boost economic activity, and create jobs, they can also have significant social consequences.

Positive Social Impacts

- **Improved Accessibility:** New roads can connect isolated communities to essential services like healthcare, education, and markets. This can improve quality of life and economic opportunities.
- **Economic Growth:** Road construction often generates jobs and stimulates local businesses. Improved transportation networks can attract investment and development.

- **Community Development:** Roads can facilitate social interaction and cultural exchange between communities.

Negative Social Impacts

- **Displacement and Resettlement:** Road construction may necessitate land acquisition, forcing residents and businesses to relocate. This can disrupt livelihoods, sever social ties, and create economic hardship.
- **Community Division:** New roads can physically divide neighborhoods, making it difficult for residents to access amenities or interact with each other.
- **Safety Concerns:** Increased traffic volume can raise safety risks for pedestrians, cyclists, and children, especially in areas without proper infrastructure.
- **Noise and Air Pollution:** Construction and increased traffic contribute to noise and air pollution, impacting the health and well-being of residents.

Community Engagement Strategies

- **Early and Transparent Communication:** Inform the community about the project's purpose, potential impacts, and timelines well in advance.
- **Public Forums and Hearings:** Provide opportunities for residents to voice their concerns and suggestions.
- **Community Advisory Committees:** Establish committees with local representatives to participate in the decision-making process.
- **Social Impact Assessments:** Conduct thorough assessments to identify potential negative impacts and develop mitigation strategies.

Mitigating Negative Impacts

- **Compensation and Resettlement Plans:** Offer fair compensation to displaced individuals and businesses, and assist with relocation.
- **Community Amenities:** Incorporate pedestrian walkways, bicycle lanes, and green spaces into road designs to promote safety and social interaction.
- **Noise Reduction Measures:** Use noise barriers and construction techniques that minimize noise pollution.
- **Traffic Management Plans:** Implement traffic calming measures and public transportation options to reduce congestion and improve safety.

Community involvement and strong mitigation strategies are crucial for maximizing positive social impacts from road projects. This minimizes negative effects, ensuring these infrastructures contribute to a more equitable and sustainable future for all stakeholders.

Environmental Criteria

Multi-Criteria Analysis (MCA) for road design demands a meticulous exploration of environmental (bio-physical) criteria, crucial for selecting an optimal road route in harmony with the natural environment.

Impacts on Water Resources dynamics such as; quantity, quality, and flow direction scrutinizing potential impacts on rivers, water sheds, and bodies, evaluating alterations and directional shifts induced by the road alignment. The criterion also considers impacts on groundwater, discerning pollution levels and water crossing structures, differentiating between bridges and culverts.

Water resources are a major environmental concern. The route assessment considers how the road might affect rivers, watersheds, and other bodies of water, including potential changes in water flow and quality. Additionally, impacts on groundwater pollution and the need for structures like bridges or culverts for water crossings are evaluated.

The route's impact on agriculture considers potential disruption to prime land and food security. Ecological factors include effects on vegetation, forests, and potential for erosion (ERA, 2013).

Engineering Criteria

During Multi-Criteria Analysis (MCA) for road design, engineering factors play a central role. A crucial aspect of this is terrain analysis, which considers geographic features, slopes, and overall topographic complexity. By examining these factors, engineers can simulate different route options and assess how variations in slope and topography will influence the final road design (ERA, 2013).

Terrain analysis helps engineers understand challenges posed by the landscape, ensuring the road design is stable, feasible, and can navigate various terrains. Another engineering factor is distance and travel time. This involves carefully examining the route's length and the resulting travel time between origin and destination (O-D) (ERA, 2013).

Weightages of Criteria for MCA

As discussed in existing literature, exemplified by Overseas Road Note 5 (Dr J Rolt et al., 2005)[2]), creating a weighting procedure devoid of subjectivity is a challenging task.

The same was also reflected in Ghanian Integrated Transport Plan –Multi-Criteria Evaluation Manual which primarily rely on professional judgment, then weights are assigned to the indicators. The weights will represent the relative importance attached to each of the sub-criteria. The final score mixing the indicators should be obtained by applying the weights stipulated by the professionals (Ghanian-Manual, 2009). Assigning weights to different criteria involves some subjectivity, but it's necessary to compare route options. These weights don't necessarily reflect the absolute value of a factor, like social impact. They help prioritize factors based on project goals (ERA, 2013).

While all route options are considered single carriageway link roads (DC-4 class), weighting adjustments are necessary based on the Ethiopian Roads Authority's Route Selection Manual. These adjustments account for the differing purposes, standards, and objectives of various road classes during the MCA process. As such, the weightages for DC-4, a road standard for simulation purpose as the Debre Birhan to Addis Ababa lay under this road classification category is picked as shown in the Table 1 Below;

Table 1. Weightings for MCA at Route Selection Stage (ERA, 2013)

Road Class			Dual Carriageway Trunk	Single Carriageway Link	Single Carriageway Collector
			DC8	DC4	DC3
Primary Criteria	Secondary Criteria	No	Max Score	Max Score	Max Score
Engineering	Road Length	1.1	9	6	5
	Terrain	1.2	5	4	3
	Earthworks	1.3	4	4	3
	Pavement	1.4	3	3	3
	Culverts & Bridges	1.5	4	4	3
	Materials	1.6	4	3	3
	Geo-hazards	1.7	4	6	5
	Cost	1.8	7	5	5
	Sub-total		**40**	**35**	**30**
Social (see also Chapter 8 for discussion)	Community Access	2.1	1	3	6
	Development	2.2	3	3	3
	Public Transport	2.3	2	4	5
	Resettlement	2.4	3	4	4
	Severance	2.5	3	2	2
	Cultural Heritage	2.6	4	4	4
	Road Safety	2.7	3	3	3
	Pollution	2.8	1	2	3
	Sub-total		**20**	**25**	**30**
Environmental (Chapter 8)	See Chapter 8		20	20	20
	Sub-total		**20**	**20**	**20**
Economic	Viability	4.1	7	7	7
	Road User Benefits	4.2	3	3	3
	Sub-total		**10**	**10**	**10**
Administrative	Sustainability	5.1	5	5	5
	Strategic Impact	5.2	5	5	5
	Sub-total		**10**	**10**	**10**
Total			**100**	**100**	**100**

Once the criteria have been defined the respective weightages has been chosen from the which is presented in Table 2 below.

Table 2. Proposed criteria weightages for MCA

SN. No.	Parameters Proposed for Comparison		
	Principal Criteria Description	**Secondary Parameters**	**Weightages**
1	**Socio-Economic**	**Community Access (Road Shape File How many Settlements will the respective road route shape files crosses)**	3
		Administrative boundaries (Administrative location of the route options (Region, Zone, Woreda, Kebele), Strategic Impact	5
		The likely number of households to be displaced **(Resettlement)**	4
		Economic structure of the area (Major economic activities such as industrial and mineral resource areas, farming, animal husbandry, trading of goods and services, etc. for each route options)	3
		Severance which will use the Land use and land cover patterns (Cultivated land (rain fed and irrigated) of the Woredas and areas traversed by each route option; Grazing areas and pasture lands of the Woredas and areas traversed by each route option; Forests, plantations, bush and shrub lands, wetlands of the, Woredas and areas traversed by each route option; **Settlement areas**, bare land and others of the Woredas and areas traversed by each route option, if possible.)	2
Total			**17**
2	**Environmental (Bio-Physical) Issues**	**Impacts on water resources** (Rivers, Water Sheds and/or bodies); (quantity, quality and flow direction); Impacts on groundwater resources (pollution and flow direction), Number of water crossing structures (Bridges vs Culverts)	6
		Prime agricultural land affected	6
		Vegetation (Flora), Forests; Erosion/ Desertification (length affected)	6
Total			**18**
3	**Engineering**	**Terrain Analysis** (Geographic, Slope and Topographic Difficulty)	4
		Distance/Travel Time	6
Total			**10**
Overall Total Score for Ranking			**45**

However, the respective weight will be changed in to decimal points when encoded to python language for ease of computing the overall scoring.

Correlating Criteria With GIS Shape Files

Accordingly, specific secondary parameters were picked from the bunch of criteria list which suit the scope of this chapter as depicted in the Table 3 herein below and matrixed with the related geospatial shape file for analysis;

Table 3. Proposed criteria and required combination of geospatial data for analysis

SN. No.	Parameters Proposed for Comparison		
	Criteria Description	**Secondary Parameters**	**Required Geo-Spatial Shape file type**
1	Socio-Economic	**Community Access (Road Shape File How many Settlements will the road shape file crosses)**	Road Shape file with Settlement Shape file
		Administrative boundaries (Administrative location of the route options (Region, Zone, Woreda, Kebele), Strategic Impact	Road Shape file with Settlement Shape file
		The likely number of households to be displaced **(Resettlement)**	Settlement shape file showing settlement Density the road may passes through
		Economic structure of the area (Major economic activities such as industrial and mineral resource areas, farming, animal husbandry, trading of goods and services, etc. for each route option)	Land use and land cover shape file
		Severance which will use the Land use and land cover patterns (Cultivated land (rain fed and irrigated) of the Woredas and areas traversed by each route option; Grazing areas and pasture lands of the Woredas and areas traversed by each route option; Forests, plantations, bush and shrub lands, wetlands of the, Woredas and areas traversed by each route option; **Settlement areas**, bare land and others of the Woredas and areas traversed by each route option, if possible.)	Land use and land cover shape file as well as Settlement shape files
2	Environmental (Bio-Physical) Issues	**Impacts on water resources** (Rivers, Water Sheds and/or bodies); (quantity, quality and flow direction); Impacts on groundwater resources (pollution and flow direction), Number of water crossing structures (Bridges vs Culverts)	River and Water bodies Shape file
		Prime agricultural land affected	Land use and land cover shape file
		Vegetation (Flora), Forests; Erosion/Desertification (length affected)	Topographic Map Shape File (Forest Cover etc)
3	Engineering	**Terrain Analysis** (Geographic, Slope and Topographic Difficulty)	Digital Elevation Model (DEM) or Topography shape file ***(Contour line data shape file)*** in conjunction with each route corridors proposed for study
		Distance/Travel Time	Road shape file length Vs. Travel time to settlements shape file

PYTHON IMPLEMENTATION FOR MCA

Defining Criteria for MCA Using Python

An algorithmic Python program which is used for defining the criteria for the MCA while adopting the aforementioned criteria selected for the simulation through MCA. Representing a category (e.g., Socio-Economic, Environmental, Engineering), and the elements within each set represent the specific criteria within that category.

Mathematically, we can represent it as follows:

Let P be the set of primary criteria:

$$P= \{C1, C2, C3\}$$

where C1, C2, C3 are the categories representing Socio-Economic, Environmental, and Engineering, respectively.

Each category Ci contains a set of specific criteria:

$$C1= \{C_{1,1}, C_{1,2}, C_{1,3}, C_{1,4}, C_{1,5}\}$$

$$C2= \{C_{2,1}, C_{2,2}, C_{2,3}\}$$

$$C3= \{C_{3,1}, C_{3,2}\}$$

where $C_{1,j}$, $C_{2,}$, and $C_{3,j}$ represent the specific criteria within each category.

So, in mathematical notation:

$$P= \{\{C_{1,1}, C_{1,2}, C_{1,3}, C_{1,4}, C_{1,5}\}, \{C_{2,1}, C_{2,2}, C_{2,3}\}, \{C_{3,1}, C_{3,2}\}\}$$

where $C_{1,1}$ through $C_{3,2}$ represent the specific criteria as mentioned in the section above where the criteria for MCA is defined.

Each of the principal criteria was sub-divided into a series of secondary criteria to enable precise comparisons to be made. This resulted in a highly detailed, and project specific, breakdown which required much information gathering and, of necessity.

Assigning Weightages to Criteria

The Python programming language is employed to define the criteria for Multi-Criteria Analysis (MCA). Additionally, it is encoded to instruct the machine on the appropriate weighting to be adopted for the MCA.

Represent it as a set of sets, where each set represents a category (Socio-Economic, Environmental, Engineering), and within each category, there are set of tuples where the first element of the tuple represents the specific criterion and the second element represents its weightage.

Mathematically, we can represent it as follows:

Let *W* be the set of weightages for primary criteria:

$$W = \{W_1, W_2, W_3\}$$

where W_1, W_2, W_3 are the categories representing Socio-Economic, Environmental, and Engineering, respectively.

Each category W_i contains a set of tuples representing specific criteria and their corresponding weightages:

$$W1 = \{(C_{1,1}, w_{1,1}), (C_{1,2}, w_{1,2}), (C_{1,3}, w_{1,3}), (C_{1,4}, w_{1,4}), (C_{1,5}, w_{1,5})\}$$

$$W2 = \{(C_{2,1}, w_{2,1}), (C_{2,2}, w_{2,2}), (C_{2,3}, w_{2,3})\}$$

$$W3 = \{(C_{3,1}, w_{3,1}), (C_{3,2}, w_{3,2})\}$$

Where, $w_{i,j}$ represents the weightage associated with criterion $C_{i,j}$.

So, in mathematical notation:

$$W = \{\{(C_{1,1}, w_{1,1}), (C_{1,2}, w_{1,2}), (C_{1,3}, w_{1,3}), (C_{1,4}, w_{1,4}), (C_{1,5}, w_{1,5})\}, \{(C_{2,1}, w_{2,1}), (C_{2,2}, w_{2,2}), (C_{2,3}, w_{2,3})\}, \{(C_{3,1}, w_{3,1}), (C_{3,2}, w_{3,2})\}\}$$

where $C_{1,1}$ through $C_{3,2}$ represent the specific criteria and $w_{i,j}$ represent the corresponding weightages as mentioned in the Python dictionary.

Data Source and Specifications

Table 4 lists data sources for the MCA criteria. While consulting these sources is crucial, most data is general and requires additional methods like remote sensing or field observations to gather project-specific details

Table 4. Summary of relevant data sources utilized

I. No.	Data Type	Source	Specification/ Format	Spatial Reference System (SRS)
1	Coding Language	Python (Version 3.11.1)	Open-source libraries available	N/A
2	Python Libraries	Geo-Pandas (version 0.14.3)	Open-source available	N/A
3	Route Alignment Options Shape file	(Landsat / Copernicus) Google Earth	Shape Files (CPG, DBF, PRJ, SHX, AutoCAD Shape Source) and KMZ	Imagery from the dates 4/10/2013
4	Contour line-Intermediate	ALOS World 3D data from JAXA 30m	etnaaloscont050m.zip etna: National Dataset alos: ALOS satellite cont: Contour line 050m: Conour line 50m, interval 100m	etnaaloscontline050m
5	Raster Dataset LandCover20m2016ESAC.zip	European space agency, ESA, Copernicus	etnaesacland020m2016.tif etna: National Dataset esac: ESA Copernicus Forest land: Land cover 020m: spatial resolution 2016: Year of observation	etnaesacland020m2016.tif
6	Main River Basins SRTM	srtm: Hydro DEM	etna: National Dataset srtm: Hydro DEM based on Shuttle Radar Topography Mission main: Main River basin of Ethiopia basn: Delineation of main basins	etnasrtmmainbasn
7	Raster Dataset PopulationDensityModel2016WLRC.zip	etna: National Dataset Ethiopia wlrc: Water and Land Resources Centre	etna: National Dataset wlrc: Water and Land Resources Centre popd: Population density modx: Own model approach	etnawlrcpopdmodx.tif
8	Shapefile SettlementdsWLRC&OSM2018	www.ema.gov.et; (government agency) www.osm.org (Open Street Map contributors); www.geonames.org; www.cde.unibe.ch; www.wlrc-eth.org	etnastlmmerg2018.zip etna: National Dataset Ethiopia stlm: Settlement Layer merg: Merged layers from different sources (cross-checking of data, removement of duplicates, hierarchical structuring of data) 2018: Year of major revisions	etnasetlmerg2018
9	Regional Boundary	etna: National Dataset Ethiopia GCS_WGS_1984	Shape Files (CPG, DBF, PRJ, SHX, AutoCAD Shape Source)	Eth_Region1
10	Towns	etna: National Dataset Ethiopia GCS_WGS_1984	Shape Files (CPG, DBF, PRJ, SHX, AutoCAD Shape Source)	Et_Towns

Data Pre-Processing

Data-processing is a task on the spatial data at the deepest level of thought possible. Data preprocessing is an improvement of the raw data set that limits undesired distortions or outliers which enhances some data features applicable for further processing and analysis tasks (Fan et al., 2021).

The first step in the data pre-processing is data extraction. Some unrelated parts of the spatial data have been neglected and the focus region of interest is concentrated. Taking out the project area from the whole part of the image is important to reduce the size of the image file to include only the Area of Interest (AOI).

For instance, the shape file data for all the three alternatives has been enveloped by taking the centre line data from google earth and extracting the particular spatial points for the study corridor and filtered in KMZ filing.

This does not only eliminate the extraneous data in the surrounding geographic environment. However, it speeds up processing due to the smaller amount of data to process (Yitayew, 2020).

This chapter explores Multi-Criteria Analysis (MCA) for comparing three proposed road routes. Data preprocessing is crucial for accurate and efficient analysis. Here, we discuss preprocessing methods used for some of the data employed in the MCA process. Similar procedures were applied to other data as well.

Route Raw Data and Potential Issues

The Raw Data obtained for the centre line of each proposed road route alignment is from Google Earth. While valuable, this data might be in a format less suitable for direct analysis with GIS and MCA. It could be in a proprietary format or lack necessary geospatial information.

This is where data preprocessing using Python libraries comes in. Tools like Geo-Pandas and potentially others might be helpful depending on the initial format of the data. Here is the preprocessing steps followed for the same:

1. **Format Conversion:** the data from its original format proprietary format from Google Earth converted to a more widely used geospatial format like Keyhole Markup Language (KMZ). This conversion streamlines the analysis process within GIS.
2. **Data Cleaning:** There might be inconsistencies in the data, such as missing coordinates or errors in the alignment. Python libraries can help identify and address these inconsistencies, ensuring data accuracy.
3. **Data Transformation:** For MCA, additional information derived from the center line data has been understood. For instance, Python has been be used to calculate the total length of each proposed route in kilometers, a crucial factor in the analysis.

Once the preprocessing has been executed by utilizing Python, the data turned tolerably clean, standardized, and enriched dataset in the form of KMZ files, ready for integration with GIS. This preprocessed data offers several benefits:

1. **Efficiency:** KMZ files are a more streamlined format for GIS analysis compared to the original data format. This allows for faster processing and analysis of route alignments within the GIS platform.
2. **Accuracy:** Data cleaning ensures the information used in the MCA reflects the actual road alignments accurately.
3. **Enriched Data:** The addition of derived data points like route length enhances the overall data available for the MCA, leading to a more comprehensive analysis.

Moreover, Settlement Data Preprocessing were also conducted prior of Integrating settlement data into the Multi-Criteria Analysis (MCA) to let python consider/identify settlements near proposed routes.

Settlement Raw Data and Challenges

Having the settlement data obtained from a government agency in the form of a shapefile. Since Shapefiles are a common geospatial data format, but directly using them in Python for complex analysis can be cumbersome.

As such Data Preprocessing with Python is needed using the library Geo-Pandas excel at handling geospatial data. Here's how Python preprocessed the settlement shapefile:

1. **Loading the Data:** Python can use Geo-Pandas to read the settlement shapefile. This creates a structured data object within Python that allows for manipulation and analysis.
2. **Data Cleaning:** The shapefile might contain inconsistencies like duplicate entries or errors in geographic coordinates. Python can identify and address these issues, ensuring data accuracy.
3. **Geometric Simplification (Optional):** Depending on the complexity of the settlement data, Python libraries like Shapely can be adopted to simplify the geometry of the settlement polygons within the shapefile. This can improve processing speed and efficiency, especially if dealing with a large number of settlements.
4. **Buffer Creation:** To identify settlements "near" a proposed route, creating buffer zones around each route alignment has been established using Python

libraries. The buffer zone defines a specific distance from the route centerline within which a settlement is considered to be potentially impacted.

Data preprocessing converts complex geospatial information from shapefiles into a format Python understands. This simplifies analysis of settlements near the proposed routes. Additionally, Python helps identify potentially impacted settlements by creating buffer zones around each route, providing vital data for the MCA process.

Correlating Criteria with GIS Data in Python

When conducting Multi-Criteria Analysis (MCA) in a GIS context, it is important to correlate criteria with spatial data in the form of GIS shape files.

The overall procedure followed through Python has been depicted as follows;

Preparing GIS Geo-Spatial Data

Make sure the GIS shape files representing the spatial features (e.g., roads, towns, rivers, administrative boundaries and any other areas of interest to be considered governing) and any associated attribute data through 3rd party GIS Software (e.g. Arc GIS).

Loading GIS Shape Files

Using the geo-pandas' library to read and manipulate GIS shape files. ensuring to load both the geometry and attribute data as show in the annotation herein below;

Mathematical Annotation:

$\Gamma(\text{routes}) = \{r_0, r_1, r_2\}$, $r_i \in \mathbb{R}^2$ (n-dimensional space)

- $\Gamma(\text{routes})$: This denotes a set named routes that represents all the available route alternatives.
- $\{r_0, r_1, r_2\}$: This curly brace notation signifies that routes is a collection of elements, where each element represents a single route. Here, r_0 corresponds to "Route_1-line.shp", r_1 to "Route_2-line.shp", and r_2 to "Route_3-line.shp".
- $r_i \in \mathbb{R}^2$ (n-dimensional space): Each route element r_i is a member of the two-dimensional Euclidean space ($\mathbb{R}^2$). This indicates that each route can be represented by a series of points or coordinates in a two-dimensional plane.

Similarly, the criteria spatial data also be loaded which is represented by the following mathematical annotation for one of the criteria (i.e., community access) for the sake of illustration;

$\Sigma(\text{settlements}) = \{s_1, s_2, ..., s_n\}$, $s_i \in \mathbb{R}^2$ (n-dimensional space)

- **Σ(settlements):** This denotes a set named settlements that represents all the settlements or community access points.
- **$\{s_1, s_2, ..., s_n\}$:** This curly brace notation signifies those settlements is a collection of elements, where each element represents a single settlement/ community access point. Here, $s_1, s_2, ..., s_n$ represent individual settlements or access points.
- **$s_i \in \mathbb{R}^2$ (n-dimensional space):** Each settlement element s_i is a member of the two-dimensional Euclidean space ($\mathbb{R}^2$). This indicates that each settlement or access point can be represented by a pair of coordinates in a two-dimensional plane (latitude and longitude).

Merge Criteria Data With Route GIS Data

After preparing the criteria data, it can be further enriched by merging it with relevant GIS data. This requires a common identifier, like a location ID, to link the datasets. For example, community access can be analyzed by merging settlement data with the route shapefile.

While the code doesn't directly involve mathematical operations, it can be interpreted as a visualization of the mathematical sets defined earlier:

- **Γ(routes)** (set of route alternatives) is visualized as lines on the plot.
- **Σ(settlements)** (set of settlements/community access points) is visualized as markers on the plot.

The plot essentially combines these two sets of geospatial data into a single visual representation, aiding in analyzing the spatial relationships between routes and criteria (e.g. settlements/access points spatial data).

Spatial Visualization

Visualize the spatial data along with the criteria information using mat-plot-lib or geo-pandas plotting functionalities. Correlating respective 'criteria column' with the column from the Data Frame containing the criteria values.

This step performs a spatial join using the geo-pandas library. It joins the settlements/community access Geodata Frame with the Route_Alternatives GeoDataFrame based on a spatial operation.

- **settlements/community access**: The first input GeoDataFrame containing settlements/community access points.[3]
- **Route_Alternatives**: The second input GeoDataFrame containing route alternatives.
- **how="left"**: This argument specifies that all settlements/community access points will be included in the output, even if they don't intersect with any route.
- **op="intersects"**: This defines the spatial operation as intersection. The output will only include settlements/community access points that intersect with at least one route alternative.

The spatial join operation combines the information from two GeoDataFrames based on their geometric relationships. In this case, it creates a new GeoDataFrame that links each settlement/community access point with the route(s) it intersects.

While the join itself doesn't involve explicit mathematical notation, it can be understood as a set operation that relates the previously defined sets:

- **Σ(settlements)**: The original set of settlements/community access points.
- **Γ(routes)**: The original set of route alternatives.

Merging criteria data with GIS data (e.g., location ID) creates a richer dataset. This process, called a spatial join, combines information from both sources. Settlements not intersecting any route (using a "left" join) will still be included but with less route-related data.

- The mathematical interpretation provides a conceptual understanding of how the join relates to the previously defined sets.

One can obtain a combined visualization of the route alignment alternatives with the criteria dataset for better understanding of the analysis process as shown in the under depicted template tiff images; while adopting the code as shown here in Figure 16 below;

Figure 16. Spatial visualization of shape files together

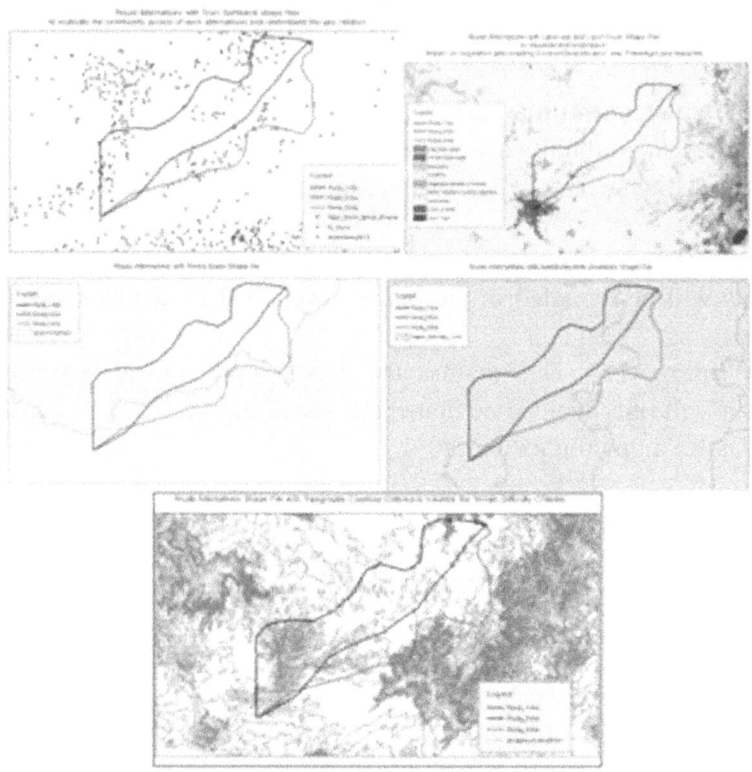

Multi-Criteria Analysis Using Python

Apply the MCA methodology to analyse the spatial features based on the criteria. one can calculate weighted scores, rankings, or any other relevant analysis. Once the code has been run it could be possible to export the merged data with the MCA results back to a new GIS shape file or other formats for further use or sharing for visualization.

Make sure to replace placeholders like 'path/to/gis_data.shp' and others with the actual paths and names in the system. Adjust the steps based on the specific data, criteria, and analysis requirements as required.

CASE STUDY OF ADDIS ABABA TO DEBRE BIRHAN ROAD PROJECT

Application of the Geospatial Approach

Road design is experiencing a significant transformation with the integration of geospatial data analysis. This approach leverages advanced technologies like Python to combine geospatial data with powerful programming capabilities. This powerful combination creates a robust framework for informed decision-making in the complex world of route selection.

At the core of the geospatial approach lies the integration of diverse geospatial datasets, encompassing satellite imagery, geographic information systems (GIS), terrain models, and road network databases. This amalgamated dataset provides an exhaustive comprehension of the Ethiopian landscape, enabling a comprehensive evaluation of factors influencing route selection. The strategic utilization of geospatial data empowers decision-makers with a visual and analytical foundation, facilitating informed choices through considerations of terrain, population density, environmental conditions, and economic variables.

The geospatial approach leverages Python's power for processing and analysing geospatial data. Python simplifies working with complex datasets and automates tasks like data manipulation, analysis, and visualization. This not only speeds up decision-making but also ensures a consistent and rigorous approach to route selection (McKinney, 2017).

The geospatial approach, combining advanced data and Python programming, offers several benefits for road design. It considers environmental, social, and engineering factors for a holistic evaluation. This data-driven approach makes decision-making more transparent, as stakeholders can understand the reasoning behind route choices. Furthermore, the geospatial approach can be adapted to different regions, promoting the development of resilient and sustainable road projects (McClain, 2022).

Analysis of Route Options

Despite that the chapter could not be expanded further showing each any every steps followed, it is reasonable only to show the generic procedure followed in the process and herein below to show how the community access has been evaluated in the code correlating with the route shape file and the same is followed for others too.

Configuration

First thing first, the configuration of Geo-Pandas Library for handling GIS shape files and Matplotlib for visualization has to be executed.

Importing

Then, importing the libraries to the workspace for recognizing the data has to be conducted.

Loading Shape Files

The first step in the geospatial analysis involves loading the relevant data as shapefiles. For instance, to evaluate community access, a socio-economic criterion, we would load the shapefiles for the road route alignments and the settlements. This allows us to analyse the level of accessibility provided by each route option to nearby communities. As such, loading the shapefiles into Geodata Frames using Geo-Pandas library as discussed earlier.

Plot Route Alignments and Settlements Together

Visualize the Route Alignment data and the respective criteria shape files as shown in the data source once they pass pre-processing stage.

Resulting in the under depicted tiff Figure 17 showing routes and criteria tag features *(i.e., settlements, town, urban, village, city, location, populated places that are indicators for community access).*

Figure 17. Python visualization of the routes and criteria shape files for first hand understanding of the measurement

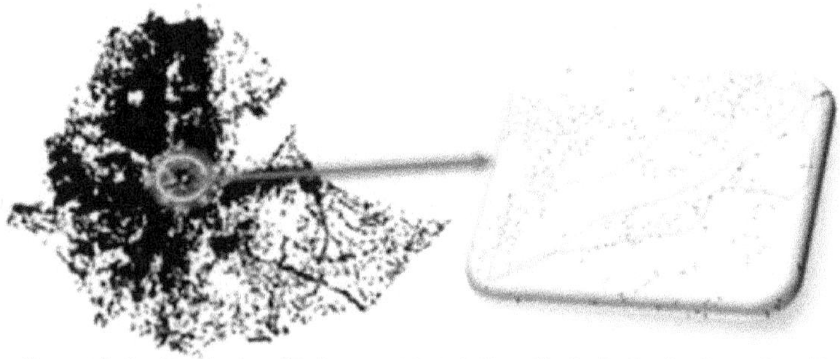

Spatially Join the Shape Files

Perform a spatial join to associate each nearby settlement points with the respective route alignments.

It will create a new Geodata Frame by combining settlement data with route information. New attributes might be added to describe intersecting routes. We repeat this process for all sub-criteria under the three main criteria (e.g., environmental, social, engineering) to prepare data for further analysis, like scoring and ranking routes. For organization, a consistent file path system and clear attribute naming conventions are essential. Finally, ensuring all data uses the same coordinate system is crucial for accurate spatial analysis.

Finally, calculating scores of the Route alterative for each criteria measurement has been computed by using the following python script. While considering the following scoring logics to be adopted in comparing the optimum route amongst the alternatives determined as shown in the following mathematical expression;

- Scoring logic for community access (assuming higher access is better).
- Assuming better connectivity is preferable for Admin Boundaries strategically.
- Scoring logic for resettlement (assuming lower impact is better).
- Scoring logic for impact on water resources (assuming lower impact is better).
- Scoring logic for impact on vegetation and deforestation (assuming lower impact is better).
- Scoring logic for travel time (assuming shorter travel times are better).
- Scoring logic for terrain analysis (assuming lower difficulty is better) gradual contour elevation change is preferred than sudden fall or rise in elevation.

Representing the functions defined for calculating scores for each criterion First represent the score calculation functions for each criterion, Here is the provided source code converted into mathematical notation:

1. For the 'Socio-Economic' criteria

Community-Access:

$$\text{community access score(access level)}= \begin{cases} 0.6 & \text{if access\_level} =' \text{High}' \\ 0.3 & \text{if access\_level} =' \text{Moderate} \\ 0 & \text{if access\_level} =' \text{Low}' \end{cases}$$

AdminBoundaries:

admin boundary score(strategical connectivity)=

$$\begin{cases} 0 & \text{if strategical\_connectivity} = \text{'Low'} \\ 0.3 & \text{if strategical\_connectivity} = \text{'Moderate} \\ 0.6 & \text{if strategical\_connectivity} = \text{'High'} \end{cases}$$

Resettlement:

$$\text{resettlement score(impact\_level)=} \begin{cases} 0.6 & \text{if impact\_level} = \text{'Low'} \\ 0.3 & \text{if impact\_level} = \text{'Moderate'} \\ 0 & \text{if impact\_level} = \text{'}H'_{igh} \end{cases}$$

2. For the ‘Environmental’ criteria:

Water Impacts:

$$\text{water\_impact\_score(impact\_level)=} \begin{cases} 0.6 & \text{if impact\_level} = \text{'Low'} \\ 0.3 & \text{if impact\_level} = \text{'Moderate} \\ 0 & \text{if impact\_level} = \text{'}H'_{igh} \end{cases}$$

PrimeAgriLand:

agricultural_impact_score(impact_level)=

$$\begin{cases} 0.6 & \text{if impact\_level} = \text{'Low'} \\ 0.3 & \text{if impact\_level} = \text{' Moderate} \\ 0 & \text{if impact\_level} = \text{'High'} \end{cases}$$

3. For the ‘Engineering’ criteria:

- Travel-Time: travel_time_score(travel_time)
- Terrain-Analysis: slope_score(slope)

This mathematical annotation represents the score calculation functions for each criterion in a more formal and concise manner.

Scoring Assumptions for each criterion is discussed in the Table 5 herein below;

Table 5. Scoring assumptions deployed

I. No.	Criteria	Assumptions
Principal Criteria: Socio-Economic		
1	Community Access	Scoring logic for community access (assuming higher access is better)
2	Admin. Boundaries	Assuming better connectivity is preferable
3	Resettlement	Scoring logic for resettlement (assuming lower impact is better)
4	Economic Structure	Ditto No. 1 and 2 (directly proportional)
5	Severance	Any route passing directly through a settlement will also create more severance (low score) than one which bypasses the settlement (high score)
Principal Criteria: Environmental		
6	Water Impacts	Scoring logic for impact on water resources (assuming lower impact is better)
7	Impact on Vegetation and resulting Erosion/Desertification	Scoring logic for impact on vegetation and deforestation (assuming lower impact is better)
8	Impact on Prime Agri. Land	
Principal Criteria: Engineering		
9	Travel Time	Scoring logic for travel time (assuming shorter travel times are better)
10	Terrain Analysis	Scoring logic for terrain analysis (assuming lower difficulty is better) gradual contour elevation change is preferred than sudden fall or rise in elevation

After applying scoring logic to each criterion, Python scripts can calculate individual criterion scores and combine them into overall scores for each route option. These scores consider predetermined weights assigned to different criteria, enabling a data-driven comparison of routes.

Applying scoring logic to create scores for each criterion we have:

- **$R = \{r_1, r_2, r_3\}$:** This set represents the three routes under consideration (Route-1, Route-2, Route-3).
- **$C = \{C_1, C_2, \ldots, Cn\}$:** This set represents all the socio-economic, environmental, and engineering criteria used for route evaluation.

Criteria Functions (f_c):

- We can assume functions f_c exist for each criterion $c \in C$. These functions take an element representing a route's data for a specific criterion and return a score based on the scoring logic defined in the original code (e.g., community_access_score).

Symbolic Representation:

Principal Criteria: Socio-Economic
$r_1 \mapsto$ [f_C$_1$(r_1), f_C$_2$(r_1), ..., f_Cn(r_1)] // Socio-economic scores for Route-1
$r_2 \mapsto$ [f_C$_1$(r_2), f_C$_2$(r_2), ..., f_Cn(r_2)] // Socio-economic scores for Route-2
$r_3 \mapsto$ [f_C$_1$(r_3), f_C$_2$(r_3), ..., f_Cn(r_3)] // Socio-economic scores for Route-3
Principal criteria: Environmental (similar structure as socio-economic)
$r_1 \mapsto$ [f_C'(r_1), f_C''(r_1), f_C'''(r_1)] // Environmental scores for Route-1
$r_2 \mapsto$ [f_C'(r_2), f_C''(r_2), f_C'''(r_2)] // Environmental scores for Route-2
$r_3 \mapsto$ [f_C'(r_3), f_C''(r_3), f_C'''(r_3)] // Environmental scores for Route-3

where, (f_C', f_C'', f_C''' represent functions for environmental criteria)

Principal criteria: Engineering (similar structure)
$r_1 \mapsto$ [f_C''''(r_1), f_C'''''(r_1)] // Engineering scores for Route-1
$r_2 \mapsto$ [f_C''''(r_2), f_C'''''(r_2)] // Engineering scores for Route-2
$r_3 \mapsto$ [f_C''''(r_3), f_C'''''(r_3)] // Engineering scores for Route-3

where, (f_C'''' and f_C''''' represent functions for engineering criteria)

The arrow ($\mapsto$) indicates a mapping between a route r_i and a list containing its scores for all criteria C. Each function f_c (or its variants for environmental and engineering criteria) represents the scoring logic for the corresponding criterion.

This approach assigns scores to each route based on various criteria using pre-defined functions. This avoids complex mathematical formulas, as scoring logic may involve conditions and qualitative assessments. Since different sub-criteria have varying importance, the overall score for each route is further broken down into its components, as shown below;

- The expression below calculates the overall score for each route (Route-1, Route-2, Route-3) using a weighted sum approach.
- Within each category (Socio-Economic, Environmental, Engineering) represented by curly braces {}, the code performs a weighted sum of individual criteria scores.
- **weight[criterion]:** This represents the weight assigned to a specific criterion within a category (e.g., weight['community_access']).

- **Route-[category]_score:** This refers to the pre-calculated score for a route on a specific category (e.g., Route-1['community_access_score']).

Using mathematical annotation;

∀ r ∈ R: // For all routes in set R
overall_score(r) = Overall score for route r
w_s * S_socio_weighted(r) = Socio-economic score (weighted)
w_e * S_env_weighted(r) = Environmental score (weighted)
w_eng * S_eng_weighted(r) = Engineering score (weighted)

where:

S_socio_weighted(r) = Weighted sum of socio-economic criteria scores
∑(w_c * S_c(r) | c ∈ C_socio)
S_env_weighted(r) = Weighted sum of environmental criteria scores
∑(w_c * S_c(r) | c ∈ C_env)
S_eng_weighted(r) = Weighted sum of engineering criteria scores
∑(w_c * S_c(r) | c ∈ C_eng)

Where,

- The overall score for each route r is a linear combination of its weighted scores for each category: Socio-Economic (S_socio_weighted(r)), Environmental (S_env_weighted(r)), and Engineering (S_eng_weighted(r)).
- The weights (w_s, w_e, w_eng) represent the relative importance of each category in the overall score.
- Within each category, the weighted score S_socio_weighted(r), S_env_weighted(r), S_eng_weighted(r)) is calculated as a sum of individual criteria scores (S_c(r)) multiplied by their corresponding weights (w_c).
- ∑ represents summation over all criteria (c) within a specific category (C_socio, C_env, C_eng).
- The overall score calculation is now a more complex weighted linear combination compared to the simple sum used previously.

Finally, Ranking Alternatives as a final remark of the multi-criteria Analysis based on the overall ranking values as;

```
ranked_options = routes.sort_values(by='overall_score',
ascending=False)
```

MCA Results and Decision-Making

The route with the highest overall score is typically chosen. However, scoring logic can differ by criteria. Some criteria Favor larger values (like distance), while others Favor smaller values (like environmental impact). Details on how each sub-criterion is scored are provided later.

Community Access

The option that provides access to the greatest number of potential road users and access to the greatest range of community services, such as education and health care, should be given the maximum score. A new rural access road would probably have the maximum score.

Administrative Boundaries Strategic Impact

Will the new route have a significant strategic impact (high score) for the province or country as a whole, for example forming part of national corridor link, or is it a side road that will benefit only a few local farmers (low score)?

Resettlement

This refers to the number of dwellings that would have to be demolished due to the road footprint and the consequent resettlement of the inhabitants. Clearly if large numbers of inhabitants will require resettlement, then this would score the minimum, and if only very small numbers are involved, then this would score the maximum.

Development Potential/Economic Structure

Will the route provide considerable development potential (high score), such as increased economic activity and urban expansion on otherwise marginal land, or does the route mainly pass through terrain that would be very difficult to develop, such as mountain or escarpment terrain (low score)?

Road Length (km)/Travel Time

This is an easily measurable item. The shortest route length should be given the maximum score, with the other routes given a lower score proportionate to their length compared to the shortest length.

Terrain

The maximum score should be given for 100% flat terrain, zero score given for 100% escarpment terrain and all other combinations scored on a pro-rata basis.

It is suggested that 100% escarpment terrain should score zero, with 100% rolling terrain scoring 66% x 4 = 2.6 and 100% mountainous terrain scoring 33% x 4 = 1.3 (ERA, 2013) which is summarized in Table 6 herein in below.

Table 6. MCA final ranking computation

ID. No.	Principal Criteria	Characteristic (Secondary Criteria)		Route-A	Score	Route-B	Score	Route-C	Score
1	**Socio-Economic**	Community Access		35.00	3.00	11.00	0.94	30.00	2.57
		Administrative Boundaries		27.00	1.45	84.00	4.52	93.00	5.00
		Resettlement		35.00	4.00	11.00	1.26	30.00	3.43
		Economic Structure		35.00	3.00	11.00	0.94	30.00	2.57
		Severance		10.00	0.40	2.00	2.00	6.00	0.67
2	**Environmental**	Impacts on water resources		1.00	6.00	3.00	2.00	5.00	1.20
		Prime agricultural land affected		56.17	3.00	67.44	2.50	28.09	6.00
		Vegetation (Flora), Forests; Erosion/ Desertification (length affected)		28.08	1.36	14.05	2.73	6.38	6.00
3	**Engineering**	Terrain Analysis (Geographic, Slope and Topographic Difficulty)	Flat (%)	40.00%	1.60	50.00%	2.00	20.00%	0.80
			Rolling (%)	16.67%	0.44	33.33%	0.88	14.29%	0.38
			Mountainous (%)	23.34%	0.31	14.29%	0.74	33.33%	1.73
			Escarpment (%)	20.00%	0.00	2.38%	0.00	32.38%	0.00
			Aggregate Score	100.00%	2.35	1.00	3.62	1.00	2.91
		Distance/Travel Time		140.42	4.80	112.40	6.00	140.45	4.80
Over All Score				**29.37**		**26.51**		**35.15**	
Percentage out of total weight allotted (45)				**65.26%**		**58.90%**		**78.11%**	
Rank				**2**		**3**		**1**	

Before interpreting the MCA result the following points should be noted:

- Road Length (km).

The shortest route length (i.e. Route B) should be given the maximum score of 6.0. The longest route length (i.e. Route C) will have a score of 6 x 112.4/140.45 = 4.8, with Route B at 6 x 112.4/140.42 = 4.8.

- Terrain.

This scores 4.0 maximum for 100% flat terrain. Considering the recommended pro-rata conversion stated above in ERA Route selection Manual 2013.

Therefore, for Route A, 40% flat terrain would have a score of 40% x 100% x 4 = 1.6, 16.67% rolling terrain would score 16.67% x 66% x 4 = 0.44 and so on.

In this case, Route C, the option with the longest length, is the preferred option scoring the most points of the Multi-Criteria-Analysis.

SUMMERY

This chapter positions geospatial data as a valuable guide for route selection. Similar to a skilled navigator considering topographies, potential impediments, and the final destination, geospatial data offers insights into land use, terrain features, and environmental factors.

The use of geospatial data mirrors careful route planning. same as skilled navigator considers the terrain, so too does this approach factor in geographic realities. This ensures the chosen route aligns with project goals and promotes the development of efficient, sustainable road networks.

Geospatial data analysis is enhanced by Python's powerful geospatial libraries like Geo-Pandas, Shapely, and Fiona. These tools, combined with Python's versatility, enable efficient data handling and complex analysis for road design. By integrating geospatial data with these libraries, we can create a unified framework that considers various criteria like terrain, environment, and socioeconomics for informed route selection.

Python's geospatial libraries offer specific functionalities. For example, some help manipulate data (Geo-Pandas) while others enable complex spatial analysis (Shapely). Additionally, compatibility and data exchange are ensured through tools for reading and writing geospatial files (Fiona). This combination empowers engineers to consider various factors (terrain, environment, socioeconomics) through a unified framework for optimal route selection.

Python's efficiency in processing geospatial data, coupled with the specialized features of its geospatial libraries, facilitates the creation of robust models that contribute to informed decision-making in the road design stage. As such, the symbiotic application of geospatial data and Python programming, leveraging the specialized features of Geo-Pandas, Shapely, and Fiona, in road design exemplifies a nuanced and technologically advanced approach, ensuring the alignment of infrastructure development with diverse criteria and sustainability imperatives.

Finally, a dataset for three different Route Alignments from Addis Ababa to Debre-Berhan Towns has been analysed to evaluate multi-criteria features of each rout alternatives and to rank their optimality using the combined weightage technique.

The multi-criteria analysis identified Route "C" as the preferred option for further development. This route, despite having the longest distance, achieved the highest overall score due to favourable scores in other criteria, like balanced terrain characteristics.

Route C emerged from the multi-criteria analysis as the best option for further development, even though it has the longest distance. This is because its score (highest overall score) closely aligns with the weight assigned to the overall assessment (45%). By considering various factors through Python's geospatial tools (Geo-Pandas, Shapely, Fiona), this analysis provides a nuanced understanding for informed decision-making. Therefore, Route C is a strong candidate for further design and construction.

While Multi-Criteria Analysis (MCA) is a proven decision-making tool, its application in road design with a vast array of social, environmental, and potentially financial factors (data permitting) is innovative. This approach transcends mere engineering feasibility, enabling a holistic evaluation of route options.

Similarly, Python is a widely used programming language. However, its integration with geospatial technologies like GIS for spatial analysis in road design is a key innovation. This allows for:

- Data-driven decision making: Python facilitates the use of real-world data on wildlife movement patterns, environmental conditions, and social demographics. This data can then be incorporated into MCA frameworks to create robust and evidence-based solutions.
- Optimization of route selection: Python scripts can be used to analyse vast amounts of geospatial data, identifying optimal road alignments that minimize habitat fragmentation, social disruption, and construction costs.
- Improved design methodologies: Python coding can be used to assist a decision all around by simulating the ecological impact of different road route alternatives which would be a drive for innovative design methods for necessary mitigation if the route is inevitable selected.

Therefore, the novelty lies not in the individual tools (MCA and Python) themselves, but in their combined and innovative application within the context of sustainable road design. This integration allows for data-driven decision making, optimization of route selection, and the exploration of advanced design methodologies – all crucial for achieving a balance between human mobility, social and ecological well-being.

REFERENCES

Althoey, F., Ansari, W. S., Sufian, M., & Deifalla, A. F. (2023). Advancements in low-carbon concrete as a construction material for the sustainable built environment. *Developments in the Built Environment, 16*. https://doi.org/https://doi.org/10.1016/j.dibe.2023.100284

Ammapa, J., Visuttiporn, P., Klaylee, J., Chayphong, S., & Iamtrakul, P. (2022). *Using GIS-Based Spatial Analysis: Comparing Pattern of Urbanization and Transportation Networks*. IEEE. doi:10.1109/ICTLE55577.2022.9902095

Asfoor, H., & Halit, P. (2017). *Using Life Cycle Cost Analysis in Pavement Design As a Decision Support Tool.*

Balkaya, C., & Tutumluer, E. (2021). *AI and Big Data in Transportation Infrastructure for Condition and Risk Assessment.*

Brucker, K., & Macharis, C. (2011). Multi-criteria analysis in transport project evaluation: An institutional approach. *European Transport \. Trasporti Europei, 47*, 3–24.

David, A., & Zeiler, M. (2004). *Designing Geodatabases: Case Studies in GIS Data Modeling*.

ERA. (2013). Route Selection Manual. Addis Ababa Ethiopia: Ethiopian Roads Authority (ERA).

Fan, C., Chen, M., Wang, X., Wang, J., & Huang, B. (2021). A Review on Data Preprocessing Techniques Toward Efficient and Reliable Knowledge Discovery From Building Operational Data. *Frontiers in Energy Research, 9*, 652801. doi:10.3389/fenrg.2021.652801

Ghana: Associated Consulting Gipps, P., Gu, K., Held, A., & Barnett, G. (2001). New technologies for transport route selection. *Transportation Research Part C: Emerging Technologies, 9*, 135-154. doi:10.1016/S0968-090X(00)00040-1

Ghanian-Manual. (2009). *Integrated Transport Plan – Ghana Multi-Criteria Evaluation Manual.*

González Del Campo, A., & Gazzola, P. (2020). Untapping the potential of technological advancements in Strategic Environmental Assessment. *Journal of Environmental Planning and Management, 63*(4), 585–603. doi:10.1080/09640568.2019.1588712

Grinberg, M. (2018). *Flask Web Development: Developing Web Applications with Python*. O'Reilly. https://books.google.com.et/books?id=cV34swEACAAJ

Heldt, B. W. T. (2022). *Exploring the use of public participation GIS in transportation planning for tourism at a Nordic destination*.

Henning Theuns. F. P. D. R. (2017). *Integrating climate change into road asset management*. World Bank. https://documents.worldbank.org/en/publication/documents-reports/documentdetail/981831493278252684/technical-report-2017

IENE. (2022). *The Problem*. Handbook wildlife traffic. https://handbookwildlifetraffic.info/1-introduction/the-problem/

Jameel, A. a. A., S. (2017). A comprehensive review of route selection methods for transportation projects. *Jordan Journal of Civil Engineering*, *11*(2).

Jani, H. J. (2018). SOCIAL IMPACT ASSESSMENT OF ROAD INFRASTRUCTURE PROJECTS. *Global Journal of Commerce, 7*. https://www.longdom.org/articles-pdfs/social-impact-assessment-of-road-infrastructure-projects.pdf

Lawhead, J. (2013). *Learning Geospatial Analysis with Python*. Packt Publishing. https://books.google.com.et/books?id=UR-8AQAAQBAJ

Li, S., Dragicevic, S., Anton, F., Sester, M., Winter, S., Coltekin, A., Pettit, C., Jiang, B., Haworth, J., Stein, A., & Cheng, T. (2015). Geospatial Big Data Handling Theory and Methods: A Review and Research Challenges. *ISPRS Journal of Photogrammetry and Remote Sensing*, *115*, 119–133. doi:10.1016/j.isprsjprs.2015.10.012

Li, S., Dragicevic, S., Castro, F. A., Sester, M., Winter, S., Coltekin, A., Pettit, C., Jiang, B., Haworth, J., Stein, A., & Cheng, T. (2018). Geospatial Big Data Handling Theory and Methods: A Review and Research Challenges. *ISPRS Journal of Photogrammetry and Remote Sensing*, *115*, 119–133. doi:10.1016/j.isprsjprs.2015.10.012

Lillesand, T., Kiefer, R. W., & Chipman, J. (2015). *Remote Sensing and Image Interpretation* (7th ed.). Wiley. https://books.google.com.et/books?id=eQXYBgAAQBAJ

McClain, B. P. (2022). *Python for Geospatial Data Analysis*. O'Reilly Media, Inc. https://www.everand.com/book/601850266/Python-for-Geospatial-Data-Analysis

McKinney, W. (2017). *Python for Data Analysis: Data Wrangling with Pandas, NumPy, and IPython*. O'Reilly Media, Incorporated. https://books.google.com.et/books?id=2BYfvgAACAAJ

Moon, W., & Longley, P. A. M. F., Goodchild, D. J., Maguire and D. W. Rhind. (2015). Geographic Information Systems and Science (3rd Edition) (Book Review). The Leading Edge (Society of Exploration Geophysicists), 31, 975-976.

Navarro, I. J., Yepes, V., & Martí, J. V. (2019). A Review of Multicriteria Assessment Techniques Applied to Sustainable Infrastructure Design. *Advances in Civil Engineering, 6134803*, 1–16. Advance online publication. doi:10.1155/2019/6134803

Raymond E. (2019). *Public–Private Partnerships for Infrastructure Development.*

Serrat, O. (2017). The Sustainable Livelihoods Approach. In O. Serrat (Ed.), *Knowledge Solutions: Tools, Methods, and Approaches to Drive Organizational Performance* (pp. 21–26). Springer Singapore. doi:10.1007/978-981-10-0983-9_5

Syamsunur, D., Ismail, A., Rahmat, R., & Karim, A, O. (2011). Knowledge-based expert system for route selection of road alignment. *Australian Journal of Basic and Applied Sciences*, *5*, 208–213.

Tischler, S. (2017). Finding the right way - a new approach for route selection procedures? *Transportation Research Procedia*, *25*, 2813–2827. doi:10.1016/j.trpro.2017.05.247

Verbytskyi, Y. (2023). DELIVERY ROUTES OPTIMIZATION USING MACHINE LEARNING ALGORITHMS. *Eastern Europe: economy, business and management.* https://doi.org/ doi:10.32782/easterneurope.38-14

Wang, L. (2014). Framework for Evaluating Sustainability of Transport System in Megalopolis and its Application. *IERI Procedia*, *9*, 110–116. doi:10.1016/j.ieri.2014.09.049

Xiao, H., & Hao, S. (2023). Public participation in infrastructure projects: An integrative review and prospects for the future research. *Engineering, Construction, and Architectural Management*, *30*(2), 456–477. doi:10.1108/ECAM-06-2021-0495

Yitayew, H. (2020). *Gis Based New Route Alignment Selection Between Dabat And Keraker Towns* [Masters, Addis Ababa University Addis Ababa Institute Of Technology School Of Civil And Environmental Engineering]. Addis Ababa. https://etd.aau.edu.et/items/47c6ce9f-40dc-497f-866e-e10bd0488830

ENDNOTES

1. Schedule I projects require an Environmental Impact Assessment (EIA) to be undertaken automatically while Schedule II projects require an Initial

Environmental Impact Assessment (IEIA) prior to deciding whether an EIA is required.

2 *Note that this is a well-established source to give an indication of how MCA works for Road Infrastructure available. As already stated, the precise format, both in criteria and weighting, varies considerably between applications.*

3 Same for other criteria spatial data

Chapter 9
Python Library for Road Network Analysis in the Case of Debre Berhan City

Tesfaye Fufa Gedefa
Space Science and Geospatial Institute, Ethiopia

Tsegaye D. Lemma
Space Science and Geospatial Institute, Ethiopia

Wondwossen Mindahun Eshetu
https://orcid.org/0000-0002-0017-4487
Space Science and Geospatial Institute, Ethiopia

Mohammad Gouse Galety
https://orcid.org/0000-0003-1666-2001
Samarkand International University of Technology, Uzbekistan

ABSTRACT

This study employs the Python library for road network analysis in the Debre Berhan Metropolitan City, focusing on determining the best route, shortest path, road types, length in km, and speed limit in km/h. Geopandas is utilized to visualize the road network data, enabling the calculation of the shortest path and extraction of road type and speed limit information. Network analysis libraries like,folium, NetworkX convert the data into a graph representation, using algorithms like Dijkstra's or A to find the shortest path based on distance or travel time. Geopandas then overlays road segments on a map, highlighting the shortest path and indicating road type, length, and speed limit, offering a robust framework for assessing and optimizing the road network in urban areas and contributing to the effective management and enhancement of transportation infrastructure.*

DOI: 10.4018/979-8-3693-1754-9.ch009

INTRODUCTION

Python is a modern, high-level language suitable for various programming tasks, whereas Geospatial refers to information located on the Earth's surface using coordinates. This can include, for example, the position of a cell phone tower, the shape of a road, or the outline of a country (Westra, 2010). Transportation networks are lifelines that support services essential to society (Gecchele et al., 2019). People everywhere need to get somewhere else and depend on automobiles, buses, trains, subways, or airplanes to reach a final destination. Most people would like to arrive at a destination in the shortest amount of time, with the fewest distances, in a safe way, and at the lowest cost with the shortest path to save time and money (Echols, 2003). Shortest Path problems are one of the most popular problems encountered in network analysis, which consists of nodes associated with each other through edges and used in decision-making studies and on which algorithms have been developed (Şirin & Erener, 2024). These transportation issues concern not only individuals but also businesses and governments. Routes and networks are interconnected features that are used for transportation and include highways, railways, city streets, rivers, transportation routes (transit, school buses, garbage collection), and utility systems (electricity, telephone, water, sewage) (Echols, 2003).

Networks are an essential part of our everyday lives, and analyzing these networks improves the movement of people, goods, services, and the flow of resources (Jamil Awel,2006). Python extends Python with robust mathematical operations that provide practical guarantees in calculations with arrays and matrices (Gedefa, T. F., Gouse, G. M., & Iticha, G. T., 2023). Network analysis examines the properties of natural and artificial networks to understand the behavior of flows within and around such networks and locational analysis (Team et al., n.d.). Network analysis in GIS is often related to finding solutions to transportation problems. One primary application of network analysis is found in transportation planning, where the issue might be to find paths corresponding to specific criteria, like finding the shortest or least cost path between two or more locations or finding all locations within a given travel cost from a specified origin (Husdal, 1999).

The Network Analyst extension for digital addressing solves problems with network traffic on streets, rivers, railroads, pipes, or any interconnected set of lines. It can find the shortest or fastest route between your origin and destination, including all the stops. It can tell whether one place is linked to another hundreds of miles away. It finds the closest facilities to a location, such as the closest hospital to a vehicle accident, and identifies service areas, such as those serviced by a bank. A network analyst can also build spatial traffic flow models (Alvi, 2012).

The primary programming language used in this data analysis is Python. Geographic Information System (GIS) mainly produces the desired input data before running Python. In the GIS environment, socio-economic data are combined with their respective spatial data, and computations of time, distance, and velocity parameters are done for further analysis in Python. Google Earth has also been employed as a supplementary data preparation tool as a reference to identify natural and artificial features like streams and unplanned parts of frequently used paths of the road network, especially in rural parts of the study area on the surface topography.

ROAD NETWORK ANALYSIS

The topology of spatial networks represents the spatial relationships, including the shape and structural configuration of the network (Dumedah & Garsonu, 2021). According to Dumedah & Garsonu, topological measures assess the network's configuration, connectedness, and robustness – and how these characteristics are distributed. The main objective of network analysis is to Find optimal school locations using spatial data (Wondwossen Mindahun & Bedasa Asefa, 2019).

Geopandas and Python library are invaluable for conducting Road Network Analysis, especially in urban areas such as Debre Berhan City's Baso Sub-City. Analysts can use Geopandas and Python to effectively assess and optimize the road network within the sub-city, considering factors such as road type, connectivity, and traffic flow. Geopandas enables the manipulation and analysis of geospatial data, making it ideal for visualizing and processing geographic information, while Python libraries such as NetworkX provide functionalities for analyzing complex networks, including road networks. In the case of Basona Werana Wereda, these tools can be utilized to evaluate the efficiency, safety, and functionality of the road network, identify areas for improvement, and make informed decisions about finding the shortest path, distance, and speed limit based on the road types for the domain experts and users.

The road network analysis can incorporate finding the shortest path, considering the distance in kilometers and the speed limit based on the road types within and outside Debre Berhan. Geopandas and Python libraries can calculate the shortest path between different locations within the city and beyond, considering the specific road types and their associated speed limits. For instance, the analysis can consider the varying speed limits based on road types such as asphalt concrete, footpaths, gravel, cobblestone, and paved roads, both within the city and in the surrounding areas. By integrating geospatial data and road network analysis, it becomes possible to assess the most efficient routes, considering the distance in kilometers and the speed limits associated with the different road types, thereby contributing to informed decision-making for the domain experts and users regarding road network optimization and management.

STUDY AREA

Debre Berhan City, the home of the North Shewa Zone Administration, is one of the cities located in the southernmost part of the North Shewa Zone in the Amhara Regional State Government. According to Ethiopia's Central Statistics Agency (CSA) population projection for 2017, the area's total population is 254,079, of which 115,515 people live in urbanized areas and 138,264 are rural. The Wereda covers a total land area of 1315.44 sq km (131544.40 ha), from which the rural portion accounts for 88.88% 90% (1169.17 sq km), and the remaining 146.27 sq km is for the Town of Debre Berhan (Wondwossen et al., 2022). The road data was collected by professionals and digitized for road network analysis, as shown in Figure 1.

Figure 1. Debre berhan city road network analysis study area

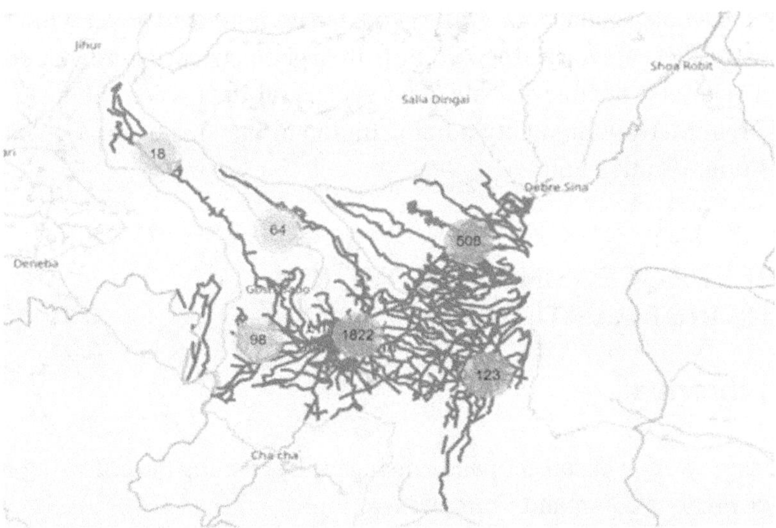

METHODOLOGY

Original Raw Data Acquisition

The starting phase of finding the primary and secondary data collection was having a close linkage with institutions. Secondary data gathering from the organizations was preferable to reduce the time, energy and resource wastage due to physical surveying data collection. Rural Wereda coordinates data collection from the Wereda Land Management and Supervision Bureau located at Debre Berhan Town.

The Precondition for Python Data Processing

ArcGIS is found to be suitable software for conducting some data entry, editing, manipulation, and mapping input data. ESRI developed ArcGIS and helped determine network problems like building network topology and datasets. It also helps to compute the socio-economic information for each network element, such as length, time taken, and speed, which are vital for the network analysis. Then, the network analyst extension helps to derive the quickest route.

The Preconditions for the core data analysis, where correction of data inconsistencies, anomalies, and absence of data are undertaken through modifications, deletions, substitutions, etc., were made sincerely during this data preprocessing. Collected spatial data from different offices and ground surveying have different formats and coordinate reference systems, and thus coordinate projections and transformations, digitization, and attribute entry are also done to bunch the reference frames of the whole spatial data. A file geodatabase is created under which different datasets are further categorized according to their geometrical nature (feature datasets and raster datasets) details of Data. The geospatial data were collected from the Debre Berhan Metropolitan City administration at the size of the Basona Werana Woreda Administrative unit.

DATA PREPROCESSING AND PYTHON LIBRARY INSTALLATION

Install Libraries

Geopandas**: we can install geopandas using pip by running the following command in your terminal or command prompt:

```
pip install geopandas
```

Matplotlib

Matplotlib is a widely used plotting library in Python and is often installed by default with other scientific packages. If not, you can install it using pip install matplotlib

Data Preprocessing

Data Acquisition

Obtain road network data in a suitable format, such as shapefiles or GeoJSON files.

Data Loading

Use geopandas to load the road network data into a GeoDataFrame:

```
import geopandas as gpdimport matplotlib.pyplot as pltr_data =
gp.read_file('RaodNet.shp')# Read the shapefiler_data
```

As the code provided reads a shapefile using geopandas, the resulting data

Figure 2. Debre Berhan City data visualization640

	OBJECTID	Name	Shape_Leng	Length_KM_	Shape_Le_1	Speed_Km_H	Time_Hr	Time_Minit	geometry
0	2	Baso footpath	1593.668314	1.593668	1593.668314	6.39	0.249400	14.964022	LINESTRING Z (563296.766 1073794.653 0.000, 56...
1	3	Baso footpath	1434.524569	1.434525	1434.524569	6.39	0.224495	13.469714	LINESTRING Z (563825.567 1072796.109 0.000, 56...
2	4	Baso footpath	1577.727754	1.577728	1577.727754	6.39	0.246906	14.814345	LINESTRING Z (565183.397 1072792.862 0.000, 56...
3	5	Baso footpath	475.969571	0.475970	475.969571	6.39	0.074487	4.469198	LINESTRING Z (567067.824 1075200.338 0.000, 56...
4	6	Baso footpath	1556.313054	1.556313	1556.313054	6.39	0.243554	14.613268	LINESTRING Z (569099.766 1073261.602 0.000, 56...
...	...	...	...	...	...	...	...	...	...
2629	2638	Baso Gravel	684.177022	0.684177	684.177022	47.00	0.014557	0.873417	LINESTRING Z (580553.343 1086376.029 0.000, 58...
2630	2639	Baso footpath	1760.485149	1.760485	834.627470	6.39	0.275506	16.530377	LINESTRING Z (579964.826 1086592.855 0.000, 57...
2631	2640	Baso Gravel	3706.406221	3.706406	228.553152	47.00	0.078860	4.731582	LINESTRING Z (579614.440 1084965.709 0.000, 57...
2632	2641	Baso footpath	4414.950265	4.414950	4414.950265	6.39	0.690916	41.454932	LINESTRING Z (556610.451 1072374.933 0.000, 55...
2633	2642	Baso Gravel	4784.753553	4.784754	1519.887418	47.00	0.101803	6.108196	LINESTRING Z (551930.675 1077819.421 0.000, 55...

2634 rows × 9 columns

can be visualized as a table with 2634 rows and 9 columns (Figure 2). Each row represents a geographic feature likely related to road networks, and each column represents a specific attribute or characteristic of the features. The table could display information such as road names, coordinates, road type, length, and other relevant data. Visualizing the data as a table allows for a structured and organized view of the geographic information, enabling further analysis and understanding of the road network dataset.

Data Exploration

```
print(r_data.head()# Display the first few rows of the data
```

The code `*print(r_data.head())*` is used to explore the loaded data and understand its structure and attributes. When executed, it will display the first few rows of the loaded data in the variable `r_data.` This is a common practice in data exploration to get a quick overview of the dataset, including the column names and the first few rows of data. By examining this output, one can gain insights into the data structure, the types of information it contains, and the format in which it is stored. This initial exploration is crucial for understanding the dataset and is often a starting point for further analysis and visualization.

Figure 3. Data type and size visualization in road network analysis

```
<class 'geopandas.geodataframe.GeoDataFrame'>
RangeIndex: 2634 entries, 0 to 2633
Data columns (total 9 columns):
 #   Column       Non-Null Count  Dtype
---  ------       --------------  -----
 0   OBJECTID     2634 non-null   int64
 1   Name         2634 non-null   object
 2   Shape_Leng   2634 non-null   float64
 3   Length_KM_   2634 non-null   float64
 4   Shape_Le_1   2634 non-null   float64
 5   Speed_Km_H   2634 non-null   float64
 6   Time_Hr      2634 non-null   float64
 7   Time_Minit   2634 non-null   float64
 8   geometry     2633 non-null   geometry
dtypes: float64(6), geometry(1), int64(1), object(1)
memory usage: 185.3+ KB
None
```

It obtains concise summary information about the loaded data in the variable r_data. When executed, it will display a summary that includes the total number of entries (rows), the number of columns, the data types of each column, and the count of non-null values. This summary provides valuable information about the dataset, such as the presence of missing values and the data types of the columns, which is essential for understanding the structure and quality of the data. This type of summary is often used in data analysis to understand the dataset's characteristics (Figure 3) quickly. The first five rows of Road Network Data displayed are given in Figure 4.

Figure 4. First five rows of road network data displaying

```
r_data.head()
✓ 0.0s
```

	OBJECTID	Name	Shape_Leng	Length_KM_	Shape_Le_1	Speed_Km_H	Time_Hr	Time_Minit	geometry
0	2	Baso footpath	1593.668314	1.593668	1593.668314	6.39	0.249400	14.964022	LINESTRING Z (563296.766 1073794.653 0.000, 56...
1	3	Baso footpath	1434.524569	1.434525	1434.524569	6.39	0.224495	13.469714	LINESTRING Z (563825.567 1072796.109 0.000, 56...
2	4	Baso footpath	1577.727754	1.577728	1577.727754	6.39	0.246906	14.814345	LINESTRING Z (565183.397 1072792.862 0.000, 56...
3	5	Baso footpath	475.969571	0.475970	475.969571	6.39	0.074487	4.469198	LINESTRING Z (567067.824 1075200.338 0.000, 56...
4	6	Baso footpath	1556.313054	1.556313	1556.313054	6.39	0.243554	14.613268	LINESTRING Z (569099.766 1073261.602 0.000, 56...

Data Cleaning

Perform necessary data cleaning steps such as handling missing values, removing duplicates, or converting data types.

```
r_data.isnull().any()
```

The above code checks for null values in each column of the r_data variable and returns a boolean value for each column, where True indicates null values in that column and False indicates no null values. In this case, the result indicates that the geometry column contains at least one null value (Figure 5).

Figure 5. Checking null value columns

```
OBJECTID       False
Name           False
Shape_Leng     False
Length_KM_     False
Shape_Le_1     False
Speed_Km_H     False
Time_Hr        False
Time_Minit     False
geometry        True
dtype: bool
```

```
# Drop any rows with NaN values in the geometryr_data = r_
data[r_data.geometry.notna()]
```

This code takes the variable r_data and filters it only to include rows where the geometry column is not null. It uses the .notna() method to check for non-null values in the geometry column and then applies this filter to the r_data variable. The result is a new version of r_data that only contains rows with non-null geometry values (Figure 6).

Figure 6. Ignoring null values from the road network data

	OBJECTID	Name	Shape_Leng	Length_KM_	Shape_Le_1	Speed_Km_H	Time_Hr	Time_Minit	geometry
0	2	Baso footpath	1593.668314	1.593668	1593.668314	6.39	0.249400	14.964022	LINESTRING Z (563296.766 1073794.653 0.000, 56...
1	3	Baso footpath	1434.524569	1.434525	1434.524569	6.39	0.224495	13.469714	LINESTRING Z (563825.567 1072796.109 0.000, 56...
2	4	Baso footpath	1577.727754	1.577728	1577.727754	6.39	0.246906	14.814345	LINESTRING Z (565183.397 1072792.862 0.000, 56...
3	5	Baso footpath	475.969571	0.475970	475.969571	6.39	0.074487	4.469198	LINESTRING Z (567067.824 1075200.338 0.000, 56...
4	6	Baso footpath	1556.313054	1.556313	1556.313054	6.39	0.243554	14.613268	LINESTRING Z (569099.766 1073261.602 0.000, 56...
...	...	...	...	...	...	...	...	...	...
2629	2638	Baso Gravel	684.177022	0.684177	684.177022	47.00	0.014557	0.873417	LINESTRING Z (580553.343 1086376.029 0.000, 58...
2630	2639	Baso footpath	1760.485149	1.760485	834.627470	6.39	0.275506	16.530377	LINESTRING Z (579964.826 1086592.855 0.000, 57...
2631	2640	Baso Gravel	3706.406221	3.706406	228.553152	47.00	0.078860	4.731582	LINESTRING Z (579614.440 1084965.709 0.000, 57...
2632	2641	Baso footpath	4414.950265	4.414950	4414.950265	6.39	0.690916	41.454932	LINESTRING Z (556610.451 1072374.933 0.000, 55...
2633	2642	Baso Gravel	4784.753553	4.784754	1519.887418	47.00	0.101803	6.108196	LINESTRING Z (551930.675 1077819.421 0.000, 55...

2633 rows × 9 columns

Data Visualization

To visualize the road network data and perform initial exploratory data analysis, we can use the matplotlib and geopandas libraries in Python. First, we read the road network shapefile using GeoPandas, which allows us to handle geographic data efficiently. Then, we can plot the road network using Matplotlib's pyplot module to represent the roads visually. This helps us gain insights into the overall structure of the road network. Additionally, we can extract various attributes from the shapefile, such as road lengths or types, and analyze those using descriptive statistics or visualizations like histograms or box plots. This initial exploratory data analysis provides a foundation for further investigation and understanding the road network dataset.

```
import geopandas as gpdimport matplotlib.pyplot as pltr_data.
plot()plt.show()
```

Figure 7. Debre berhan city road network topology

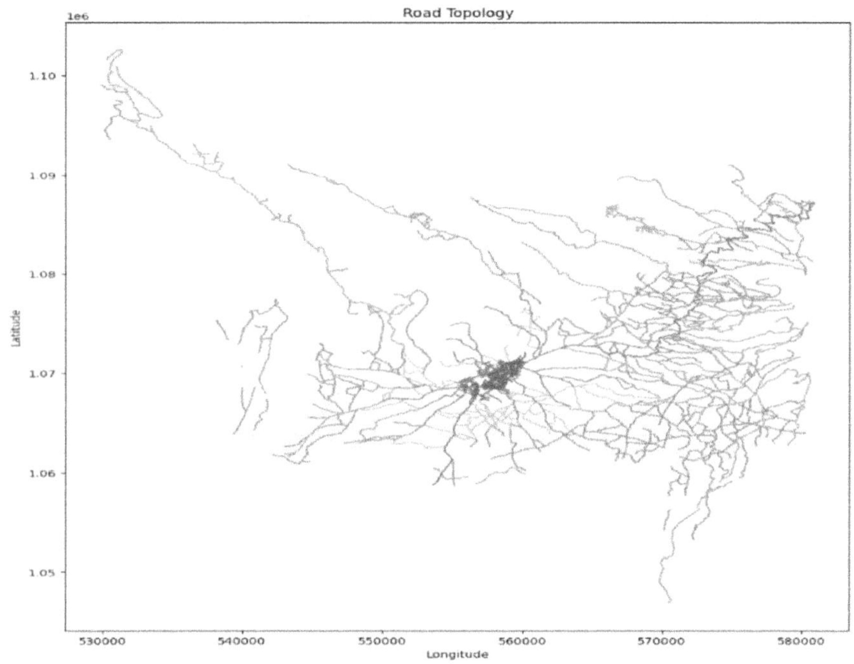

The provided code utilizes the geopandas and matplotlib libraries to visualize the road network data from a shapefile. The code first reads the shapefile using `gpd.read_file()` and stores it in the GeoDataFrame `r_data`. Then, a base map is created using `plt.subplots()`, with a specified figure size. The road network data is plotted on the map using `r_data.plot()`, with specified linewidth and color map parameters. The plot's title is set as 'Road Topology' using `plt.title()`. Finally, the plot is displayed using `plt.show()`. The output of this code will be a visualization of the road network topology, where each road segment is displayed based on its geometry and attributes present in the shapefile. Debre Berhan City road network topology is given in Figure 7.

```
import matplotlib.pyplot as pltfig, ax = plt.
subplots(figsize=(10, 10)) # set the figure sizer_data.
plot(column='Name', legend=True, ax=ax)plt.xlabel('')plt.show()
```

Figure 8. Debre berhan city road types

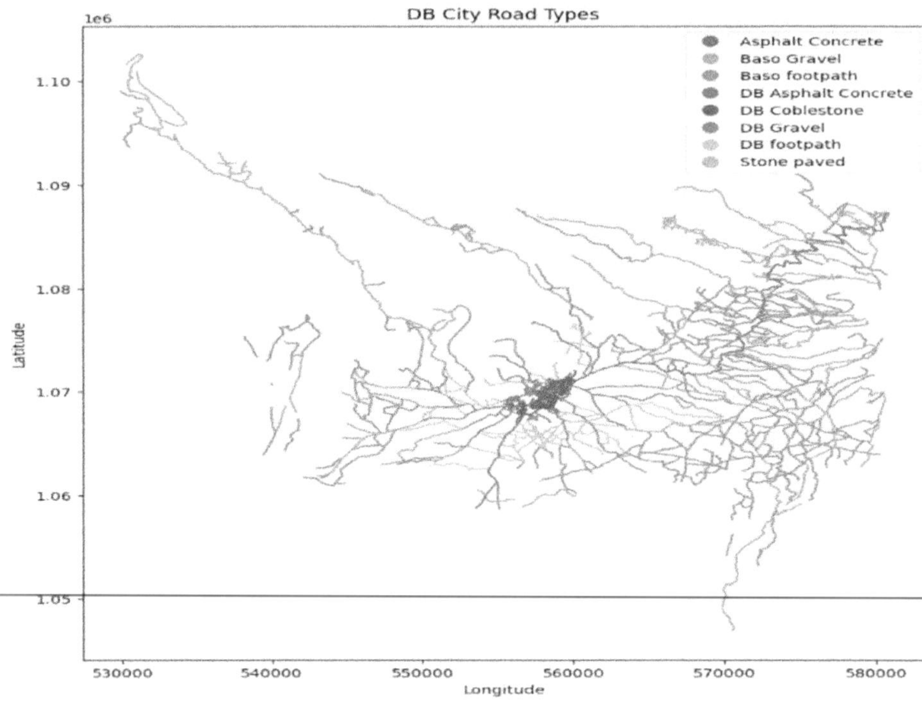

The provided code uses the matplotlib library to plot the road network data from the GeoDataFrame `r_data`. It creates a figure and axis object with a specified size using `plt.subplots(figsize=(10, 10))`. The road network data is plotted on the map with different colors for each road segment based on the 'Name' column using `r_data.plot(column='Name', legend=True, ax=ax)`. The legend is displayed to indicate the color-coding of road names. The x-axis label is an empty string using `plt.xlabel('')` to remove any label. Finally, the plot is shown using `plt.show()`. This code generates a visual representation of the road network data, where each road segment is colored based on its name, allowing for easy identification and analysis of different roads in the dataset, as shown below in figure 8.

```
import matplotlib.pyplot as pltfig, ax = plt.
subplots(figsize=(10, 10)) # set the figure sizer_data.
plot(column='Length_KM_', legend=True, ax=ax)plt.xlabel('x')#
r_data.plot(column='Speed_Km_H', legend=True, ax=ax)# r_
data.plot(column='Shape_Leng', legend=True, ax=ax)# r_
data.plot(column='Time_Hr', legend=True, ax=ax)# r_data.
```

```
plot(column='Time_Minit', legend=True, ax=ax)plt.xlabel('')plt.
show()
```

The above code uses the matplotlib library to plot the road network data from the GeoDataFrame `r_data`. It creates a figure and axis object with a specified size using `plt.subplots(figsize=(10, 10))`. The road network data is plotted on the map with different colors for each road segment based on the 'Length_KM_' column using `r_data.plot(column='Length_KM_', legend=True, ax=ax)`. The legend is displayed to indicate the color-coding of road lengths. The x-axis label is set as 'x' using `plt.xlabel('x')` to provide a label for the x-axis. Finally, an empty string is used for another x-axis label (`plt.xlabel('')`) to remove additional labels. The resulting plot provides a visual representation of the road network data, where each road segment is colored based on its length and can be analyzed in terms of varying lengths across different areas (Figure 9).

```
import matplotlib.pyplot as pltfig, ax = plt.
```

Figure 9. Debre Berhan City road network length in km

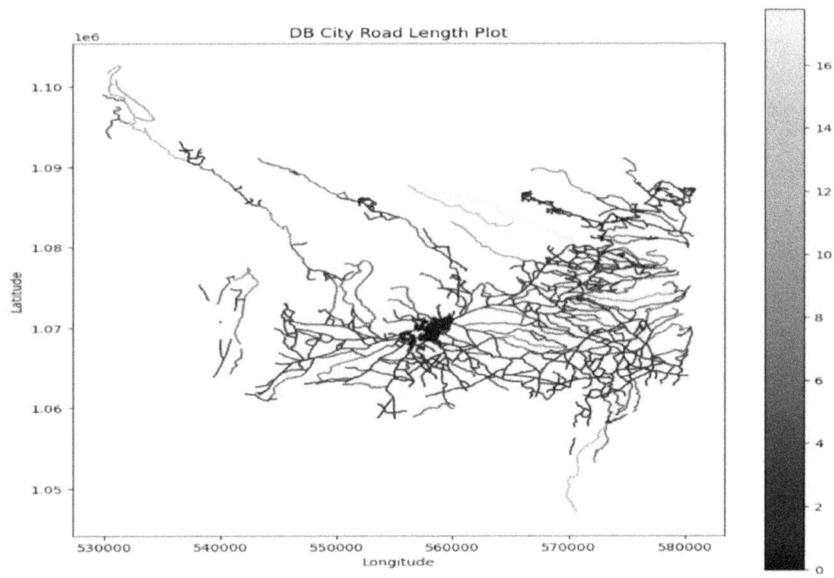


```
subplots(figsize=(10, 10)) # set the figure sizer_data.
plot(column='Time_Minit', legend=True, ax=ax)# Add legend
for shortest and longest distances# shortest_distance = r_
```

```
data['Length_KM_'].min()# longest_distance = r_data['Length_
KM_'].max()# ax.text(0.95, 0.9, f"Shortest Distance: {shortest_
distance} km", transform=ax.transAxes)# ax.text(0.95, 0.85,
f"Longest Distance: {longest_distance} km", transform=ax.
transAxes)plt.xlabel('Length(M)')plt.show("off")
```

This code creates a plot using the matplotlib library. It sets the figure size to 10x10 and then uses the r_data variable to plot a column named 'Time_Minit.' The legend is included in the plot. Additionally, the code sets the plot's title to "Road Length and Speed in km/h" and labels the x-axis as 'Longitude' and the y-axis as 'Latitude'. Finally, the plot is displayed using plt.show(). Debre Berhan City Road network Visualization Based on Speed in Km/h is given in Figure 10.

```
import matplotlib.pyplot as pltimport matplotlib.cm as cm#
Assuming you have a DataFrame "r_data" containing road
data with 'Name' and 'Length_KM' columnsfig, ax = plt.
```

Figure 10. Debre Berhan city road network visualization based on speed in Km/h

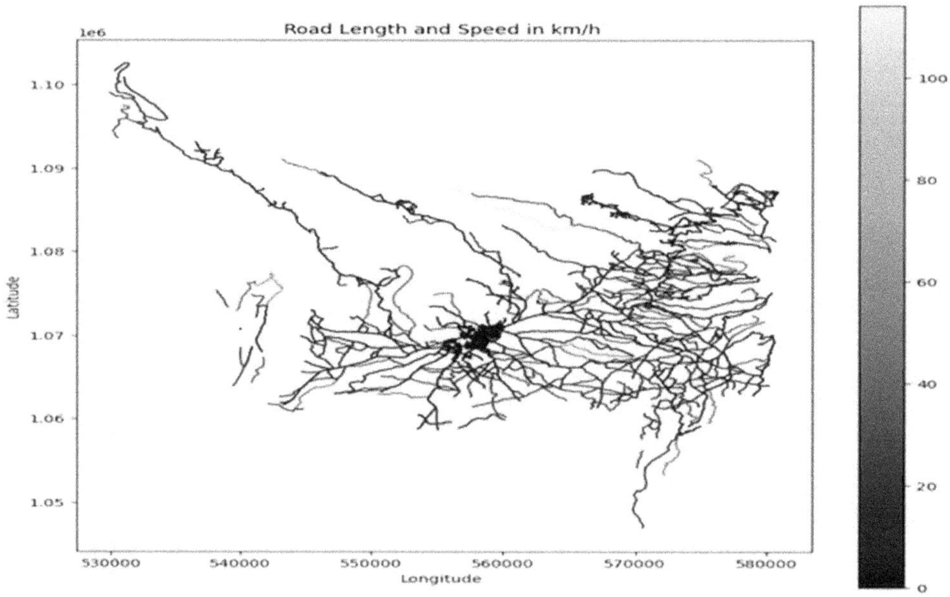


```
subplots(figsize=(10, 10))# Get unique road typesroad_types
= r_data['Speed_Km_H'].unique()# Create colormap# cmap =
```

```
cm.get_cmap('BrBG')# Plot each road type separately with a
different colorfor i, road_type in enumerate(road_types):road_
data = r_data[r_data['Speed_Km_H'] == road_type]ax.barh(road_
data['Name'], road_data['Speed_Km_H'])plt.title('Road
Types vs Speed(Km/h)', fontsize=16)plt.xlabel('Speed (KM/
H)',fontsize=12) # Set the x-axis labelplt.ylabel('Road
Types',fontsize=12) # Set the y-axis labelplt.show()
```

This code creates a horizontal bar plot using the matplotlib library. It first sets the figure size to 10x10 and then retrieves the unique road types from the 'Speed_Km_H' column in the r_data DataFrame. It then creates a colormap using matplotlib.cm. Using a horizontal bar plot, the code then plots each road type separately with a different color. The title of the plot is set to 'Road Types vs Speed(Km/h)' and the x-axis label is set to 'Speed (KM/H)' while the y-axis label is set to 'Road Types'. Finally, the plot is displayed using plt.show(). Debre Berhan City Road Types vs Speed in km/h is given in Figure 11.

```
import numpy as npmin_speed_kmh = r_data['Speed_Km_H'].min() #
Convert to kilometers per hourmax_speed_kmh = r_data['Speed_
Km_H'].max() # Convert to kilometers per hourprint("Minimum
Speed:", min_speed_kmh, "km/h")print("Maximum Speed:", max_
speed_kmh, "km/h")#displays:
```

Minimum Speed: 0.0 km/h

Figure 11. Debre berhan city road types vs. speed in km/h

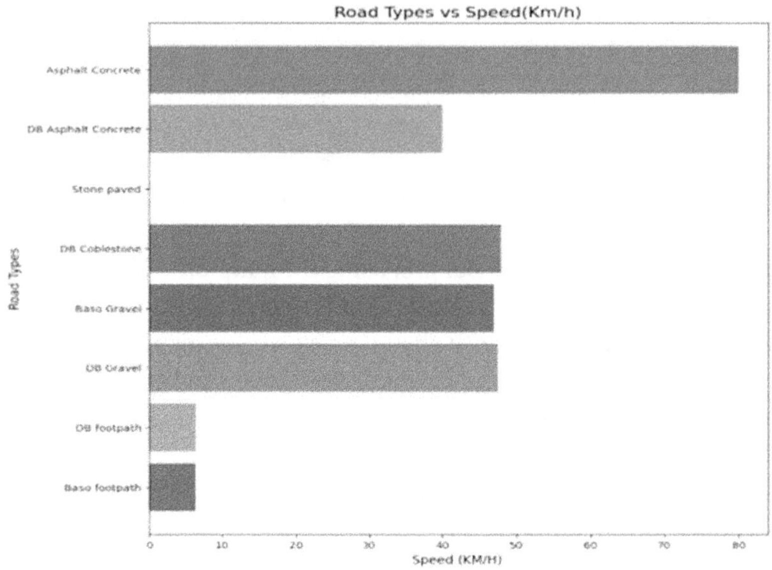

Maximum Speed: 80.0 km/h

```
import geopandas as gpdimport foliumfrom folium.plugins
import MarkerClusterfrom IPython.display import display#
Read the shapefileroad_data = gpd.read_file('RaodNet.shp')#
Reproject the GeoDataFrame to a new CRSroad_data = road_
data.to_crs(crs='EPSG:4236')# Drop any rows with NaN values
in the geometryroad_data = road_data[road_data.geometry.
notna()]# Create a folium map centered at the mean of centroid
coordinatesm = folium.Map(location=[road_data.geometry.
centroid.y.mean(), road_data.geometry.centroid.x.mean()],
zoom_start=12)# Add the shapefile as a GeoJSON layer to the
mapfolium.GeoJson(road_data).add_to(m)# Add the shapefile
name as markers at the centroid locationsmarker_cluster
= MarkerCluster().add_to(m)for idx, row in road_data.
iterrows():folium.Marker([row.geometry.centroid.y, row.
geometry.centroid.x], popup=row['Name']).add_to(marker_
cluster)# Save the map to an HTML file or display itm.
save('map_with_shapefile_overlay.html')display(m)
```

The above code uses geopandas and folium libraries to visualize road network data. First, it reads the shapefile and reprojects the data to a new CRS. Rows with NaN values in the geometry column are dropped for data integrity. A folium map is created, centered at the mean of centroid coordinates. The shapefile is added as a GeoJSON layer on the map, allowing for interactive visualization. Additionally, markers are added at centroid locations with corresponding names as popup content. The resulting map represents the road network data and facilitates exploration and analysis of its spatial distribution and attributes with attribute values represented on the map with numbers (i.e., 60,48,13, etc., see Figure 12).

CONCLUSION

In conclusion, this study utilized the Python library to analyze, integrate, and visualize road network information within the Debre Berhan Metropolitan City. The focus was on demonstrating road network analysis capabilities, including determining the best route between destinations, identifying the shortest path, and displaying road types with their respective lengths in kilometers and speed limits in kilometers per hour. Geopandas, a Python library for geospatial data manipulation, analysis, and visualization, played a central role in loading, preprocessing, and visualizing the road network data. The integration of Geopandas with network analysis libraries such as NetworkX facilitated the conversion of data into a graph representation and the application of algorithms like Dijkstra's or A* to find the shortest path based on distance or travel time. Visualizing results through Geopandas allowed for overlaying road segments on a map, highlighting the shortest path, and representing

Figure 12. Debre Berhan City road network visualization on the open street map

road type, length, and speed limit through color, labels, or markers. The study also recommended the analysis of road traffic using real-time data and emphasized the robust framework provided by integrating Geopandas and the Python library for road network analysis to assess and optimize urban road networks. By considering factors such as the shortest path, distance, and speed limits based on road types, this approach offers valuable insights into the characteristics and performance of different road types, ultimately contributing to effective management and enhancement of the city's transportation infrastructure.

Future Work

I recommend exploring integrating real-time road traffic data into the existing framework. This would provide valuable insights into the dynamic nature of road networks, ultimately leading to more accurate and responsive transportation planning and management. Additionally, I suggest integrating the developed framework with advanced GIS software, such as ArcGIS or QGIS, to leverage their sophisticated extensions for enhanced spatial analysis, visualization, and decision-making capabilities. Lastly, I recommend expanding the analysis to include multi-modal transportation networks, incorporating data on public transit, pedestrian pathways, and cycling routes to develop integrated transportation systems that cater to diverse mobility needs within the city. These future works can further enhance the capabilities of the existing framework and contribute to more comprehensive and sophisticated approaches to urban transportation infrastructure analysis and management, not only for Debre Berhan City but also for other cities in Ethiopia.

REFERENCES

Dumedah, G., & Garsonu, E. K. (2021). Characterising the structural pattern of urban road networks in Ghana using geometric and topological measures. *Geo : Geography and Environment*, *8*(1), e00095. doi:10.1002/geo2.95

Echols, N. (2003). *Finding the Fastest Route.*

Galety, M. G., Thiagarajan, R., Sangeetha, R., Vignesh, L. B., Arun, S., & Krishnamoorthy, R. (2022, April). Personalized music recommendation model based on Machine Learning. In *2022 8th International Conference on Smart Structures and Systems (ICSSS)* (pp. 1-6). IEEE. 10.1109/ICSSS54381.2022.9782288

Gecchele, G., Ph, D., Ceccato, R., & Gastaldi, M. (2019). Road Network Vulnerability Analysis. *Journal of Transportation Engineering. Part A, Systems*, *145*(7), 1–10. doi:10.1061/JTEPBS.0000252

Husdal, J. (1999). *Network analysis – raster versus vector – A comparison.* Husdal. https://husdal.com/1999/10/11/network-analysis-raster-versus-vector-a-comparison-study/

Pradhan, M. R. (2023). PIRAP: Medical Cancer Rehabilitation Healthcare Center Data Maintenance Based on IoT-Based Deep Federated Collaborative Learning. *International Journal of Cooperative Information Systems.*

Şirin, A., & Erener, A. (2024). Python-based evaluation of road network constraints for electric scooters and bicycles. *Izmit Example.*, *9*(1), 34–48.

Team, D., Investigator, P., Islamia, J. M., Delhi, N., Coordinator, P., & Writer, C. (n.d.). *Network Analysis in GIS.*

Westra, E. (2010). *Python Geospatial Development.*

Wondwossen. (2022). *GPS SUPPORTED ASSESSMENT INTEGRATED WITH GIS SERVICE AREA ANALYSIS OF HEALTHCARE INSTITUTIONS OF BASONA WERANA IN COMPARISON WITH DEBRE BERHAN Wondwossen MINDAHUN.*, *48*(4), 209–217.

Wondwossen Mindahun & Bedasa Asefa. (2019). *Location Allocation Analysis for Urban Public Services Using GIS Techniques : A Case of Primary Schools in. July.* https://doi.org/ doi:10.5923/j.ajgis.20190801.03

Chapter 10
Impacts of Climate Changes on Traffic Flows Using Geospatial Data Analysis:
Shifting the Traffic Flow to the Next Level

Jimbo Henri Claver
Samarkand International University of Science and Technology, Uzbekistan

Bouetou Thomas
University of Yaounde, Cameroon

Nagueu Djambong Lionel Perin
University of Yaounde, Cameroon

Tchoua Paul
University of Ngoundere, Cameroon

ABSTRACT

This chapter explores the complex interplay between climate change and the accuracy of traffic flow predictions, focusing on the crucial use of geospatial data analysis. The potential effects of extreme weather events, such as heavy precipitation, storms, and heat waves, on traffic patterns should be considered to improve the robustness of traffic management systems. In this study, the authors demonstrate the effectiveness of geospatial data analysis in considering climatic and environmental variables to improve the accuracy of traffic flow forecasts. By integrating data into predictive models, we provide tangible evidence of the impacts of climate change on urban traffic patterns. The results obtained from data and simulations on machine learning models such as Lasso regression, random forest, XGboost and LTSM gave us very good results. prediction performance on the random forest with a correlation coefficient of 0.94; an RMSE of 265 and a MAE of 279 thus demonstrating its effectiveness for predicting traffic flow.

DOI: 10.4018/979-8-3693-1754-9.ch010

INTRODUCTION

Rapid urbanization in underdeveloped countries such as Cameroon, India, Niger, Angola and Thailand, combined with the effects of climate change and changing modes of transport, has led to significant challenges in road traffic management (United Nations, n.d.). Accurately predicting traffic flow is essential to ensure smooth traffic flow, reduce congestion and improve transportation infrastructure planning. However, climate change has introduced additional complexity to this task, requiring a thorough understanding of its interaction with traffic flow forecasts. Learning, a branch of machine learning, is used to estimate corn yield forecasts in Cameroon using geospatial climate data. Geospatial climate data includes parameters such as temperature, precipitation, wind speed, and sun exposure, which are key factors influencing crop growth and yield (The TIME, n.d.). This study aims to predict corn yield in Cameroon using geospatial climatic data using two methods, LSTM and GRU, to determine which is the appropriate model for a good prediction of corn yield using temperature as geospatial climatic parameters, wind speed, precipitation, and solar radiation. Our study focuses on comprehensively exploring the complex interplay between climate change and the accuracy of traffic flow forecasts, emphasizing the crucial use of geospatial data analysis.

The underlying problem lies in understanding how climate variations affect traffic patterns, which challenges existing forecasting systems. Extreme weather events, such as heavy precipitation, storms, and heat waves, can significantly affect traffic patterns, requiring adequate consideration of traffic management tools. The complexities associated with traffic forecasting arise from the nature of the traffic domain, which includes the constraints imposed by the physical traffic infrastructure, such as road network capacities, traffic regulations and management policies, and the behavior of individual agents (road users), as well as exogenous factors, such as calendar (i.e. traffic time). day, day of the week, etc.), weather, accidents and incidents, events, and road closures, to name a few (Essien et al., 2019). Traffic data science has evolved by expanding many data sources to develop predictive models. Early studies investigated the importance of geospatial weather data on traffic flow parameters influencing driving behavior, travel demand, travel mode, road safety, and traffic flow characteristics. Additionally, research has shown over the years that precipitation reduces traffic capacity and operating speeds, thereby increasing traffic congestion and loss of road network productivity.

For example, the authors (Essien et al., 2018) reported that precipitation intensity affected urban traffic characteristics by reducing traffic speed by 4-9%. In contrast, rush hour traffic jams correlate significantly with the temperature intensity. Despite the importance of weather as a traffic indicator, most traffic forecast models used in ITS assume weather conditions are clear, thus missing important sources of environmental data that could allow a more precise assessment of the state of the traffic network. Despite the importance of weather as a traffic indicator, most traffic forecast models used in ITS assume clear weather conditions, thereby missing important sources of environmental data that could enable a more precise assessment of the state of the traffic network (Agarwal, 2005). Geospatial data analysis has proven to be a practical approach to improve the accuracy of traffic flow forecasts by considering climate variables. Integrating geospatial data such as weather data and environmental characteristics into predictive models makes it possible to capture the complex interactions between climate change and urban traffic flows. Therefore, this study aims to demonstrate the effectiveness of geospatial data analysis in considering climatic and environmental variables to improve the accuracy of traffic flow forecasts. By providing tangible evidence of the impacts of climate change on urban traffic patterns, this research will contribute to improving traffic management systems in underdeveloped countries.

STATE OF THE ART

Recent studies have investigated the impact of including non-traffic-related input datasets for forecasting urban traffic parameters, many (if not all) of which have been found to improve forecast accuracy (Essien et al., 2019a), (Essien et al., 2019). For example, a deep bidirectional LSTM model was proposed (Jia et al., 2017), which was trained using precipitation and temperature datasets and traffic flow features. Some studies have shown improved prediction accuracy compared to reference datasets (i.e., traffic only). Similar results were obtained in studies that included non-traffic input data. This can be explained by the fact that data-driven prediction of traffic parameters typically relies on predictive analytics techniques applied to historical data observations to extract patterns that can be used to predict future observations. However, even the most accurate predictive models will perform poorly predictively in unusual or non-recurring situations, such as events or incidents that cannot be inferred from historical observations.

Typical examples of non-recurring or stochastic events/incidents include accidents, lane closures, and sporting and public events. Since such events may be sudden, unexpected and/or rare, developing robust predictive models to enable accurate traffic forecasting in these circumstances is necessary. Existing studies have presented research efforts to leverage social media data for traffic parameter prediction, for example, in Jia et al. (2017a), Jia et al. (2017b), He et al. (2013), where an optimization technique based on linear regression is used for road traffic forecasting. Deep learning (DL) approaches have been used in image and speech recognition, natural language processing (NLP), and intelligent gamification.

However, they have also been used in short-term traffic forecasting (Essien & Giannetti, 2019), (Kim et al., 2017) (Poonia et al., 2018).

Deep learning refers to techniques for learning complex features from high-level data hierarchically using stacked architectures and multilayers [15]. Other authors have studied the prediction of short-term traffic parameters essential for modern urban traffic management and control systems. They used a deep-learning urban traffic forecasting model that combines information extracted from tweet messages with traffic and weather information. The predictive model adopts a bidirectional long short-term memory (LSTM) stacked autoencoder (SAE) architecture for multi-stage traffic flow prediction, trained using tweets, traffic and weather (Essien & Petrounias, 2020) (Goodfellow & Bengio, 2015) (Nilsson, 2015).

METHODOLOGY

This section presents the methodology used to predict traffic flow or volume to allow traffic managers to better manage road networks. By anticipating traffic variations, they can adjust traffic signals, implement reversible lanes, optimize recommended routes, etc. This helps reduce congestion, improve traffic flow, and reduce travel times. This section presents the methodology for predicting traffic flow or volume to allow traffic managers to manage road networks better. By anticipating traffic variations, they can adjust traffic signals, implement reversible lanes, optimize recommended routes, etc. This helps reduce congestion, improve traffic flow, and reduce travel times. We used geospatial data containing the climatic and environmental parameters for this prediction.

This section presents the basic concepts of our three models, Random Forest, XGboost, Lasso regression, and LSTM, and their uses in our work. We will start with a description of the Random Forest, then with the XGboost, then with the Lasso Regression, and finally with the LSTM

Figure 1. Traffic flow prediction steps

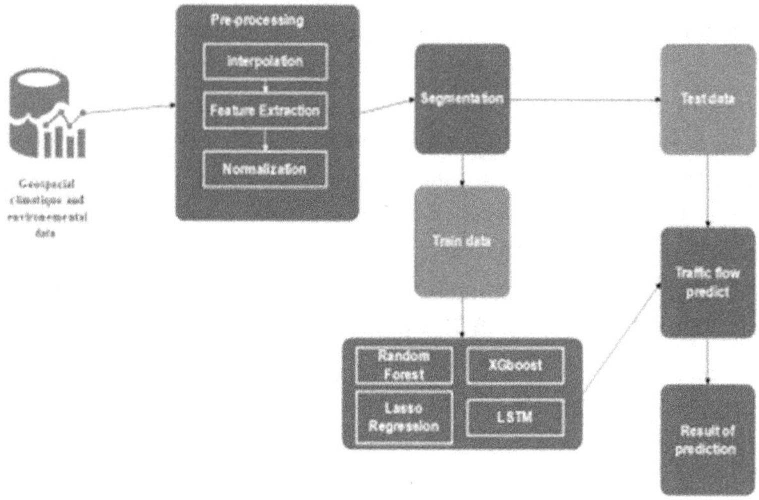

Random Forest

Random Forest is a machine learning algorithm for classification, regression, and other tasks. It is an ensemble model that combines several decision trees to improve the accuracy and robustness of predictions. How this model works is explained below:

Creation of Decision Trees

A set of decision trees is constructed, with each tree trained on a random subsection of the training data. Random selection of data is carried out with replacement (bootstrap sampling).

Construction of Trees

Each tree is constructed by selecting, at each node, the best split from a random subset of the features. Trees are often shallow decision trees to avoid overfitting.

Majority Vote

When predicting, each tree gives an answer, and the final answer is determined by a majority vote (classification) or an average (regression) of the individual tree responses.

This model also uses parameters to work, and as a parameter, we have:

$$\boldsymbol{H}(T) = \sum_{i-1}^{C} p_i * log_2(pi)$$

- **Entropy (classification):** The Entropy measures the uncertainty in a data set. The lower it is, the more homogeneous the data.

Where H(T) is the entropy of the set C, the number of classes, and is the probability of belonging to class i.

- **Gini Criterion (classification):** The Gini criterion measures the probability of misclassification. The lower it is, the more homogeneous the data is.

$$\boldsymbol{G}(T) = \sum_{i-1}^{C} p_i * (1 - p_i)$$

- **Mean Square Error (Regression):** The mean square error measures the dispersion of the predicted values compared to the actual values.

$$MSE = \frac{1}{N} \sum_{i-1}^{N} (y_i - \ddot{y}_i) \quad (1)$$

where N is the number of observations, yi is the actual value, and ÿi is the predicted value.

- **Weighted Average (regression):** The predicted value for the regression is the weighted average of the values predicted by each tree.

XGboost

XGBoost (eXtreme Gradient Boosting) is a machine learning algorithm that belongs to the family of gradient boosting algorithms. It is trendy for its power, flexibility, and ability to produce well-performing models in various supervised learning tasks. Here's how XGBoost works, followed by some critical mathematical formulas associated with its optimization. Ce modèle fonctionne comme suit:

- **Initialization:** The model starts with a simple prediction, usually the mean of the target variable in the case of regression or the majority class in the case of classification.

- **Creation of Trees (Boosting):** A decision tree is added to the model to correct prediction errors of the current model. Trees are added sequentially, and each tree is fitted to the residuals (errors) of the existing model.
- **Optimization of the Cost Function:** XGBoost optimizes a regularized cost function, the sum of the loss function and the regularization terms. The following formula defines this cost function

$$\mathcal{L}(\theta) = \sum_{i-1}^{n} l(y_i, \ddot{y}_i) + \sum_{k-1}^{k} ©(f_k,) \quad (2)$$

Where l is the loss function, is the model prediction, and ÿi is the predicted value, $\Omega(f_k)$ Is the regularization term for the tree k, K is the number of trees, and *(θ)* represents the model parameters.

- **Loss Function:** The loss function l depends on the type of problem (regression, classification, etc.) and measures the gap between the model's prediction and the true value.
- **Regularization Terms:** XGBoost uses two regularization terms: L1 regularization (Lasso) to promote parameter parsimony and L2 regularization (Ridge) to avoid overfitting.
- **Updated Predictions:** The model's predictions are updated by adding the predictions of each new tree weighted by a learning rate.
- **Stopping the Process:** The tree-building process is repeated until a stopping criterion is reached, such as the maximum number of trees or no significant improvement.

Lasso Regression

Lasso Regression, or L1 regularization, is a linear regression technique incorporating a penalty term to the cost function, encouraging sparsity in the model by driving some of the coefficient values to zero. This helps in feature selection by eliminating irrelevant features. The mathematical formula for Lasso Regression can be expressed as follows:

- **Objective Function (Cost Function):** The objective function for Lasso Regression is a combination of the least squares cost function and the L1 regularization term:

$$J(\theta) = \frac{1}{2m} \sum_{i-1}^{m} (h_\theta \left(x^{(i)}\right) - y^{(i)})^2 + \lambda \sum_{j-1}^{n} \left|\theta_j\right| \tag{3}$$

Here, $J(\theta)$ is the objective function (cost function), m is the number of training examples, $h\theta_{(x}(i^{))}$ is the predicted value for the i-th example, $y(i^{)}$ ^{i}s the actual value for the i-th example, n *is* the number of features, θj i_s the j-th coefficient and λ is the regularization parameter, also known as the penalty parameter.

- L1 **Regularization Term:**

$$\lambda \sum_{j-1}^{n} \left|\theta_j\right| \tag{4}$$

λ controls the strength of the regularization, $\sum_{j-1}^{n} \left|, j\right|$ is the sum of the absolute values of the coefficients. This term penalizes the absolute values of the coefficients, encouraging sparsity.

- Optimization Objective:

The goal is to minimize the objective function $J(\theta)$ concerning the model parameters θ. This involves adjusting the coefficients to balance fitting the training data well and keeping the magnitude of the coefficients small.

- **Optimization Algorithm**

Standard optimization algorithms, such as gradient descent or coordinate descent, can be used to minimize the objective function. The L1 regularization term introduces a subdifferential, used in the optimization process to handle cases where the absolute value is not differentiable.

- **The Effect of L1 Regularization**

Lasso Regression tends to produce sparse models by forcing some coefficients to be exactly zero. This property makes it useful for feature selection in scenarios where not all features are relevant to the prediction task.

LSTM

Long Short-Term Memory (LSTM) is a recurrent neural network (RNN) architecture designed to address the vanishing gradient problem in traditional RNNs. LSTMs are particularly effective in capturing long-range dependencies in sequential data, making them well-suited for tasks such as time series prediction, natural language processing, and speech recognition. Here's an overview of how LSTMs operate, along with their mathematical formulas:

- **Forget Gate:** Determines which information from the previous cell state should be discarded or kept. The sigmoid activation function is applied to the output of the forget gate.

$$f_t = \sigma(w_f.[h_{t-1}, x_t] + b_f) \tag{5}$$

- **Input Gate:** Decides which new information should be stored in the cell state. Sigmoid activation is applied to the output of the input gate. Hyperbolic tangent (tanh) activation is applied to the candidate new cell state values.

$$i_t = \sigma(w_i * [h_{t-1}, x_t] + b_i) \tag{6}$$

$$\ddot{}C_t = \sigma(w_f * [h_{t-1}, x_t] + b_f) \tag{7}$$

- **Update Cell State:** The old cell state is multiplied by the forget gate output, and the new candidate values are multiplied by the input gate output. These results are added to update the cell state.

$$C_t = f_t * C_{t-1} + i_t * \ddot{}C_t \tag{8}$$

- **Output Gate:** Determines the output based on the updated cell state. Sigmoid activation is applied to the output of the output gate. The updated cell state is passed through the tanh function.

$$o_t = \sigma(w_o * [h_{t-1}, x_t] + b_o) \tag{9}$$

$$h_t = o_t * \tanh(C_t) \quad (10)$$

Data Collection

The data we collected for this study are geospatial climatic and environmental data. The geospatial climate data used are real-time precipitation, temperature, weather data and the description of this weather for the last 5 years from 2012 to 2017. 2022. These geospatial climate and environmental data are obtained from an open github repository and are in CSV format (road_Interstate_Traffic_Volume.csv). the quantity of these climatic data obtained is 33750 x 15 parameters or 506250 data.

Data Training

Data training will be carried out on the LSTM, random forest, linear regression and decision tree data models which have been previously pre-processed to adapt them to our different models. The preprocessing was carried out in four steps, namely interpolation, feature extraction, segmentation and normalization. Data Interpolation: Interpolation is the process of finding the value between several known:

1. data points. The purpose of interpolation is to correct for data that is not measured or not recorded by finding the median value between two values.
2. Feature Extraction: Feature extraction is a pro- cess of finding the largest value of each variable, so that the data becomes monthly and annual data with the maximum value of each variable.
3. 3. Normalization: Normalization is the process of converting data into normal form. Normaliza- tion is needed when the data is very large, very small, or has different units. In this process, the Min-Max normalization is carried out by scaling in the range of zero to one [17].

$$Z = \frac{((X - min(.)))}{((max(.) - min(.)))} \quad (11)$$

where X are the data to be normalized, max is the highest data in the column, and min are the lowest data in the column.

1. Segmentation: Segmentation is the process of separating and grouping raw data into data needed by the system. In this study, the data that has become

monthly and annual will be separated into training data which will be used in our LTSM, GRU model and test data which will be used to evaluate the capacity of the model to generalize and make precise predic- tions on new data not seen before.

LTSM, Random Forest, Lasso Regression and XGBOOST Training

After the preprocessing phase of our data, 27009 x 21 parameters or 405135 training data sets were produced to be used as input in the LSTM, Random Forest, Lasso Regression and Xgboost and 6741x 21 parameters or 101115 sets data was produced to be used as a test for our models.

Traffic Flow Prediction

For this prediction step, the Random forest architecture will be used to predict the traffic flow, with the training data used in the input layer. The architecture of the Random Forest is presented in the following figure 2 designed in the re- current neural network method can be seen in Figure 2 below:

Figure 2. Random forest architecture for traffic flow prediction

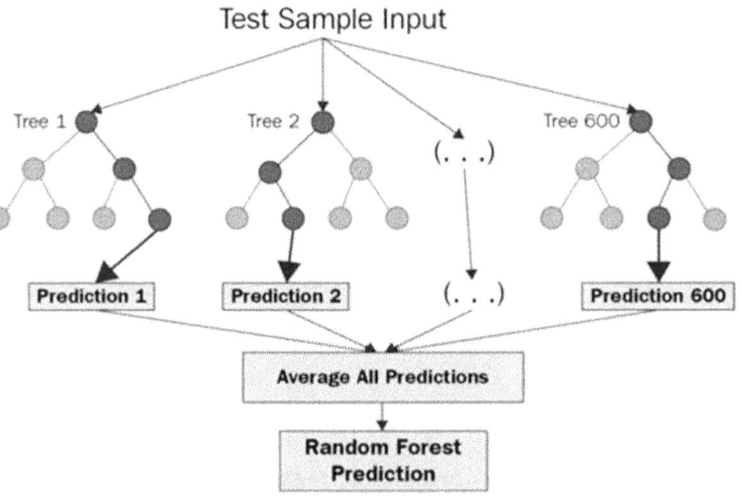

As can be seen in Figure 2 regarding the MLP architecture used in this study to predict corn yield, where there are 50 neurons in the climate and agricultural parameters input layer. There are 50 neurons in the hidden layer and in the output layer there is one neuron which represents the predictions output. The testing process uses the weight of the training results to produce an output value in the form of a corn yield prediction value for each variable.

Prediction Results

This study produced a road traffic flow forecast based on climatic and environmental parameters. The prediction results can be seen in Figures 3, 4 and 5 below:

Figure 3. Result of random forest training

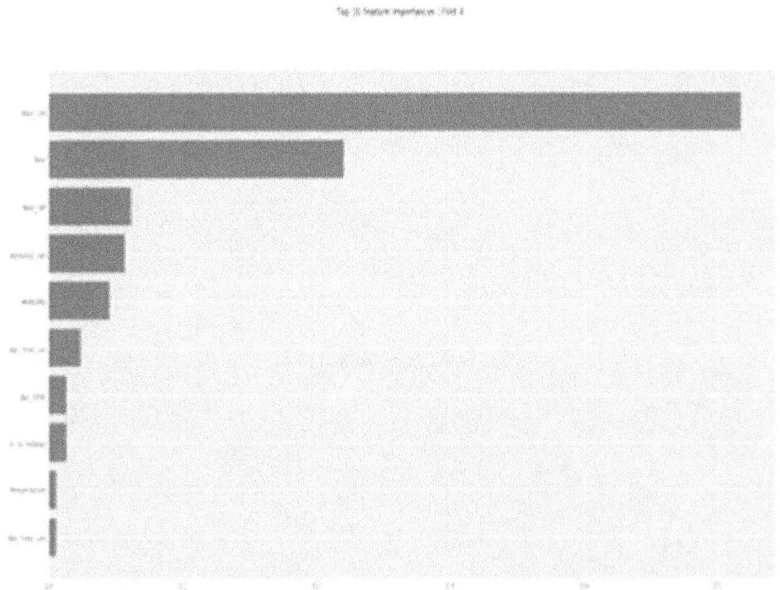

Figure 4. Result of XGboost training

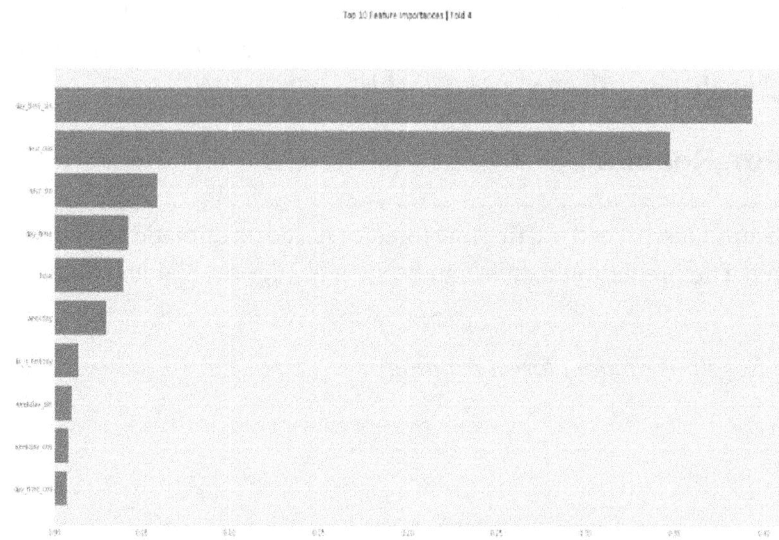

Figure 5. Result of LSTM training

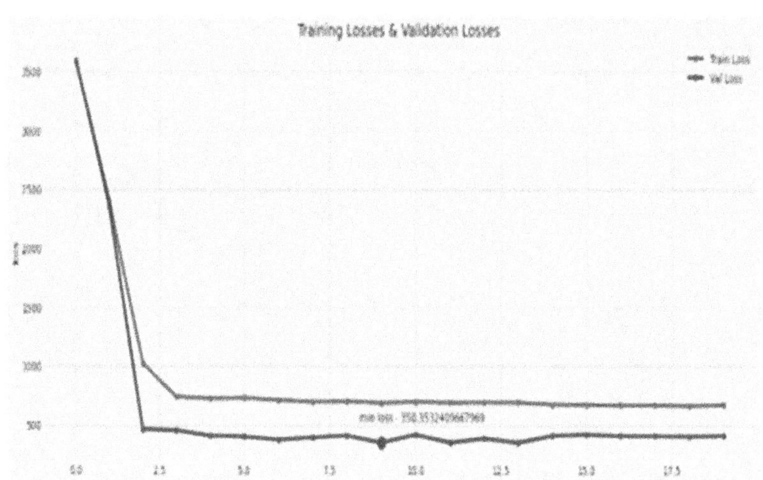

RESULTS AND DISCUSSION

After training our model and making predictions, we performed the optimization tests on ours. At this stage, an analysis of each parameter tested was carried out. Before testing our model, training was carried out using pre-processed data. Our models were trained on Random Forest, XGboost, Lasso Regression, and LSTM architectures. In addition, this training was carried out on 4 Folds, except the LTSM, which was carried out on 300 epochs.

Optimization Model Influence Test

The fold performance test is performed to determine which of our models is best suited to our forecast. The test accuracy results of our models can be seen in Table 1 below:

Table 1. Performance test of the LSTM, Random Forest, XGboost and Lasso Regression models.

Models	Correlation coefficient	MAE	RSME
Random Forest	0.94	279	265
XGboost	0.94	289	465
Lasso Regression	0.94	610	820
LSTM	0.36	1228	1565

As a metric to measure the performance of our models, we used prediction results on metrics such as the correlation coefficient, mean absolute error (MAE), and root mean square error (RMSE) below 279 and 265 for Random Forest, 289 and 465 for XGboost, 610 and 820 for Lasso Regression, finally 1228 and 1565 for LSTM with correlation coefficients of 0.94 for the Random Forest, 36 for the LSTM which means that the characteristics of the first three models are strongly correlated and that of the LSTM is weakly correlated. Given that these models have the same correlation coefficient except that of the LSTM, and we cannot only limit ourselves to this metric, this is why we focused on the two other metrics, the MAE and the RMSE. After comparing these models, Random Forest was found to give better results, which means that the Random Forest model is more accurate in predicting traffic flow.

CONCLUSION

Ultimately, our work aimed to highlight the importance of geospatial data analysis for understanding the complex interplay between climate change and the accuracy of traffic flow forecasts. Extreme weather events, such as heavy precipitation, storms and heat waves, must be considered to improve the robustness of traffic management systems. The integration of climatic and environmental data into predictive models, such as Lasso Regression, random forest, XGboost, and LTSM, has made it possible to obtain promising prediction performances, particularly with random forest, which has demonstrated effective high with a correlation coefficient of 0.941, RMSE of 265 and MAE of 279 to predict traffic flow. These results highlight the importance of an integrated and innovative approach to address emerging challenges linked to climate change in urban traffic management. The recommendations from this study can contribute to developing more resilient and adaptive traffic management systems aligned with climate change's challenges.

REFERENCES

Agarwal, M., Maze, T., & Souleyrette, R. (2005). Impacts of weather on urban freeway traffic flow characteristics and facility capacity. In: *Proceedings of the 2005 mid-continent transportation research symposium*. IEEE.

Essien, A., Chukwukelu, G., & Giannetti, C. (2019). *A Scalable Deep Convolutional LSTM Neural Network for Large-Scale Urban Traffic Flow Prediction using Recurrence Plots. In: 2019 IEEE AFRICON*. IEEE Xplore. doi:10.1109/AFRICON46755.2019.9134031

Essien, A., & Giannetti, C. (2019). A deep learning framework for Univariate time series prediction using convolutional LSTM stacked autoencoders. In: 2019 IEEE international conference on INnovations in intelligent SysTems and applications (INISTA). IEEE. doi:10.1109/INISTA.2019.8778417

Essien, A., Petrounias, I., Sampaio, P., & Sampaio, S. (2018). The impact of rainfall and temperature on peak and offpeak urban traffic. In: *International Conference on Database and Expert Systems Applications*, (pp. 399–407). Springer, Cham. 10.1007/978-3-319-98812-2_36

Essien, A., Petrounias, I., Sampaio, P., & Sampaio, S. (2019). Deep-PRESIMM: Integrating Deep Learning with Microsimulation for Traffic Prediction. In: *IEEE International Conference on Systems, Man, and Cybernetics* (pp. 1–6). IEEE Xplore. 10.1109/SMC.2019.8914604

Essien, A., Petrounias, I., Sampaio, P., & Sampaio, S. (2019). Improving Urban Traffic Speed Prediction Using Data Source Fusion and Deep Learning. In: *2019 IEEE International Conference on Big Data and Smart Computing (BigComp),* (pp. 1–8). IEEE. 10.1109/BIGCOMP.2019.8679231

Goodfellow, I., & Bengio, Y. (2015). *Deep learning*. MIT Press.

He, J. & Shen, W. (2013). *Divakaruni et al: Improving traffic prediction with tweet semantics.* AAAI.org.

Jia, Y., Wu, J., Ben-Akiva, M., Seshadri, R., & Du, Y. (2017). Rainfall-integrated traffic speed prediction using deep learning method. *IET Intelligent Transport Systems*, *11*(9), 531–536. doi:10.1049/iet-its.2016.0257

Jia, Y., Wu, J., & Xu, M. (2017). Traffic flow prediction with rainfall impact using a deep learning method. *Journal of Advanced Transportation*, *2017*, 1–10. doi:10.1155/2017/6575947

Kim, S., Hong, S., Joh, M., & Song, S. (2017). DeepRain: ConvLSTM network for precipitation prediction using multichannel radar data. ArXiv preprint. 1711.

Nilsson, N. J. (2015). *An introduction to machine learning*.

Poonia, P., Jain, V. K., & Kumar, A. (2018). Short term traffic flow prediction methodologies: A review. *Mody University International Journal of Computing and Engineering Research.*, *2*, 37–39.

The TIME. (n.d.). Le changement climatique est un facteur aggravant du trafic d'êtres humains. *Le Temps*. https://www.letemps.ch/monde/changement-climatique-facteur-aggravant-trafic-detres-humains

United Nations. (n.d.). *L'urbanisation rapide augmente les risques liés au climat pour des milliards d'individus.* United Nations Framework Convention on Climate Change. https://unfccc.int/fr/news/l-urbanisation-rapide-augmente-les-risques-lies-au-climat-pour-des-milliards-d-individus

Chapter 11
Deep Learning Approach to Estimate the Maize Yield Prediction Using Data From Cameroon:
Shifting the Maize Yield Production to the Next Level

Jimbo Henri Claver
Samarkand International University of Science and Tecchnology, Uzbekistan

Nagueu Djambong Lionel Perin
University of Yaounde, Cameroon

Bouetou Thomas
University of Yaounde, Cameroon

Tchoua Paul
University of Ngoundere, Cambodia

ABSTRACT

Corn cultivation plays a crucial role in the Cameroon's food production, providing an important source of food and income for many farmers. However, climatic variability and unstable agricultural conditions can have a significant impact on the yield of agricultural products in general and that of maize in particular; accurately predicting these yields in different regions of Cameroon remains a difficult process due to the uncertain evolution of climatic data.. This is where deep learning comes in, a powerful approach to analyzing large amounts of data and generating predictive models. This study aims to estimate maize yield forecasts in Cameroon using geospatial climate data parameters such as temperature, precipitation, wind speed and sun exposure as well as agricultural data. The study results, based on performance evaluations of o GRU model, have 24751 in 200 epochs for GRU, a mean absolute percentage error (MAPE) of 237%, and a root mean square error (RMSE) of 518 for GRU, which demonstrates the effectiveness of the deep learning approach in predicting corn yield.

DOI: 10.4018/979-8-3693-1754-9.ch011

INTRODUCTION

Cameroon, a country in Central Africa, relies heavily on agriculture for its economy and food security (World Bank Open Data, n.d.) . Maize cultivation plays a crucial role in the country's food production, providing many farmers with an essential source of food and income (FAOSTAT,n.d.). However, climate variability and unstable agricultural conditions often affect maize yield, which can significantly impact rural populations' food security and livelihoods. In this context, accurate corn production forecasts become essential for farmers and policymakers to make informed decisions and optimize agricultural resources (Kenneth et al., 2019). The consequences of inaccurate crop yield prediction can be dire, leading to food shortages and economic instability. Many technologies, including machine learning and deep learning, are now available to predict crop yields. Machine learning offers many algorithms and techniques to predict various phenomena accurately (Kamilaris et al., 2018). Many aspects must be considered in order to produce good quality corn crops, such as climate (Beck et al., 2018). The climate is the weather conditions in an area at a given time and is influenced by several factors, such as temperature and humidity (Thaddaeus et al., 2022).

Climatic variability and unstable agricultural conditions can significantly impact the yield of agricultural products in general and that of maize in particular; accurately predicting these yields in different regions of Cameroon remains a complex process due to the uncertain evolution of climatic data. Poor prediction of agricultural yield can lead to food shortages, particularly in regions where agriculture is the primary source of livelihood, but it also causes economic instability, with repercussions on farmers' incomes, overall economic growth, and the country's trade balance. In this article, we use deep learning, a branch of machine learning, to estimate corn yield forecasts in Cameroon using geospatial climate data. Geospatial climate data includes parameters such as temperature, precipitation, wind speed, and sun exposure, key factors influencing crop growth and yield. This study aims to predict corn yield in Cameroon using geospatial climatic data using two methods, LSTM and GRU, to determine which is the appropriate model for a good prediction of corn yield using temperature as geospatial climatic parameters, wind speed, precipitation, and solar radiation. The potential benefits of our findings could revolutionize the way farmers and policymakers approach crop yield prediction, offering a more accurate and reliable tool for decision-making.

STATE OF ART

In this era, many technologies are capable of predicting the climate in agriculture, one of which is machine learning technology (Aravind et al., 2022). Machine learning technology is one of the automatic lessons intended to enable a machine to have the ability to predict, analyze, and recognize a pattern (Ying et al., 2019). Machine learning has many methods, such as K-means, recurrent neural networks, and artificial neural networks (Jiahuan et al., 2022). Several previous studies have examined the machine learning technology used to predict climate in the agricultural sector. Using recurrent neural network methods, deep learning has been used to predict climate and estimate agricultural products based on Koppen classification. The parameters are precipitation, temperature, sun exposure, farming, and soil management (Novia et al., 2021). Previous research has also studied Machine learning methods for crop yield and climate forecasting with an assessment of the impact of change in agriculture with parameters temperature, precipitation, humidity, and wind speed and with the neural networks semi-parametric as the method used (Andrew et al., 2018).

Previous research has also studied Machine learning methods for crop yield and climate forecasting with an assessment of the impact of change in agriculture with parameters temperature, precipitation, humidity, and wind speed and with the neural networks semi-parametric as the method used 12. Several other studies have also made predictions using neural networks and machine learning methods to estimate crop yields using climate data (Kodimalar et al., 2019) (Rayda et al., 2021) (Vinushi et al., 2020). Many previous studies have examined the use of technological learning to predict climate and estimate agricultural yields. However, no research has been conducted on deep learning using the Gated Recurrent Network and Long-Short-Term method in Cameroon. Memory and application of several parameters such as humidity, temperature, precipitation, and solar radiation (exposure to the Sun) combined with local agricultural data to estimate corn yield forecast. The Gated Recurrent Network is an improved version of LSTM that performs learning by reviewing previous information. LSTM is a recurrent neural network method capable of overcoming the weaknesses of RNN so that it can improve performance and increase accuracy (Naveen et al., 2019). On this basis, this method is used in which climatic conditions are closely related to past climatic conditions.

METHODOLOGY

The subject of this research is the geospatial climate data of Cameroon obtained from the latitude and longitude coordinates of the satellite application of NASA (National Aeronautics and Space Administration), agricultural data (maize production and tons of pesticides used), the results of corn production and the Recurrent Neural Network Methods for the prediction process. In this study, several steps will be carried out using climate data to estimate the forecast of agricultural yields, mainly corn. The process of these steps is elaborated in Figure 1.

Data Collection

The data we collected for this study are geospatial and agricultural climate data from the official NASA satellite sites https://power.larc.nasa.gov/data-access-viewer/ and that of the FAO (https://www.fao.org/faostat/en/#data) for agricultural data that will be used for corn yield prediction. The geospatial climate data used are rainfall data, average temperature, duration of exposure to the Sun (solar radiation), and wind speed, and the agricultural data is data on corn production and use of pesticides for the last 12 years from 2010 to 2022 The data is obtained from monthly and annual geospatial climate data from the Cameroon geophysical station through the official NASA website. PowerPointDaily and FAOSTATdata are data in a file (.csv). The quantity of climatic data obtained is 4748 x 6 climatic parameters = 28,488 data, and that of agricultural data is 12 x 6 agricultural parameters.

Figure 1. Corn yield prediction steps

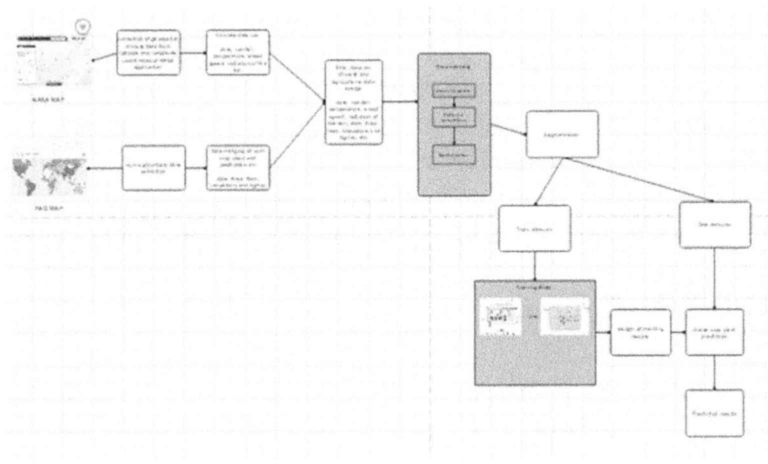

Data Training

The data training will be carried out on the LSTM and GRU models, which have been preprocessed to make this data suitable for our different models. The preprocessing was carried out in four stages: interpolation, feature extraction, segmentation, and normalization.

1. Data Interpolation: Interpolation is the process of finding the value between several known
2. data points. The purpose of interpolation is to correct for data that is not measured or not recorded by finding the median value between two values.
3. Feature Extraction: Feature extraction is a pro- cess of finding the largest value of each variable so that the data becomes monthly and annual data with the maximum value of each variable.
4. Normalization: Normalization is converting data into standard form. Normalization is needed when the data is extensive, very small, or has different units. In this process, the Min-Max normalization is carried out by scaling from zero to one (Nilsson et al., 2015).

$$Z = \frac{(X - \min(.))}{(\max(.) - \min(.))} \tag{1}$$

X is the data to be normalized, max is the highest data in the column, and the minimum is the lowest data in the column.

5. Segmentation: Segmentation separates and groups raw data into data the system needs. In this study, the data that has become monthly and annual will be separated into training data, which will be used in our LTSM, GRU model, and test data, which will be used to evaluate the capacity of the model to generalize and make precise predictions about new data not seen before.

LTSM AND GRU Training

After the preprocessing phase of our data, 3506 x 8 climatic and agricultural parameters, or 28,048 data sets, were produced to be used as input in the LSTM and GRU training process, and 877 x 8 climatic and agricultural parameters, or 7016 Data sets, were produced to be used as test data for our models.

Maize Crop Yield Prediction

For this prediction step, the MLP architecture will be used in the recurrent neural networks method to predict the corn yield, with the input data used in the input layer, namely precipitation, average temperature, duration of exposure to sunlight, wind speed, and Yield the MLP architecture designed in the recurrent neural network method can be seen in Figure 2.

Figure 2. MLP architecture for corn yield prediction

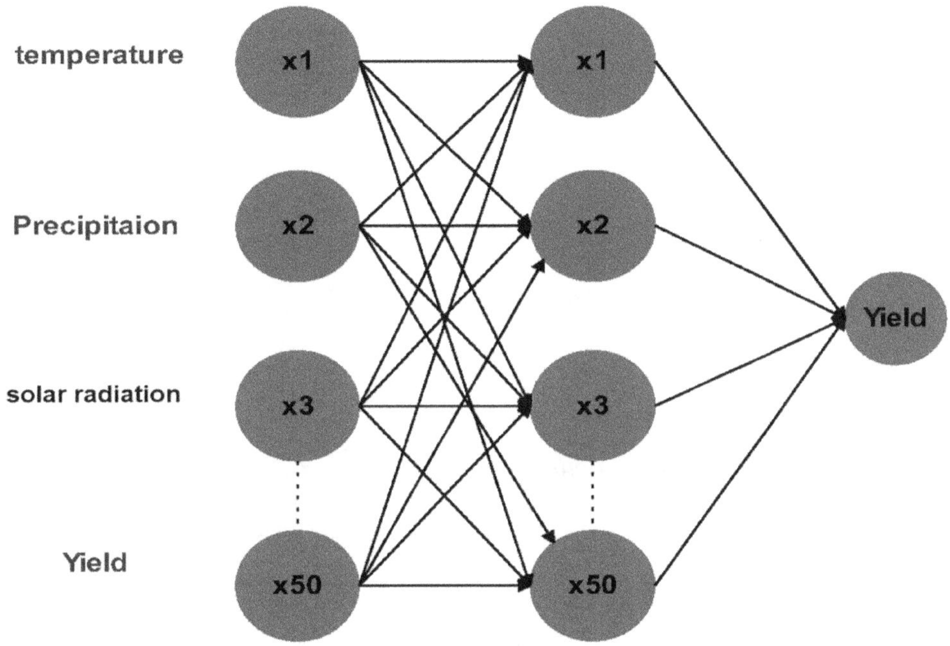

As can be seen in Figure 2 regarding the MLP architecture used in this study to predict corn yield, where there are 50 neurons in the climate and agricultural parameters input layer. There are 50 neurons in the hidden layer; in the output layer, one neuron represents the prediction output. The testing process uses the weight of the training results to produce an output value in the form of a corn yield prediction value for each variable.

Prediction Results

This study produced a corn yield forecast based on four climatic parameters: rainfall data, average temperature, duration of exposure to the Sun and wind speed, and an agricultural parameter, yield. The prediction result can be seen in Figures 3 and 4.

Figure 3. LSTM corn yield prediction

	Date	Prédictions LSTM	Données Réelles
3506	2019	16718.763672	17754
3507	2019	17271.986328	17754
3508	2019	17559.941406	17754
3509	2019	17317.242188	17754
3510	2019	17503.349609	17754
...	...	...	...
4378	2021	18336.974609	17705
4379	2021	17653.253906	17705
4380	2021	18201.833984	17705
4381	2021	17596.888672	17705
4382	2021	18163.960938	17705

[877 rows x 3 columns]

Figure 4. GRU corn yield prediction

	Date	Prédictions GRU	Données Réelles
3506	2019	16843.314453	17754
3507	2019	17223.080078	17754
3508	2019	17416.332031	17754
3509	2019	17402.382812	17754
3510	2019	17340.591797	17754
...	...	...	...
4378	2021	17972.798828	17705
4379	2021	17503.722656	17705
4380	2021	17862.517578	17705
4381	2021	17465.078125	17705
4382	2021	17899.130859	17705

[877 rows x 3 columns]

RESULTS AND DISCUSSION

After training our model and making predictions, we carried out the tests to optimize our model and the tests on the number of epochs. At this stage, an analysis of each parameter tested was carried out. Before testing our model, the training process was carried out using data that had been preprocessed. Our model was trained on the LSTM and GRU architectures preprocessed on climatic and agricultural data. In addition, this training was carried out on two optimizers, namely Adam and RMSprop, with several epochs, including 50, 100, and 200, and also used 0.001 for the learning speed.

Optimization Model Influence Test

The performance test on the optimizers is carried out to determine which of our two models is more suitable for our forecast. These optimizers are used to update weights during training; as optimizers, we have Adam optimizers and RMSprop optimizers. The test accuracy results of our two models can be seen in Tables 1 and 2.

Table 1. Performance test of the LSTM model on the Adam and RMSprop optimizers

LSTM	Loss	MAPE	RSME
Adam optimizer	11882580	279%	629
RMSprop optimizer	1186899	323%	734

Table 2. Performance test of the GRU model on the Adam and RMSprop optimizers

GRU	Loss	MAPE	RSME
Adam optimizer	124751	237%	518
RMSprop optimizer	11434247	325%	755

After testing, it was found that the Adam optimization method has a lower loss than RMSprop. The graph of the RMSprop and Adam test results in the two models can be seen in Figures 5 and 6 for the LTSM, figures 7 and 8.

Figure 5. Result of Adam loss on MAPE and RMSE in the LSTM

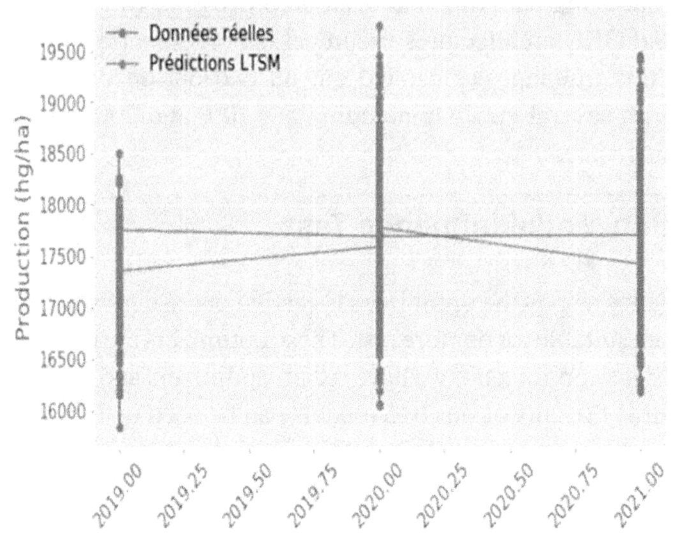

Figure 6. Result of RMSprop loss on MAPE and RMSE in the LSTM

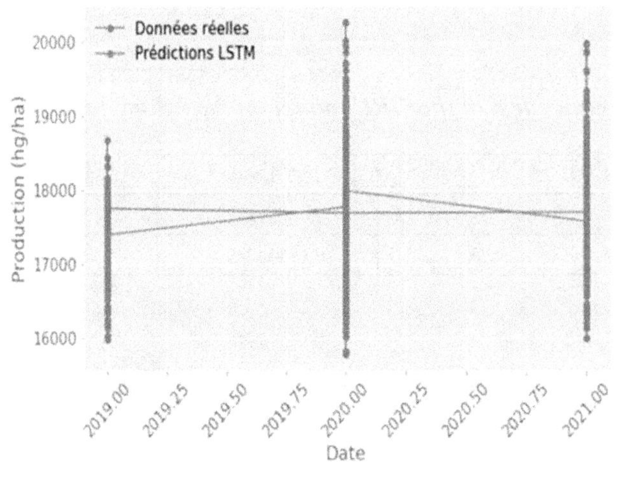

Figure 7. Result of Adam loss on MAPE and RMSE in the GRU

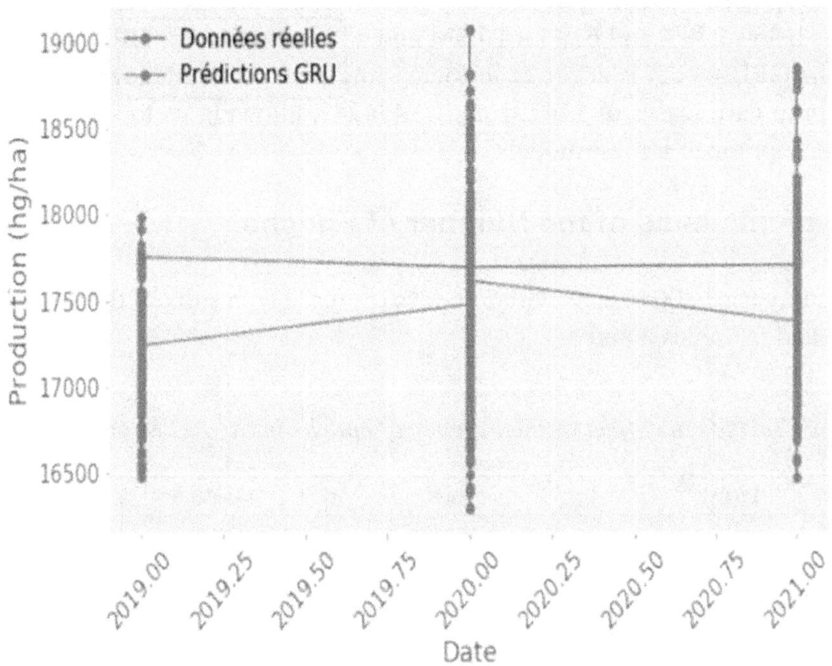

Figure 8. Result of RMSprop loss on MAPE and RMSE in the GRU

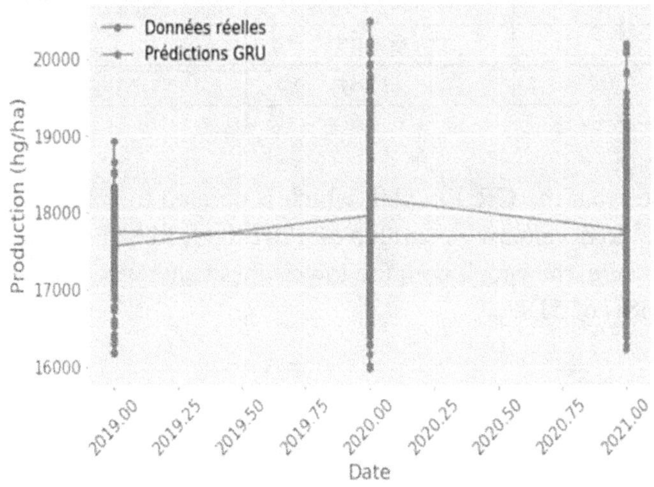

As a metric to measure the performance of our models, we used the Mean Absolute Error by percentage (MAPE) and the Root Mean Square Error (RMSE) prediction results below 279% and 629 for LTSM and 237% and 518 for GRU, which means that the GRU model is accurate in forecasting corn yield. Meanwhile, Adam and RMSprop gave losses of 124751 and 11434247, allowing us to say that Adam is the best optimizer for our models.

Test of Influence of the Number of Epochs

This study used 200 epochs with low-loss results. The results of the epoch test are presented in Tables 3 and 4.

Table 3. Test for the effect of the number of epochs in the LTSM model

LSTM	Loss	MAPE	RSME
50 epochs	12644395	390%	882
100 epochs	12539294	318%	721%
200 epochs	11882580	279%	629

Table 4. Test for the effect of the number of epochs in the GRU model

GRU	Loss	MAPE	RSME
50 epochs	12073044	342%	781
100 epochs	11336415	310%	712
200 epochs	124751	237%	518

It was found that the GRU model, which is curated to LSTM, using epoch 50, has the highest loss value with a value of 12073044, MAPE of 342%, and RSME of 781. Meanwhile, the epoch with the lowest loss value was 200, with a MAPE of 237% and RMSE of 518.

CONCLUSION

This research highlights the effectiveness of the deep learning approach, particularly with the GRU model, in accurately predicting maize yield in Cameroon. Based on geospatial climate data and agricultural parameters, the results reveal a minimum loss of 124751 after 200 epochs, a mean absolute percentage error (MAPE) of 237%, and a root mean square error (RMSE) of 518. This performance demonstrates the reliability of the approach in providing robust predictions, making it an essential tool for optimizing crop management and mitigating risks related to changing climate conditions. These results, integrated into the decision-making process, are intended to strengthen food security and the sustainability of agriculture in Cameroon.

REFERENCES

Amaratunga, V., & Wickramasinghe, L. (2020, July). *Artificial Neural Network to Estimate the Paddy Yield Prediction Using Climatic Data, 2020*, 11.

Beck, H., Zimmermann, N. E., McVicar, T., & Vergopolan, N. (2018). *Present and Future K¨oppen-Geiger climate Classification Maps at 1-Km Resolution.*

Ben Ayed, R., & Hanana, M. (2021, April). Artificial Intelligence to Improve the Food and Agriculture Sector. *Journal of Food Quality, 2021*, 7. doi:10.1155/2021/5584754

Crane-Droesch, A. (2018). *Machine learning methods for crop yield prediction and climate change impact assessment in agriculture*. IOP publishing.

Galety, M. G., Thiagarajan, R., Sangeetha, R., Vignesh, L. B., Arun, S., & Krishnamoorthy, R. (2022, April). Personalized music recommendation model based on Machine Learning. *In 2022 8th International Conference on Smart Structures and Systems (ICSSS)* (pp. 1-6). IEEE. 10.1109/ICSSS54381.2022.9782288

Gouse, D. G. M., Haji, C. M., & Saravanan, D. (2018). Improved reconfigurable based lightweight crypto algorithms for IoT based applications. *Journal of Advanced Research in Dynamical & Control Systems*, *10*(12), 186–193.

Huang, Y., Zhao, H., & Huang, X. (2019). A Prediction Scheme for Daily Maximum and Minimum Temperature Forecasts Using Recurrent Neural Network and Rough set. *IOP Conference Series: Earth and Environmental Science ICAESEE.* IEEE. 10.1088/1755-1315/237/2/022005

Idrees, B. A., Naif, S. S., Galety, M. G., & Arulkumar, N. (2022). An optimized algorithm for selecting stable multipath routing in MANET using proficient multipath routing and glowworm detection techniques. In *Sustainable Advanced Computing: Select Proceedings of ICSAC 2021* (pp. 511-523). Singapore: Springer Singapore. 10.1007/978-981-16-9012-9_41

Jason E. (2022). *An introduction to machine learning for classification and prediction.*

Kamilaris, A. (2018). Deep learning in agriculture: A survey. *Computers and Electronics in Agriculture*.

Kenneth, J. (2019). *Boote et al: Simulation of maize evapotranspiration: An inter-comparison among 29 maize models*. Elsevier.

Lu, J., Xiong, R., Tian, J., & Wang, C. (2022). *Battery degradation prediction against uncertain future conditions with recurrent neural network enabled deep learning*. Elsevier. doi:10.1016/j.ensm.2022.05.007

Naveen, K. (2019). *Impact of Climate Change on Agriculture Production and Its Sustainable Solutions*. Environmental Sustainabilit.

Palanivel, K., & Surianarayanan, C. (2019, May-June). An Approach for Prediction of Crop Yield using Machine Learning and Big Data Techniques, IOP publishing [IJCET]. *International Journal of Computer Engineering and Technology*, *10*(03), 110–118. doi:10.34218/IJCET.10.3.2019.013

World Bank Open Data. (n.d.). *CM*. World Bank Open Data. https://donnees.banquemondiale.org/country/CM

Compilation of References

Agarwal, M., Maze, T., & Souleyrette, R. (2005). Impacts of weather on urban freeway traffic flow characteristics and facility capacity. In: *Proceedings of the 2005 mid-continent transportation research symposium.* IEEE.

Airlangga, P., & Rachman, A. A. (2022). Deployment of Location Mapping Results Based on Application Programming Interface. *NEWTON: Networking and Information Technology*, *1*(3), 111–118. doi:10.32764/newton.v1i3.1922

Ajagbe, S. A., Oladipupo, M. A., & Balogun, E. O. (2020). Crime belt monitoring via data visualization: A case study of folium. International Journal of Information Security. *Privacy and Digital Forensic*, *4*(2), 35–44.

Alam, M. M., Torgo, L., & Bifet, A. (2022). A survey on spatio-temporal data analytics systems. *ACM Computing Surveys*, *54*(10s), 1–38.

Althoey, F., Ansari, W. S., Sufian, M., & Deifalla, A. F. (2023). Advancements in low-carbon concrete as a construction material for the sustainable built environment. *Developments in the Built Environment, 16*. https://doi.org/https://doi.org/10.1016/j.dibe.2023.100284

Amaratunga, V., & Wickramasinghe, L. (2020, July). *Artificial Neural Network to Estimate the Paddy Yield Prediction Using Climatic Data*, *2020*, 11.

Ammapa, J., Visuttiporn, P., Klaylee, J., Chayphong, S., & Iamtrakul, P. (2022). *Using GIS-Based Spatial Analysis: Comparing Pattern of Urbanization and Transportation Networks*. IEEE. doi:10.1109/ICTLE55577.2022.9902095

Anselin, L., Li, X., & Koschinsky, J. (2022). GeoDa, from the desktop to an ecosystem for exploring spatial data. *Geographical Analysis*, *54*(3), 439–466.

Anselin, L., Syabri, I., & Kho, Y. (2006). GeoDa: An introduction to spatial data analysis. *Geographical Analysis*, *38*(1), 5–22. doi:10.1111/j.0016-7363.2005.00671.x

Asfoor, H., & Halit, P. (2017). *Using Life Cycle Cost Analysis in Pavement Design As a Decision Support Tool.*

Balkaya, C., & Tutumluer, E. (2021). *AI and Big Data in Transportation Infrastructure for Condition and Risk Assessment.*

Balla, D. & Mester, T. (2017). *Possibilities of spatial data visualization with web technologies for cognitive interpretation*. Research Gate.

Balla, D. (2017). *Possibilities of spatial data visualization with web technologies for cognitive interpretation*. Research Gate.

Bao, W., Gong, A., Zhang, T., Zhao, Y., Li, B., & Chen, S. (2023). Mapping population distribution with high spatiotemporal resolution in Beijing using baidu heat map data. *Remote Sensing (Basel)*, *15*(2), 458. doi:10.3390/rs15020458

Barman, D., Saha, R., Bhowmick, T., Bagui, A., & Dutta, G. (2022). Role of GIS, remote sensing and agro advisory in conservation agriculture. In *Conservation Agriculture and Climate Change* (pp. 233–248). CRC Press.

Beck, H., Zimmermann, N. E., McVicar, T., & Vergopolan, N. (2018). *Present and Future K¨oppen-Geiger climate Classification Maps at 1-Km Resolution*.

Bekker, M. (2023). Depends on how you count them: The value of general propensity choropleth maps for visualising databases of protest incidents. *Journal of Maps*, *19*(1), 2064778. doi:10.1080/17445647.2022.2064778

Ben Ayed, R., & Hanana, M. (2021, April). Artificial Intelligence to Improve the Food and Agriculture Sector. *Journal of Food Quality*, *2021*, 7. doi:10.1155/2021/5584754

Brown, J. E. (1971). Introduction to computers and their development. *Journal of the American Oil Chemists' Society*, *48*(5), 185–190. doi:10.1007/BF02883750

Brucker, K., & Macharis, C. (2011). Multi-criteria analysis in transport project evaluation: An institutional approach. *European Transport \. Trasporti Europei*, *47*, 3–24.

Camara, G. S., Camboim, S. P., & Bravo, J. V. M. (2021). Using Jupyter Notebooks for Viewing and Analysing Geospatial Data: Two Examples for Emotional Maps and Education Data. *The International Archives of the Photogrammetry, Remote Sensing and Spatial Information Sciences*, *46*, 17–24. doi:10.5194/isprs-archives-XLVI-4-W2-2021-17-2021

Campbell, J. B., & Wynne, R. H. (2011). *Introduction to remote sensing*. Guilford press.

Capata, A. (2016). *AIS Data Visualization for Maritime Spatial Planning (MSP)*. Research Gate.

Chen, C., Judge, J., & Hulse, D. (2022). PyLUSAT: An open-source Python toolkit for GIS-based land use suitability analysis. *Environmental Modelling & Software*, *151*, 105362. doi:10.1016/j.envsoft.2022.105362

Choropleth Map in Folium (n.d.). Python. https://python-visualization.github.io/folium/latest/reference.html #module-folium.features

Cho, S. K., Kim, Y. T., & Choi, J. Y. (2020). Design and implementation of mobile 3D visualization service system on the integrated underground geospatial information map. *Journal of The Korea Society of Computer and Information*, *25*(12), 173–186.

Coppock, J. T., Grelot, J. ., Yuju, H., Kanakubo, T., Morrison, J. L., & Rystedt, B. (1994). *Thematic Mapping from Satellite Imagery Visualization in Modern Cartography.*

Crane-Droesch, A. (2018). *Machine learning methods for crop yield prediction and climate change impact assessment in agriculture.* IOP publishing.

David, A., & Zeiler, M. (2004). *Designing Geodatabases: Case Studies in GIS Data Modeling.*

Dumedah, G., & Garsonu, E. K. (2021). Characterising the structural pattern of urban road networks in Ghana using geometric and topological measures. *Geo : Geography and Environment, 8*(1), e00095. doi:10.1002/geo2.95

Eberhard, K. (2023). The effects of visualization on judgment and decision-making: a systematic literature review. In Management Review Quarterly (Vol. 73, Issue 1). Springer International Publishing. doi:10.1007/s11301-021-00235-8

Echols, N. (2003). *Finding the Fastest Route.*

ERA. (2013). Route Selection Manual. Addis Ababa Ethiopia: Ethiopian Roads Authority (ERA).

Esmaili, R. B. (2021). *Earth Observation Using Python: A Practical Programming Guide* (Vol. 75). John Wiley & Sons. doi:10.1002/9781119606925

Essien, A., & Giannetti, C. (2019). A deep learning framework for Univariate time series prediction using convolutional LSTM stacked autoencoders. In: 2019 IEEE international conference on INnovations in intelligent SysTems and applications (INISTA). IEEE. doi:10.1109/INISTA.2019.8778417

Essien, A., Petrounias, I., Sampaio, P., & Sampaio, S. (2019). Deep-PRESIMM: Integrating Deep Learning with Microsimulation for Traffic Prediction. In: *IEEE International Conference on Systems, Man, and Cybernetics* (pp. 1–6). IEEE Xplore. 10.1109/SMC.2019.8914604

Essien, A., Chukwukelu, G., & Giannetti, C. (2019). *A Scalable Deep Convolutional LSTM Neural Network for Large-Scale Urban Traffic Flow Prediction using Recurrence Plots. In: 2019 IEEE AFRICON.* IEEE Xplore. doi:10.1109/AFRICON46755.2019.9134031

Essien, A., Petrounias, I., Sampaio, P., & Sampaio, S. (2018). The impact of rainfall and temperature on peak and offpeak urban traffic. In: *International Conference on Database and Expert Systems Applications*, (pp. 399–407). Springer, Cham. 10.1007/978-3-319-98812-2_36

Essien, A., Petrounias, I., Sampaio, P., & Sampaio, S. (2019). Improving Urban Traffic Speed Prediction Using Data Source Fusion and Deep Learning. In: *2019 IEEE International Conference on Big Data and Smart Computing (BigComp),* (pp. 1–8). IEEE. 10.1109/BIGCOMP.2019.8679231

Fairbairn, D., Andrienko, G., Andrienko, N., Buziek, G., & Dykes, J. (2001). Representation and its relationship with cartographic visualization. *Cartography and Geographic Information Science, 28*(1), 13–28. doi:10.1559/152304001782174005

Faisal, G., Sreelakshmi, S., & Chandra, S. S. V. (2023, April). Crop yield prediction for smart agriculture with climatic parameters using random forest. In *International Conference on Advances in Computing and Data Sciences* (pp. 367-376). Cham: Springer Nature Switzerland. 10.1007/978-3-031-37940-6_30

Fan, C., Chen, M., Wang, X., Wang, J., & Huang, B. (2021). A Review on Data Preprocessing Techniques Toward Efficient and Reliable Knowledge Discovery From Building Operational Data. *Frontiers in Energy Research*, *9*, 652801. doi:10.3389/fenrg.2021.652801

Fan, H., Zhao, Z., & Li, W. (2021). Towards measuring shape similarity of polygons based on multiscale features and grid context descriptors. *ISPRS International Journal of Geo-Information*, *10*(5), 279.

Farkas, G. (2020). Possibilities of using raster data in client-side web maps. *Transactions in GIS*, *24*(1), 72–84. doi:10.1111/tgis.12588

Feng, X., & Koch, J. (2024). Combining vector and raster data in regionalization: A unified framework for delineating spatial unit boundaries for socio-environmental systems analyses. *International Journal of Applied Earth Observation and Geoinformation*, *128*, 103745.

Fernando, J. (2018). *The Spatial-Perceptual Design Space: A New Comprehension for Data Visualization.*

Fernando, J. (2018). The Spatial-Perceptual Design Space: A New Comprehension for Data Visualization. Research Gate.

Fiorinia, M. (2016). *AIS Data Visualization for Maritime Spatial Planning*. MSP.

Folium API for Choropleth Map (n.d.). Python. https://python-visualization.github.io/folium/latest/getting_started.html

Folium API for Click for Marker (n.d.). Python. https://python-visualization.github.io/folium/latest/reference.html#folium.features.ClickForMarker

Folium API for Heat Map (n.d.). Python. https://python-visualization.github.io/folium/latest/reference.html#folium.plugins.HeatMap

Folium API for Map (n.d.). Python. https://python-visualization.github.io/folium/latest/reference.html

Folium API for Map Layer Control (n.d.). Python. https://python-visualization.github.io/folium/latest/reference.html#folium.map.LayerControl

Folium API for Marker (n.d.). Python. https://python-visualization.github.io/folium/latest/reference.html#folium.map.Marker

Folium API for Subplots in Figure (n.d.). Python. https://python-visualization.github.io/folium/latest/advanced_guide/subplots.html

Folium API for Tile Layer (n.d.). Python. https://python-visualization.github.io/folium/latest/reference.html#folium.raster_layers.TileLayer

Fornell, D. (2015). Understanding How Big Data Will Change Healthcare. In *Imaging Technology News* (*Vol. 55*, Issue 7, pp. 26–28). https://login.ezproxy.library.ualberta.ca/login?url=https://search.ebscohost.com/login.aspx?direct=true&db=rzh&AN=110150176&site=ehost-live&scope=site

Galety, M. G., Thiagarajan, R., Sangeetha, R., Vignesh, L. B., Arun, S., & Krishnamoorthy, R. (2022, April). Personalized music recommendation model based on Machine Learning. In *2022 8th International Conference on Smart Structures and Systems (ICSSS)* (pp. 1-6). IEEE. 10.1109/ICSSS54381.2022.9782288

Garcia, C., Rabadi, G., Abujaber, D., & Seck, M. (2023). Supporting Humanitarian Crisis Decision Making with Reliable Intelligence Derived from Social Media Using AI. *Journal of Homeland Security and Emergency Management*, *20*(2), 97–131.

Garcia, E., Peyman, M., Serrat, C., & Xhafa, F. (2023). Join Operation for Semantic Data Enrichment of Asynchronous Time Series Data. *Axioms*, *12*(4), 349.

Garg, P. (2023). Understanding Geospatial Data. In Emerging Trends, Techniques, and Applications in Geospatial Data Science (pp. 1-14). IGI Global.

Gecchele, G., Ph, D., Ceccato, R., & Gastaldi, M. (2019). Road Network Vulnerability Analysis. *Journal of Transportation Engineering. Part A, Systems*, *145*(7), 1–10. doi:10.1061/JTEPBS.0000252

Geetha, K., Srivani, A., Gunasekaran, S., Ananthi, S., & Sangeetha, S. (2023, September). Geospatial Data Exploration Using Machine Learning. In *2023 4th International Conference on Smart Electronics and Communication (ICOSEC)* (pp. 1485-1489). IEEE. 10.1109/ICOSEC58147.2023.10275920

Gete, Z. (2000). Implications of land use and land cover dynamics for mountain resource degradation in the Northwestern Ethiopian highlands. *Mountain Research and Development*, *21*(2), 184–191. doi:10.1659/0276-4741(2001)021[0184:IOLUAL]2.0.CO;2

Getting Started with Folium (n.d.). Python. https://python-visualization.github.io/folium/latest/getting_started.html

Ghana: Associated Consulting Gipps, P., Gu, K., Held, A., & Barnett, G. (2001). New technologies for transport route selection. *Transportation Research Part C: Emerging Technologies*, *9*, 135-154. doi:10.1016/S0968-090X(00)00040-1

Ghanian-Manual. (2009). *Integrated Transport Plan – Ghana Multi-Criteria Evaluation Manual.*

González Del Campo, A., & Gazzola, P. (2020). Untapping the potential of technological advancements in Strategic Environmental Assessment. *Journal of Environmental Planning and Management, 63*(4), 585–603. doi:10.1080/09640568.2019.1588712

Goodfellow, I., & Bengio, Y. (2015). *Deep learning*. MIT Press.

Gouse, D. G. M., Haji, C. M., & Saravanan, D. (2018). Improved reconfigurable based lightweight crypto algorithms for IoT based applications. *Journal of Advanced Research in Dynamical & Control Systems, 10*(12), 186–193.

Grinberg, M. (2018). *Flask Web Development: Developing Web Applications with Python*. O'Reilly. https://books.google.com.et/books?id=cV34swEACAAJ

Gupta and Bagchi. (2024a). Python Language Basics. In *Essentials of Python for Artificial Intelligence and Machine Learning* (pp. 87–126). Springer Nature Switzerland. doi:10.1007/978-3-031-43725-0_3

Gupta and Bagchi. (2024b). Data Visualization with Python. In *Essentials of Python for Artificial Intelligence and Machine Learning* (pp. 237–282). Springer Nature Switzerland. doi:10.1007/978-3-031-43725-0_7

He, J. & Shen, W. (2013). *Divakaruni et al: Improving traffic prediction with tweet semantics*. AAAI.org.

Heat Map in Folium (n.d.). Python. https://python-visualization.github.io/folium/latest/user_guide/plugins/heatmap.html

Heldt, B. W. T. (2022). *Exploring the use of public participation GIS in transportation planning for tourism at a Nordic destination.*

Helmus, J. J., & Collis, S. M. (2016). The Python ARM Radar Toolkit (Py-ART), a library for working with weather radar data in the Python programming language. *Journal of Open Research Software, 4*(1), 4. doi:10.5334/jors.119

Henning Theuns. F. P. D. R. (2017). *Integrating climate change into road asset management*. World Bank. https://documents.worldbank.org/en/publication/documents-reports/documentdetail/981831493278252684/technical-report-2017

Howari, F. M., & Ghrefat, H. (2021). Geographic information system: spatial data structures, models, and case studies. In *Pollution Assessment for Sustainable Practices in Applied Sciences and Engineering* (pp. 165–198). Butterworth-Heinemann.

Huang, B. (2010). *An integration of GIS, virtual reality and the Internet for visualization, analysis and exploration of spatial data.* Research Gate.

Huang, B., Jiang, B., & Li, H. (2010). *An integration of GIS, virtual reality and the Internet for visualization, analysis and exploration of spatial data.*

Huang, Y., Zhao, H., & Huang, X. (2019). A Prediction Scheme for Daily Maximum and Minimum Temperature Forecasts Using Recurrent Neural Network and Rough set. *IOP Conference Series: Earth and Environmental Science ICAESEE.* IEEE. 10.1088/1755-1315/237/2/022005

Hu, N., Meng, X., Liu, H., & Li, Y. (2022, June). Exploring Geospatial Data Visualization Based on Python. In *Proceedings of the 2022 2nd International Conference on Control and Intelligent Robotics* (pp. 858-862). 10.1145/3548608.3559322

Husdal, J. (1999). *Network analysis – raster versus vector – A comparison.* Husdal. https://husdal.com/1999/10/11/network-analysis-raster-versus-vector-a-comparison-study/

Idrees, B. A., Naif, S. S., Galety, M. G., & Arulkumar, N. (2022). An optimized algorithm for selecting stable multipath routing in MANET using proficient multipath routing and glowworm detection techniques. In *Sustainable Advanced Computing: Select Proceedings of ICSAC 2021* (pp. 511-523). Singapore: Springer Singapore. 10.1007/978-981-16-9012-9_41

IENE. (2022). *The Problem.* Handbook wildlife traffic. https://handbookwildlifetraffic.info/1-introduction/the-problem/

Ilie, G. A., Vasile, G., d'Urso, G., & Boldo, D. (2011). *Spaceborne SAR tomography: application in urban environment* [Doctoral dissertation, GIPSA-LAB].

Indian State GeoJSON Data. (2020). Kaggle. https://www.kaggle.com/datasets/sauravmishra1710/indian-state-geojson-data

Jameel, A. a. A., S. (2017). A comprehensive review of route selection methods for transportation projects. *Jordan Journal of Civil Engineering, 11*(2).

Jamil, A., Padubidri, C., Karatsiolis, S., Kalita, I., Guley, A., & Kamilaris, A. (2024). GAEA: A Country-Scale Geospatial Environmental Modelling Tool: Towards a Digital Twin for Real Estate. In *Environmental Informatics* (pp. 177–199). Springer.

Jan, A. F., & Sachar, A. (2022). Use of GIS in Road Infrastructure Development and Management. *International Journal of Innovative Research in Engineering & Management, 9*(2), 338–342.

Jani, H. J. (2018). SOCIAL IMPACT ASSESSMENT OF ROAD INFRASTRUCTURE PROJECTS. *Global Journal of Commerce, 7.* https://www.longdom.org/articles-pdfs/social-impact-assessment-of-road-infrastructure-projects.pdf

Jason E. (2022). *An introduction to machine learning for classification and prediction.*

Jia, J., Sun, H., Jiang, C., Karila, K., Karjalainen, M., Ahokas, E., Khoramshahi, E., Hu, P., Chen, C., Xue, T., Wang, T., Chen, Y., & Hyyppä, J. (2021). Review on active and passive remote sensing techniques for road extraction. *Remote Sensing (Basel), 13*(21), 4235. doi:10.3390/rs13214235

Jia, Y., Wu, J., Ben-Akiva, M., Seshadri, R., & Du, Y. (2017). Rainfall-integrated traffic speed prediction using deep learning method. *IET Intelligent Transport Systems*, *11*(9), 531–536. doi:10.1049/iet-its.2016.0257

Jia, Y., Wu, J., & Xu, M. (2017). Traffic flow prediction with rainfall impact using a deep learning method. *Journal of Advanced Transportation*, *2017*, 1–10. doi:10.1155/2017/6575947

Joosten, T. (2021). *Detecting criminal activity in rural areas: A study into the application of satellite imagery in drugs detection*. [Master Thesis, Eindhoven University of Technology].

Kamilaris, A. (2018). Deep learning in agriculture: A survey. *Computers and Electronics in Agriculture*.

Kamraju, M. (2023). Overview of Geospatial Science and Technology. In *Temporal and Spatial Environmental Impact of the COVID-19 Pandemic* (pp. 1–6). Springer Nature Singapore. doi:10.1007/978-981-99-1934-5_1

Kaswan, K. S., Dhatterwal, J. S., & Balamurugan, B. (2023). *Python for Beginners*. Chapman and Hall/CRC. doi:10.1201/9781003202035

Kathar, S. P., Nagne, A. D., Awate, P. L., & Bhosle, S. (2023, December). Comparative Study of Supervised Classification for LULC Using Geospatial Technology. In International Conference on Soft Computing and its Engineering Applications (pp. 79-93). Cham: Springer Nature Switzerland.

Kenneth, J. (2019). *Boote et al: Simulation of maize evapotranspiration: An inter-comparison among 29 maize models*. Elsevier.

Kim, S., Hong, S., Joh, M., & Song, S. (2017). DeepRain: ConvLSTM network for precipitation prediction using multichannel radar data. ArXiv preprint. 1711.

Kirilenko, A. P. (2022). Geographic Information System (GIS) Making Sense of Geospatial Data. In *Applied Data Science in Tourism: Interdisciplinary Approaches, Methodologies, and Applications* (pp. 513–526). Springer International Publishing.

Koperski, K2019). Spatial Data Mining. Progress and Challenges Survey paper.

Koperski, K., Adhikary, J., & Han, J. (2019). *Spatial Data Mining: Progress and Challenges Survey.*

Kumar, S., & Roy, U. B. (2023). A technique of data collection: web scraping with Python. In *Statistical Modeling in Machine Learning* (pp. 23–36). Academic Press. doi:10.1016/B978-0-323-91776-6.00011-7

Kumar, T. S., Jothilakshmi, S., James, B. C., Prakash, M., Arulkumar, N., & Rekha, C. (2023). HHO-based vector quantization technique for biomedical image compression in cloud computing. *International Journal of Image and Graphics*, *23*(03), 2240008. doi:10.1142/S0219467822400083

Kwon, O. (2021). *Fostering Spatial Visualization Ability through Web-Based Virtual-Reality Program and Paper-Based Program.*

Kwon, O. (2021). *Fostering Spatial Visualization Ability through Web-Based Virtual-Reality Program and Paper-Based Program.* Research Gate.

Lansley, G., & Cheshire, J. (2016). An Introduction to Spatial Data Analysis and Visualisation in R. 1–121.

Lanucara, S., Praticò, S., Pioggia, G., Di Fazio, S., & Modica, G. (2024). Web-Based Spatial Decision Support System for Precision Agriculture: A tool for delineating dynamic management unit zones (MUZs). *Smart Agricultural Technology, 8*, 100444. doi:10.1016/j.atech.2024.100444

Lawhead, J. (2013). *Learning Geospatial Analysis with Python.* Packt Publishing. https://books.google.com.et/books?id=UR-8AQAAQBAJ

Lemenkova, P. (2020). Python libraries matplotlib, seaborn and pandas for visualization geospatial datasets generated by QGIS. *Analele Stiintifice Ale Universitatii "Alexandru Ioan Cuza" Din Iasi - Seria Geografie, 1*(1), 13–32.

Lemenkova, P. (2020). Python libraries matplotlib, seaborn and pandas for visualization geospatial datasets generated by QGIS. *Analele stiintifice ale Universitatii" Alexandru Ioan Cuza" din Iasi-seria Geografie, 64*(1), 13-32.

Li, J. P., Haq, A. U., Din, S. U., Khan, J., Khan, A., & Saboor, A. (2020). Heart Disease Identification Method Using Machine Learning Classification in E-Healthcare. *IEEE Access : Practical Innovations, Open Solutions, 8*, 107562–107582. doi:10.1109/ACCESS.2020.3001149

Lillesand, T., Kiefer, R. W., & Chipman, J. (2015). *Remote Sensing and Image Interpretation* (7th ed.). Wiley. https://books.google.com.et/books?id=eQXYBgAAQBAJ

Li, S., Dragicevic, S., Anton, F., Sester, M., Winter, S., Coltekin, A., Pettit, C., Jiang, B., Haworth, J., Stein, A., & Cheng, T. (2015). Geospatial Big Data Handling Theory and Methods: A Review and Research Challenges. *ISPRS Journal of Photogrammetry and Remote Sensing, 115*, 119–133. doi:10.1016/j.isprsjprs.2015.10.012

Liu, P. (2015). A survey of remote-sensing big data. *Frontiers in Environmental Science, 3,* 45.

Lu, J., Xiong, R., Tian, J., & Wang, C. (2022). *Battery degradation prediction against uncertain future conditions with recurrent neural network enabled deep learning.* Elsevier. doi:10.1016/j.ensm.2022.05.007

McClain, B. P. (2022). *Python for Geospatial Data Analysis.* O'Reilly Media, Inc. https://www.everand.com/book/601850266/Python-for-Geospatial-Data-Analysis

McClain, B. P. (2022). *Python for Geospatial Data Analysis.* O'Reilly Media, Inc.

McKinney, W. (2017). *Python for Data Analysis: Data Wrangling with Pandas, NumPy, and IPython.* O'Reilly Media, Incorporated. https://books.google.com.et/books?id=2BYfvgAACAAJ

McKitrick, M. K., Schuurman, N., & Crooks, V. A. (2023). Collecting, analyzing, and visualizing location-based social media data: Review of methods in GIS-social media analysis. *GeoJournal*, *88*(1), 1035–1057.

Mirzakarimova, G. M. (2023). Remote sensing data: international experiences and applications. *Italy" actual problems of science and education in the face of modern challenges" 14*(1).

Mohan, P., Neelakandan, S., Mardani, A., Maurya, S., Arulkumar, N., & Thangaraj, K. (2023). Eagle strategy arithmetic optimisation algorithm with optimal deep convolutional forest based fintech application for hyper-automation. *Enterprise Information Systems*, *17*(10), 2188123. doi:10.1080/17517575.2023.2188123

Moon, W., & Longley, P. A. M. F., Goodchild, D. J., Maguire and D. W. Rhind. (2015). Geographic Information Systems and Science (3rd Edition) (Book Review). The Leading Edge (Society of Exploration Geophysicists), 31, 975-976.

Navalgund, R. R., Jayaraman, V., & Roy, P. S. (2007). Remote sensing applications: An overview. *Current Science*, 1747-1766.

Navarro, I. J., Yepes, V., & Martí, J. V. (2019). A Review of Multicriteria Assessment Techniques Applied to Sustainable Infrastructure Design. *Advances in Civil Engineering*, *6134803*, 1–16. Advance online publication. doi:10.1155/2019/6134803

Naveen, K. (2019). *Impact of Climate Change on Agriculture Production and Its Sustainable Solutions*. Environmental Sustainabilit.

Nilsson, N. J. (2015). *An introduction to machine learning*.

Palanivel, K., & Surianarayanan, C. (2019, May-June). An Approach for Prediction of Crop Yield using Machine Learning and Big Data Techniques, IOP publishing [IJCET]. *International Journal of Computer Engineering and Technology*, *10*(03), 110–118. doi:10.34218/IJCET.10.3.2019.013

Philippou, A. (2022). *Data visualisation of environmental risks in real estate properties* [Bachelor's thesis, University of Twente].

Poonia, P., Jain, V. K., & Kumar, A. (2018). Short term traffic flow prediction methodologies: A review. *Mody University International Journal of Computing and Engineering Research.*, *2*, 37–39.

Pradhan, M. R. (2023). PIRAP: Medical Cancer Rehabilitation Healthcare Center Data Maintenance Based on IoT-Based Deep Federated Collaborative Learning. *International Journal of Cooperative Information Systems*.

Pradhan, M. R. (2023). PIRAP: Medical Cancer Rehabilitation Healthcare Center Data Maintenance Based on IoT-Based Deep Federated Collaborative Learning. *International Journal of Cooperative Information Systems*.

Pundt, H. (2010). *Visualization of spatial data for field-based GIS*. Research Gate.

Rajamani, S. K., & Iyer, R. S. (2023). Use of Python Modules in Ecological Research. In *Perspectives on the Transition Toward Green and Climate Neutral Economies in Asia* (pp. 182–206). IGI Global. doi:10.4018/978-1-6684-8613-9.ch011

Raymond E. (2019). *Public–Private Partnerships for Infrastructure Development.*

Raza, A., Shahid, M. A., Safdar, M., Tariq, M. A. R., Zaman, M., & Hassan, M. U. (2023, October). Exploring the Impact of Digital Farming on Agricultural Engineering Practices. In *Biology and Life Sciences Forum* (Vol. 27, No. 1, p. 10). MDPI. 10.3390/IECAG2023-15386

Reddy, C. S., Jha, C. S., Diwakar, P. G., & Dadhwal, V. K. (2015). Nationwide classification of forest types of India using remote sensing and GIS. *Environmental Monitoring and Assessment, 187*(12), 1–30. doi:10.1007/s10661-015-4990-8 PMID:26615560

Rey, S. J., & Anselin, L. (2007). PySAL: A Python library of spatial analytical methods. *The Review of Regional Studies, 37*(1), 5–27. doi:10.52324/001c.8285

Rey, S., Arribas-Bel, D., & Wolf, L. J. (2023). *Geographic data science with python.* CRC Press. doi:10.1201/9780429292507

Rogowitz B. (2021). *Data visualization: the end of the rainbow.* Research Gate.

Rogowitz, B. (2021). *Data visualization: the end of the rainbow.* Research Gate.

Román, A., Heredia, S., Windle, A. E., Tovar-Sánchez, A., & Navarro, G. (2024). Enhancing Georeferencing and Mosaicking Techniques over Water Surfaces with High-Resolution Unmanned Aerial Vehicle (UAV) Imagery. *Remote Sensing, 16*(2), 290.

Segarra, J., Buchaillot, M. L., Araus, J. L., & Kefauver, S. C. (2020). Remote sensing for precision agriculture: Sentinel-2 improved features and applications. *Agronomy (Basel), 10*(5), 641. doi:10.3390/agronomy10050641

Serrat, O. (2017). The Sustainable Livelihoods Approach. In O. Serrat (Ed.), *Knowledge Solutions: Tools, Methods, and Approaches to Drive Organizational Performance* (pp. 21–26). Springer Singapore. doi:10.1007/978-981-10-0983-9_5

Shadare, A. E., Musa, S. M., & Akujuobi, C. (2016). *Data visualization. December.*

Shaito, M. (2021). *Map Visualization using Spatial and Spatio-Temporal Data: Application to COVID-19 Data.* Research Gate.

Sharif, A., Zhang, A. M., Shih, A., Wobbrock, J. O., & Reinecke, K. (2022, October). Understanding and improving information extraction from online geospatial data visualizations for screen-reader users. In *Proceedings of the 24th International ACM SIGACCESS Conference on Computers and Accessibility* (pp. 1-5). 10.1145/3517428.3550363

Shuo, Z., Jingyu, Z., Zhengxiang, Z., & Jianjun, Z. (2021). Identifying the density of grassland fire points with kernel density estimation based on spatial distribution characteristics. *Open Geosciences, 13*(1), 796–806. doi:10.1515/geo-2020-0265

Singh, R. (2023). Remote Sensing and GIS Applications in Soil Conservation. In *Soil and Water Conservation Structures Design* (pp. 367–413). Springer Nature Singapore.

Şirin, A., & Erener, A. (2024). Python-based evaluation of road network constraints for electric scooters and bicycles. *Izmit Example.*, *9*(1), 34–48.

Sishodia, R. P., Ray, R. L., & Singh, S. K. (2020). Applications of remote sensing in precision agriculture: A review. *Remote Sensing (Basel)*, *12*(19), 3136. doi:10.3390/rs12193136

Sládek, D. (2023, May). Applicability of Python-based geospatial analysis for operational assessment of weather station deployment. In *2023 International Conference on Military Technologies (ICMT)* (pp. 1-8). IEEE. 10.1109/ICMT58149.2023.10171287

Sreelakshmi, S., & Vinod Chandra, S. S. (2023, January). Landslide classification using deep convolutional neural network with synthetic minority oversampling technique. In *International Conference on Distributed Computing and Intelligent Technology* (pp. 240-252). Cham: Springer Nature Switzerland. 10.1007/978-3-031-24848-1_17

Sreelakshmi, S., & Chandra, S. V. (2024). Visual saliency-based landslide identification using super-resolution remote sensing data. *Results in Engineering*, *21*, 101656. doi:10.1016/j.rineng.2023.101656

Srikanth, L., & Srikanth, I. (2020, January). A case study on kernel density estimation and hotspot analysis methods in traffic safety management. In *2020 international conference on communication systems & networks (COMSNETS)* (pp. 99-104). IEEE.

Sunitha, G., Abouali, A., Galety, M. G., & Sriharsha, A. V. (2023). Dynamic Programming with Python. In *Advanced Applications of Python Data Structures and Algorithms* (pp. 102–122). IGI Global. doi:10.4018/978-1-6684-7100-5.ch005

Sunitha, G., Sriharsha, A. V., Yalgashev, O., & Mamatov, I. (2023). Interactive Visualization with Plotly Express. In *Advanced Applications of Python Data Structures and Algorithms* (pp. 182–206). IGI Global. doi:10.4018/978-1-6684-7100-5.ch009

Syamsunur, D., Ismail, A., Rahmat, R., & Karim, A, O. (2011). Knowledge-based expert system for route selection of road alignment. *Australian Journal of Basic and Applied Sciences*, *5*, 208–213.

Talal, M., Raza, A., Safdar, M., Al Ansari, M. S., Ali, S. K., & Sattar, J. (2024). Optical Sensing for Precision Agriculture. In Agriculture and Aquaculture Applications of Biosensors and Bioelectronics (pp. 210-237). IGI Global. doi:10.4018/979-8-3693-2069-3.ch011

Team, D., Investigator, P., Islamia, J. M., Delhi, N., Coordinator, P., & Writer, C. (n.d.). *Network Analysis in GIS*.

The TIME. (n.d.). Le changement climatique est un facteur aggravant du trafic d'êtres humains. *Le Temps*. https://www.letemps.ch/monde/changement-climatique-facteur-aggravant-trafic-detres-humains

Thomas, B. (2014). *Spatial augmented reality—A tool for 3D data visualization*. Research Gate.

Thomas, B. (2014). Spatial augmented reality — A tool for 3D data visualization. Reseearch Gate.

Tile Providers: OpenStreetMap (n.d.). OpenStreet Map. https://www.openstreetmap.org/

Tile Providers: Stamen Maps (n.d.). Stamen Maps. https://maps.stamen.com/

Tile Providers: Thunderforest (n.d.). Thunder Forest. https://www.thunderforest.com/

Tile Providers: xyzservices (n.d.). ZYX Services. https://martinfleischmann.net/xyzservices-a-unified-source-of-xyz-tile-providers-in-python/

Tischler, S. (2017). Finding the right way - a new approach for route selection procedures? *Transportation Research Procedia*, *25*, 2813–2827. doi:10.1016/j.trpro.2017.05.247

Tran, B. H., Aussenac-Gilles, N., Comparot, C., & Trojahn, C. (2020). Semantic integration of raster data for earth observation: An RDF dataset of territorial unit versions with their land cover. *ISPRS International Journal of Geo-Information*, *9*(9), 503.

United Nations. (n.d.). *L'urbanisation rapide augmente les risques liés au climat pour des milliards d'individus.* United Nations Framework Convention on Climate Change. https://unfccc.int/fr/news/l-urbanisation-rapide-augmente-les-risques-lies-au-climat-pour-des-milliards-d-individus

Upadhye, S., Neelakandan, S., Thangaraj, K., Babu, D. V., Arulkumar, N., & Qureshi, K. (2023). Modeling of Real Time Traffic Flow Monitoring System Using Deep Learning and Unmanned Aerial Vehicles. J. *Mobile Multimedia*, *19*(2), 477–496.

Verbytskyi, Y. (2023). DELIVERY ROUTES OPTIMIZATION USING MACHINE LEARNING ALGORITHMS. *Eastern Europe: economy, business and management.* https://doi.org/doi:10.32782/easterneurope.38-14

Vishwa Kiran, S., Kaur, I., Thangaraj, K., Saveetha, V., Kingsy Grace, R., & Arulkumar, N. (2023). Machine learning with data science-enabled lung cancer diagnosis and classification using computed tomography images. *International Journal of Image and Graphics*, *23*(03), 2240002. doi:10.1142/S0219467822400022

Vuckovic, M., Schmidt, J., Ortner, T., & Cornel, D. (2021). Combining 2D and 3D visualization with visual analytics in the environmental domain. *Information (Basel)*, *13*(1), 7.

Walker, J. D., Letcher, B. H., Rodgers, K. D., Muhlfeld, C. C., & D'Angelo, V. S. (2020). An interactive data visualization framework for exploring geospatial environmental datasets and model predictions. *Water (Basel)*, *12*(10), 2928. doi:10.3390/w12102928

Wang, D. (2017). *Spatial Temporal Data Visualization In Emergency Management: A view from data-driven decision.* Research Gate.

Wang, L. (2014). Framework for Evaluating Sustainability of Transport System in Megalopolis and its Application. *IERI Procedia*, *9*, 110–116. doi:10.1016/j.ieri.2014.09.049

Wang, Z., & Trauth, K. M. (2020). Development of GIS-based Python scripts to calculate a water surface profile on a landscape for wetlands decision-making. *Journal of Hydroinformatics*, *22*(3), 628–640. doi:10.2166/hydro.2020.167

Wei, S., Zhang, T., Yu, D., Ji, S., Zhang, Y., & Gong, J. (2024). From lines to Polygons: Polygonal building contour extraction from High-Resolution remote sensing imagery. *ISPRS Journal of Photogrammetry and Remote Sensing*, *209*, 213–232.

Westra, E. (2010). *Python Geospatial Development.*

Wondwossen Mindahun & Bedasa Asefa. (2019). *Location Allocation Analysis for Urban Public Services Using GIS Techniques : A Case of Primary Schools in. July*. https://doi.org/doi:10.5923/j.ajgis.20190801.03

Wondwossen. (2022). *GPS SUPPORTED ASSESSMENT INTEGRATED WITH GIS SERVICE AREA ANALYSIS OF HEALTHCARE INSTITUTIONS OF BASONA WERANA IN COMPARISON WITH DEBRE BERHAN Wondwossen MINDAHUN.*, *48*(4), 209–217.

World Bank Open Data. (n.d.). *CM.* World Bank Open Data. https://donnees.banquemondiale.org/country/CM

Wu, N. (2023, November). *Construction and Application of Financial Management Network Teaching System in Colleges and Universities under Python Language.* In 4th International Conference on Modern Education and Information Management, ICMEIM 2023, Wuhan, China. 10.4108/eai.8-9-2023.2340077

Xiao, H., & Hao, S. (2023). Public participation in infrastructure projects: An integrative review and prospects for the future research. *Engineering, Construction, and Architectural Management*, *30*(2), 456–477. doi:10.1108/ECAM-06-2021-0495

Yitayew, H. (2020). *Gis Based New Route Alignment Selection Between Dabat And Keraker Towns* [Masters, Addis Ababa University Addis Ababa Institute Of Technology School Of Civil And Environmental Engineering]. Addis Ababa. https://etd.aau.edu.et/items/47c6ce9f-40dc-497f-866e-e10bd0488830

Zhang, H., Tang, G., Xiong, L., Yang, X., & Li, F. (2024). Geomorphology-oriented theoretical framework and construction method for value-added DEM. *Journal of Geographical Sciences*, *34*(1), 165–184.

About the Contributors

Mohammad Gouse Galety has worked in CS. His current position is CS professor at Samarkand International University of Technology in Uzbekistan. He has two Ph. D.s in web mining and CS image processing and two master's degrees in computing. First, I have a master's in computer applications and then computer science. In computer and information science, he studies web mining, IoT, and AI. Four patents, five books, and 50 Springer and Scopus-indexed journal publications and international conference proceedings are his (co)authorship. IEEE and ACM Fellow. He has worked in various organizations, including Sree Vidyanikethan Degree College, Emeralds Degree College, Brindavan College of Engineering, Kuwait Educational Centre, Ambo University, Debre Berhan University, Lebanese French University, Iraq, and Catholic University in Erbil, Iraq, since 1999. UG and PG students take computer science and engineering courses.

Arul Kumar Natarajan currently serves as an Assistant Professor in the Department of Computer Science at the Samarkand International University of Technology in Uzbekistan. He earned his Doctor of Philosophy degree from Bharathidasan University, India 2017. Concurrently, he is engaged in postdoctoral research in Generative AI for Cybersecurity at the Singapore Institute of Technology, Singapore. Throughout his 14-year teaching career, Dr. Arul has held esteemed positions at various institutions, including Christ University, Bishop Heber College in India, and Debre Berhan University in Ethiopia. Dr. Arul has made significant contributions to academia, specializing in cybersecurity and artificial intelligence, evidenced by his portfolio of scholarly works. He has authored 32 international publications indexed in Scopus and 03 international publications and has delivered 34 conference presentations. Additionally, he has edited and published two books with IGI Global Publisher, USA. In addition to his academic pursuits, Dr. Arul is a prolific innovator, holding 17 patents in India and 01 granted patent in the United Kingdom across diverse domains, including communication and computer science. He demonstrates proficiency in Networking and Cybersecurity, having completed the CCNA Routing and Switching Exam from CISCO and the Networking Fundamentals exam from Microsoft. He maintains a sincere interest in GenAI for Cybersecurity.

Tesfaye Fufa Gedefa holds a Bachelor's Degree in Information Technology from Jima University (2012) and a Master of Science in Software Engineering from Adama Science and Technology University, transitioned from lecturing at Debre Berhan University to his current role as an Associate Researcher II at the Space Science and Geospatial Institute. His research journey encompasses diverse fields such as artificial intelligence, network security, cloud computing, big data, image processing, data science, and disaster management. With a keen interest in extending his expertise, Tesfaye plans to delve into new sensor development and calibration, big data analytics, data assimilation, and integration, alongside image fusion techniques for urban climate monitoring, disaster risk reduction, and food security. He is committed to applying space technology across multidisciplinary domains, leveraging Python programming to amplify the impact of his research endeavors. Through international collaborations and a dedication to advancing knowledge, Tesfaye aims to make meaningful contributions at the intersection of technology and societal challenges.

Tsegaye Demsis Lemma is a current PhD candidate in Remote Sensing, Department of Remote Sensing Research and Development, Entoto Observatory and Research Center, Ethiopian Space Science and Technology Institute (ESSTI) affiliated with Addis Ababa University. Formerly serve as an Associate Researcher and satellite control engineer in Satellite Research, Development and Operation Directorate, ESSTI. He is an extremely motivated and sociable person, who has a passion for aerospace and remote sensing. his current work deals with the use of Earth Observation for drought and flood forecasting in Awash basin, Ethiopia. Mr. Tsegaye has a good personal skill in social, individual and team works. Participating in different satellite driven products for disaster risk reduction, urban climate, agriculture and drought.... Respect others culture, self-esteemed. He is a hard worker struggling challenges by new idea, and showing good performance in a given task, as well as he has a very good academic achievements so far. he has a plan to extend his research work (new sensor development and calibration, Big data, data assimilation, and integration(optical and SAR) and image fusion for vegetation mapping, urban climate monitoring, change detection and disaster risk reduction and food security)

Ehyaudin Musema Ahmed (PhD Candid.) a young Engineer who has embarked on a scholarly journey culminating in the achievement of a MSc. in School of Civil and Environmental Engineering, Construction Technology and Management Sub-Stream [2020] from Addis Ababa University, Addis Ababa Institute of Technology

(AAiT). With a penchant for both Engineering and innovative technology, his academic pursuits and research passion will left an indelible mark on the landscape. Upon the successful completion of his graduate studies, Ing. Ahmed emerged with a research motivated author, penning a couple of grant proposals a couple of research papers and contributing to international conference proceedings indexed by esteemed platforms such as SCI, Springer, Scopus, DOAJ, and WoS through a book chapter in the IGI Global. His intellectual prowess extends to the realms of innovation, that would work for his credit in the near future. Additionally, he has shared his insights through the creation of curricula modules, manuals and syllabus with the consideration of the applied university concept under the new educational transformation map of Ethiopia for Debre Berhan University Institute of Technology (IoT). The tapestry of Ing. Ahmed 's professional journey began in 2015, weaving through the fabric of various national higher education institutions. From the educational landscapes of Ethiopia at institutions like Debre Markos University, under the Institute of Technology, College of Engineering to serving at Debre Birhan University, his influence transcended also to the road construction industry participating In multi-million federal road construction projects under the Ethiopian RSDP-IV and V starting from the year 2016 in different regions of Ethiopia. Presently, Ing. Ahmed dons the role of Senior Lecturer and Module Coordinator for the Course Modern Construction Technology and Building Information Modeling (BIM) of the Construction Technology and Management Department at Debre Birhan University, Technology Institute in Debre Birhan, Ethiopia. In this capacity, he imparts knowledge to undergraduate students, offering a spectrum of courses in Construction Technology and Management. Ing. Ahmed has carved a niche in the research landscape, with a focal point on innovative Construction Technology, Building Information Modeling (BIM), Immersive Construction Technology (AR,VR and MR) 3D Printing, Geospatial Analysis for Design and Construction. Ing. Ahmed 's narrative is one of intellectual curiosity, global engagement, and a dedication to the advancement of knowledge in the dynamic realms of the Technology institute under the blanket of Construction Technology Management.

Mubashir Ahmed is a Pakistani Agricultural Engineer, M.Sc. (Hons), leveraging Python and deep learning to revolutionize seasonal monsoon prediction. Impacting agriculture, water management, and sustainability, establishing prominence in the field.

Mohammed Saleh Al Ansari has been working as an associate professor in the College of Engineering, Department of Chemical Engineering at the University of Bahrain. He earned his bachelor's degree in chemical engineering from King Saud University in Riyadh in 1983. In 1987, he received his MSc degree in desalination engineering, Desalination Technology, Glasgow. In 1998, he received his Ph.D. in corrosion and protection science and engineering. He has over 35 years of experience in teaching, research, and industry.

Jimbo H. Claver (Professor, PhD Supervisor, HoD, and JSPS Fellow) received in 2000 the Ph.D. degree in Applied Mathematics and Statistics from the Moscow State University named Lomonosov and he did the post-doctorate studies with the Institute of Statistical Mathematics in Tokyo, Japan. Currently, he is Full-Professor and Head of the Department of Applied Mathematics with the Samarkand International University of Technology (SIUT). His previous positions include: Head of the Department of Mathematics with the American International University (Kuwait), Professor of Applied Mathematics with the American University of Afghanistan, Associate Professor of Applied Mathematics with the Nara Institute of Science and Technology, Nara (Japan), Visiting Professor of Applied Mathematics with Waseda University (Japan), Assistant Professor of Mathematics and Statistics with Lakeland College, Tokyo (Japan), Senior Lecturer with the University of York (United Kingdom), Lecturer with Sino British College-Shanghai Institute of Science and Technology (China), visiting research fellow with Swiss Banking Institute (Zurich) and visiting fellow Morehouse College, Atlanta, Georgia (USA). His research focuses on system modelling, predictive analysis, system control, stochastic processes, applied probability, applied statistics, data analysis, quantitative finance, and actuary. In addition, he serves as editor and reviewer of several international journals, he is member of the International Mathematical Union (IMU), Japan Mathematical Society (JMS), Japan Statistical Society (JSS) and Moscow Mathematical Society (MMS). He has published four books and numerous scientific papers in pure, applied mathematics, applied statistics, quantitative finance, quantitative risk management and data analysis in peer-reviewed journals and conference proceedings. Also, he supervised countless number of MSc, PhD, and MBA students. JSPS: Japan Society for the Promotion of Science.

Deepthi Das, an Associate Professor and Associate Dean at the School of Sciences, CHRIST (Deemed to be University) in Bangalore, India, is a dedicated educator with over two decades of teaching experience. With a profound commitment to fostering academic excellence and ethical values among students, Dr. Deepthi creates inclusive learning environments that promote critical thinking and social responsibility. Holding a Ph.D. in Computer Science and Engineering from CMR University, Bangalore, she actively integrates innovative teaching methods to deliver dynamic educational experiences. Dr. Deepthi's research interests primarily focus on Artificial Intelligence and Machine Learning, with notable contributions to customer churn prediction in the motor insurance sector. Alongside her academic responsibilities, she has served in various leadership roles, including Head of the Department of Statistics and Data Science and Programme Coordinator for Undergraduate Programmes. Dr. Deepthi's dedication to education, research, and leadership acumen make her a valuable asset to the academic community.

Soressa Beyene obtained his M.S.c degree in Computer Science from Hacalu Hundessa Campus and B.S.c in Information Science from Madda Walabu University, Ethiopia. He is a lecturer at the College of Informatics, Bule Hora University. His research interests include Information retrieval, machine learning, Artificial Intelligence, Data Science, and Natural Language Processing(NLP).

Nagueu Djambong Lionel Perin is a passionate researcher working in the exciting field of data science and Artificial Intelligence. Currently enrolled in a doctoral thesis within the prestigious computer science department at the University of Yaoundé I in Cameroon, he is engaged in an in-depth and innovative exploration on the use of geospatial climate data for the prediction of climate changes in the agricultural sector. His academic journey began with obtaining his bachelor's degree at the University of Yaoundé 1 where he quickly developed a strong interest in data science and artificial intelligence. My thirst for knowledge led me to pursue a master's program in Artificial Intelligence at this University, where I had the opportunity to further immerse myself in challenging research projects. His PhD thesis, entitled "Use of geospatial climate data for the prediction of climate change in the agricultural sector" focuses on the climate change that the majority of farmers encounter during their food production. His goal is to make a significant contribution to the understanding of climate change by developing innovative solutions to help these farmers better produce their food in order to grow not only their economies but also the state's economy. He uses methodologies based on Machine Learning and Deep Learning to deepen his investigations.

Abdul Malik did M.Sc.(Hons). in Agriculture - Plant Breeding and Genetics from University of Agriculture, Faisalabad. Throughout his academic journey, He has achieved notable successes that have strengthened his foundation in plant breeding and genetics, agriculture, farming, bioinformatics and biotechnology . His Master's research reflects his dedication to create a change in polyploidy of ginger by using colchicine chemical with different doses to make it grow under Pakistani environment. Additionally, his active participation in a conferences and seminars, such as International Wheat Conference an abstract published title "Wheat Blast-A Future Threat to Wheat Production and Possible Strategies to Cope with this challenge" [2022].

A. Manimaran pursued a Bachelor of Science from Bharathidasan University, Tamil Nadu, in 2006, a Master of Computer Applications from Anna University in 2009, and a Ph.D. from Bharathidasan University in 2018. He is an assistant professor in the Department of Mathematics at VIT-AP University, Andhra Pradesh. He has published 25 research papers in reputed international journals and conferences. His primary research focuses on machine learning, deep learning, NLP, and data science. He has 10 years of teaching experience and 4 years of research experience.

Mohammed Nuru Mohammed is a Senior Lecturer of the Information Technology Department, College of Computing at the Debre Birhan University (DBU). Mohammed has a strong portfolio in research, teaching, student advising and research-project management in the field of Information Technology. He has also certified in Block chain Developer, Cloud Application Developer and CISCO CCNA Instructor trainer. He has also published more than five peer-reviewed articles in scientific journals, and conference proceedings. Mohammed has been granted the role of Team Member for the project EDU-ABCM (101082957) in the organization Debre Birhan University (DBU). He received awards in recognition of his research and teaching-learning process from DBU (2014). His research interest areas are Artificial Intelligence and Digital Privacy. In addition to the main roles of teaching-learning process, he prepared and managed research-project training to write successful grant proposals in partnership with other scholars and researchers. Mohammed was awarded a Bachelor of Science in Information Technology (2012) from Hawassa University and Master of Science in Information Technology (2016) from Jimma University, Ethiopia. From Feb 2016 – December 2018 he worked as a Course Coordinator of Information Technology, at Debre Berhan University. From Dec. 2018- Sep 20, 2022, he worked as Associate Dean for Research and Community Service at College of Computing, Debre Birhan University. From Feb 20, 2022- Dec 2023, he has been working as System Administrator. From Jan 2024- present, he has been working as a Research Extension Officer of the DBU's Vice President for Research and Community Services and Technology Transfer Office.

Tchoua Paul is Associate Professor, leading researcher with the department applied mathematics, University of Ngaoundere, Cameroon. His area of research focuses on applied mathematics and mathematical modelling. He has supervised several PhD students in various fields of applied mathematics he has also established a strong research group in his city.

K. G. Suma is working as an Associate Professor in the Department of Computer Science and Engineering, VIT-AP, India. She has received her Ph.D. degree from Anna University Chennai, India in 2017. Her research areas include Image

Processing, Medical Imaging, Machine Learning and Soft Computing. She carried several administrative roles in academics. She has won Young Women Engineer from IET in the year of 2020. She has published several Scopus and SCI indexed Journal publications, Book chapters, and attended national and international conferences and won prizes. She has 2 National/ International patents.

Gurram Sunitha is a Professor at Department of AI & ML, School of Computing, Mohan Babu University, Tirupati, A.P., India. She received her Ph.D. degree in Computer Science and Engineering from S.V.University, Tirupati. She has 23 years of experience in academia. Her research interests include Data Mining, Spatio-Temporal Analytics, Machine Learning and Artificial Intelligence. She has 13 patents and 7 books (International & National, Edited & Authored) published to her credit. She has published around 80+ research papers in various International Refereed Journals (SCIE/Scopus/Web of Science) and International Conferences indexed in Springer, IEEEXplore etc. She has been serving as reviewer for several reputed journals; She has been acting as Technical Program/Organizing/ Advisory/ Reviewer Committee Member and has been serving on the program committees and co-chaired various International Conferences across the world.

Bouetou Thomas is a full professor, leading researcher and head of department of computer science at the University of Yaoundé 1, Cameroon. His area of research focuses on applied mathematics and computing. He has supervised several PhD students in various fields of applied mathematics he is opened to research collaboration and international research to support the development of Africa.

Chinthapatla Pranay Varna is pursuing his B.Tech. degree in Computer Science and Engineering at Sree Vidyanikethan Engineering College, Tirupati, India. His research interests include Machine Learning, Artificial Intelligence, Deep learning and Predictive analysis. These areas of interest indicate a strong passion for understanding and developing advanced algorithms and models to solve complex problems, make data-driven decisions, and uncover patterns and insights from large datasets.

Desalegn Aweke Wako is a senior lecturer, and he is currently working as an assistant professor at the College of Informatics, Bule Hora University. His areas of research interests include artificial intelligence, data science, big data analytics, machine learning, and computer vision.

Index

H

I

L

M

O

P

R

S

T

V

W

X

www.ingramcontent.com/pod-product-compliance
Lightning Source LLC
LaVergne TN
LVHW080635011026
849160LV00014B/1227